Macmillan Building and Surveying Series

Series Editor: Ivor H. Seeley

Emeritus Professor, The Nottingham Trent University

(*continued overleaf*)

Series Standing Order

If you would like to receive future titles in this series as they are published, you can make use of our standing order facility. To place a standing order please contact your bookseller or, in case of difficulty, write to us at the address below with your name and address and the name of the series. Please state with which title you wish to begin your standing order. (If you live outside the United Kingdom we may not have the rights for your area, in which case we will forward your order to the publisher concerned.)

Customer Services Department, Macmillan Distribution Ltd
Houndmills, Basingstoke, Hampshire, RG21 6XS, England.

Building Services

George Hassan

PhD, MCIBSE, MIMechE, MIMarE
Formerly
Anglia Polytechnic University

MACMILLAN

© George Hassan 1996

First published 1996 by
MACMILLAN PRESS LTD
Houndmills, Basingstoke, Hampshire RG21 6XS
and London
Companies and representatives
throughout the world

ISBN 0–333–53704–1

A catalogue record for this book is available from the British Library

Printed and bound in Great Britain by
Antony Rowe Ltd
Chippenham, Wiltshire

10 9   8   7   6   5   4   3   2   1
05 04 03 02 01 00 99 98 97 96

SCIENCE

Talent alone cannot make a writer. There must be a man behind the book. [Representative Men, 'Goethe']

Ralph Waldo Emerson (1803–82)

Contents

Preface

To carry out their chosen profession in a satisfactory manner, all practitioners associated with the building function must have some knowledge of and be familiar with the state of the art of 'building services'.

This book will be of use to students of all ages, levels and abilities engaged in studying for professional qualifications on any grade of course within the built environment. The author hopes that it will also assist those who have been given the opportunity of making a career change into the building services sector.

The useful life of building services can be in excess of 20 years and students must expect to encounter both 'old ' and 'new' technology, often functioning together in the same premises. Such conditions make it a requirement that they should understand and appreciate the limitations of both the 'old' and the 'new' plant, so that they can exercise their function in operating, or altering, any part of any installation effectively. Clearly, practitioners must first appreciate the basic design parameters of the system and, to help in this, I have prefaced the introduction of most of the main topics with a short history of their development, and have also given the reasons for arranging the separate parts of each installation in the manner described.

I am confident that this book will prove to be of use to all architects and their assistants, to builders, sanitary and heating engineers, as well as to electrical and mechanical engineering consultants, building and quantity surveyors, and building managers and inspectors. Throughout the text, I have dealt with the topics from first principles, without assuming any prior knowledge of the subject on the part of the reader. Analytical work has been included only when the topic cannot be covered satisfactorily in any other way, such as in the chapter dealing with pipe sizing.

I also recommend this book as a 'first read' for engineers, who will discover that when describing and classifying the various thermodynamic processes currently in use in the built environment, I have categorised them using some of the descriptive terminology associated with applied thermodynamics.

One of the main functions of building services consultants, building managers, inspectors and engineers is to deal with energy in the built environment. Such energy is available in electrical and in non-electrical forms, and is processed in separate stages, each of which requires a separate field of knowledge, expertise and equipment. I have found it to be both logical and convenient to describe the principal aspects of these stages over four chapters, each of which may be described as being 'energy based'. Chapters 3, 4 and 5 deal with the processing of energy in its non-electrical forms, while Chapter 6 deals with the processing of energy in its electrical form. Hence, for energy in its non-electrical form: Chapter 3 deals with its collection and storage, Chapter 4 deals with its release and absorption, and Chapter 5 deals with its distribution, emission and control. And for energy in its electrical form: Chapter 6 deals with its generation, transmission, distribution, emission and control.

Chelmsford, Essex   Dr George Hassan

Acknowledgements

The author and publishers would like to thank the following for permission to reproduce copyright material:

The Chartered Institution of Building Services Engineers for tables 3.2, 7.1 to 7.11 and 9.1.

Woods of Colchester Ltd for table 2.1.

Every effort has been made to trace all the copyright holders but if any have been inadvertently overlooked the publishers will be pleased to make the necessary arrangement at the earliest opportunity.

1 Natural and Mechanical Ventilation

1.1 Natural ventilation

It is a well-known fact that hot air rises and that its place is then taken up by colder air. This natural effect occurs because the density (mass/unit volume) of the rising column of hot air is less than that of the falling column of cold air. This process in its continuously operating mode is encountered in any naturally vented enclosure and also in any chimney that is functioning properly. It is referred to as the 'stack effect'.

The provision of openings in the walls of early buildings containing fires served to: (a) aid combustion; (b) dispose of the smoke; (c) provide natural ventilation.

1.2 Stack effect

An interesting case of the effect of rising warm air can be observed in the open stairwells of multi-storey office blocks, particularly those where the stairs are built adjacent to, but away from, an outside fenestrated wall, see figure 1.1. Whenever a door leading to the foyer and/or landing is opened, a vigorous current of air will start to rise to the highest part of the building.

It is possible that an over-zealous heating consultant might prescribe radiators to be fixed at position A on each floor. The reader might find this understandable because radiators are often placed under windows, to reduce the effect of the cold downdraught. In this case, however, the provision of the radiators in the building is equivalent to providing a fire at the base of a chimney (the stairwell) and it will be found that the common circulation areas turn out to be very draughty, because of the rapidly rising column of warm air. Very often provision has to be made for a receptionist/telephonist in the foyer of the building and, when this is the case, it is better to provide a heated enclosure, rather than attempt to heat all of the foyer. Generally, the extent of heating in circulation

1

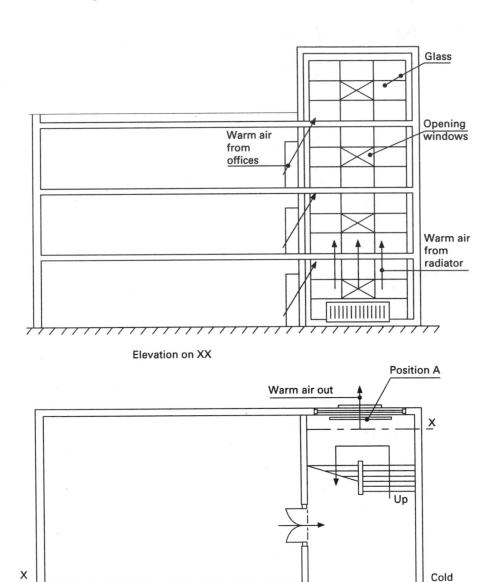

Figure 1.1   Effect of rising warm air in a stairwell

areas is lower than in the office section of the building as most people are only passing through.

The current architectural fashion for roofing-in the space between adjacent multi-storey buildings, using large areas of glass to create an enclosure of light and space, is referred to as an 'atrium', and it is within such constructions that the effect of rising columns of warm air may readily be observed. In particular, in the summer, outside air is allowed to flow into the building at ground level, usually through a number of open doors fixed across opposite sides of the building. When walking through these doors, one is aware of a flow of air, approaching a breeze, which cools the enclosure. The air is normally exhausted through louvres in the roof, which may be opened and closed automatically.

In winter, in order to restrict the flow of cold air into the space, the entrance is usually limited to the use of a single revolving door, or an airlock having sliding doors. In this way the heating effect of the solar radiation passing through the glass walls and roof is conserved. In these energy-conscious days, it is likely that in order to reduce the amount of fossil fuel being burnt, structures which are referred to as 'energy efficient' will become the norm, and it is likely that, in cold and temperate climates, passive heating of dwellings and workplaces will be encouraged.

A passive system implies that the flow of energy is by natural means, that is by natural convection, conduction and/or radiation. The stack effect can be used to transfer convective energy from a solar heated wall, to the working space, see figure 1.2, and a similar effect can be obtained using a conservatory instead of a close coupled sheet of glass as in figure 1.3.

1.3 Use of ventilating towers

The reader may be interested to learn of the way in which the stack effect is used in ventilating towers, constructed in the Iranian desert. Bahadori [1.1] gives details of the means by which natural cooling is achieved in such areas, where at night clear skies are a frequent occurrence, giving rise to a highly radiative environment. Figure 1.4 shows a vertical cross-section taken through such a ventilating tower, and the following operating scenarios are possible:

(1) *At night with no wind blowing*
The tower behaves as a chimney, drawing warmed air from the living space which is replaced by cool outside air. During this process the tower walls eventually become cool.
(2) *At night with a wind blowing*
In this case the velocity of the wind often reverses the flow of air in the tower and cold air is directed into the dwelling space cooling the walls.

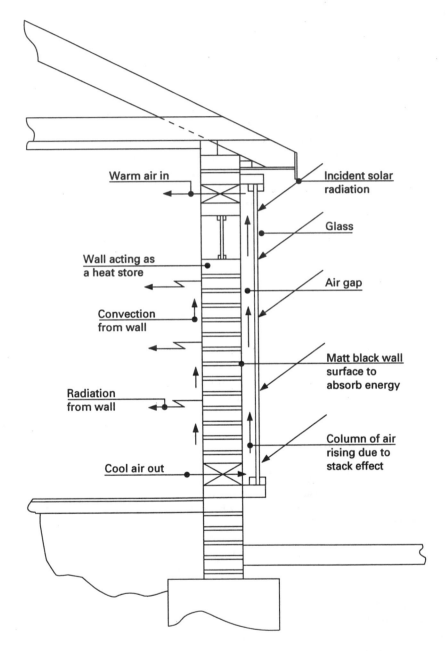

Figure 1.2 Use of the stack effect in a passive heating system

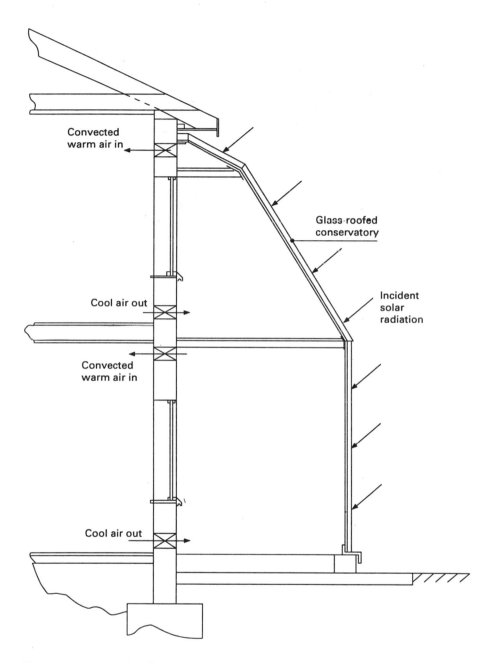

Figure 1.3 A passive heating system using a conservatory

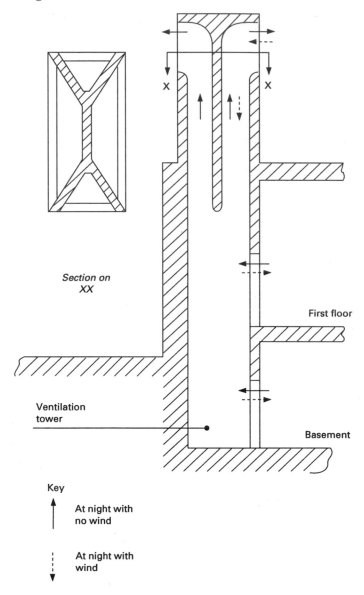

Section on
XX

X

X

First floor

Ventilation
tower

Basement

Key

At night with
no wind

At night with
wind

Figure 1.4   Operation of a ventilating tower

(3) *During the day with no wind blowing*
The warm air in contact with the cool tower walls becomes more dense,
descending and cooling the living space.
(4) *During the day with a wind blowing*
The result is the same as in (3), but the effect is greater.

The ventilating tower uses both the stack effect and the velocity of the wind to achieve ventilation. Many village halls, churches and schools have ridge ventilators on their roofs, together with low-level adjustable openings in their walls. The ridge ventilators use the kinetic energy of the wind to help the existing stack effect. These roof ventilators are open to the wind on all sides and are constructed so that the flow of air produces a drop in pressure inside the ventilator, inducing a flow of warm stale air from the underside of the roof to the outside air. This process is brought about by the 'venturi effect', which forms the basis of many static roof ventilators and also has many other applications in technology.

1.4 Venturi effect

This effect can be explained using the principle of the conservation of energy and by applying the appropriate equation of continuity.

Because air is compressible, the analysis of the flow regime is complicated owing to a possible change in density, as it passes through the tapered pipe shown in figure 1.5. In the case of a ventilator, any change in density depends on a change in temperature and, because the temperature remains constant, it can be assumed that the density also does not change. Because of this and in order to make the analysis simpler, the air may be assumed to behave as an incompressible fluid. The fluid can be said to possess energy which may be recognised in various forms:

(1) potential energy (energy of position);
(2) kinetic energy (energy of motion);
(3) flow work energy (energy needed to move the fluid).

The use of the concept of the conservation of energy implies that the sum of items (1), (2) and (3) equals a constant (ignoring losses between points 1 and 2 in figure 1.5 and any effects due to magnetism or capillarity).

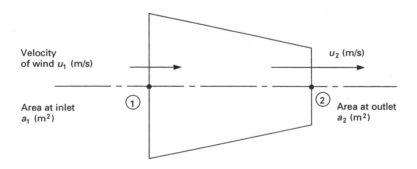

Figure 1.5   Tapered pipe

In this case there is no change in the energy of position above or below a datum level, as the tapered pipe is horizontal and item (1) is therefore zero. The energy equation then becomes:

item (2) + item (3) = a constant

From this equation it is clear that any change in the magnitude of item (2) must be at the expense of an equivalent change in item (3), and vice versa. Using the equation of continuity of flow for an incompressible fluid we can write:

volume flow rate (Q) = area (a) × the velocity of the fluid (u)

As there are no tee pieces we can write:

$$a_1 \times u_1 = a_2 \times u_2$$

which when transposed gives

$$u_2 = a_1 \times u_1/a_2$$

As the ratio a_1/a_2 is greater than 1, then u_2 is greater than u_1

The magnitude of item (2) is a function of the velocity squared (u_2^2) therefore, if u_2 is increased, item (2) will increase, and because of the reasoning given earlier, item (3) must decrease.

The magnitude of item (3) is proportional to the pressure and inversely proportional to the density of the fluid; also as in this case the density is assumed to be constant, it is clear that the value of the pressure at the outlet must be less than the atmospheric pressure acting at the inlet. It follows that any opening connected to the outlet of the tapered pipe will tend to draw fluid into the system. Figure 1.6 shows the application of this effect to a static roof ventilator and figure 1.7 the application to a chimney cowl. In both cases the operation of the device will be unimpaired by changes in the direction of the wind, and its effectiveness will depend on the magnitude of the wind velocity.

1.5 Effect of wind on buildings

Research carried out in wind tunnels using scale models of buildings by Jackman [1.2] gives details of the way in which pressures on the roof and walls of tall buildings vary with the direction of the wind, see figure 1.8.

The values given show that it would be possible to ventilate any floor of an office block by using these pressure differences. In practice, with much badly sealed fenestration, the occupants of tall office buildings know only too well the effect that wind can have on the papers on their desks, even when all window openings are shut! However, there will come a

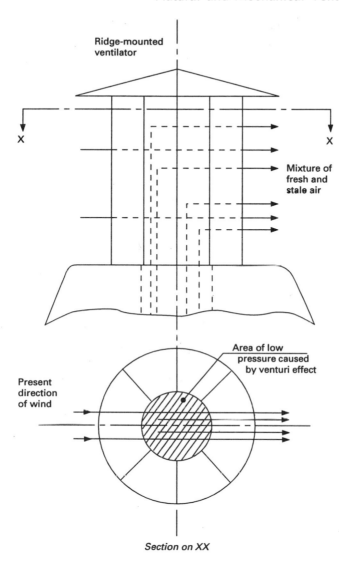

Ridge-mounted
ventilator

X

X

Mixture of
fresh and
stale air

Area of low
pressure caused
by venturi effect

Present
direction
of wind

Section on XX

Figure 1.6   Application of the venturi effect to a static roof ventilator

time when the majority of fenestration will be double glazed, coated and
properly sealed and, when this is the case, the users of these buildings
will require some carefully controlled ventilation.

A possible solution to this need could be the provision of adjustable
openings in every wall in the building; the opening and closing of these
openings would be automatically controlled by the computer dealing with
energy management, using input signals received from pressure sensing
devices placed on the outside of each wall. In this way natural ventilation

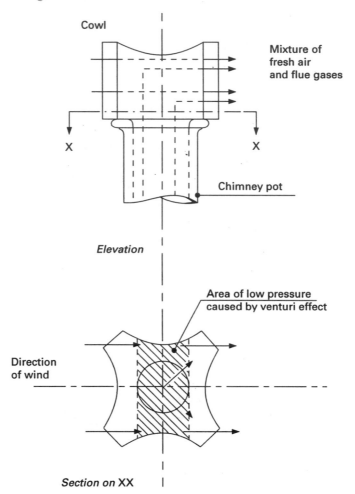

Figure 1.7   Application of the venturi effect to a chimney cowl

would be accomplished, using the pressure differences across the walls (and roof) of the building, whatever the wind direction. A suitable computer program could vary the area of the openings, depending on the magnitude of the pressure difference across the walls/roof, so that it should never again be necessary for those who work in multi-storey offices to be troubled with an impromptu migration of paper from the desk to the opposite wall of the building!

For those who wish to investigate the effect of the pressure differences across buildings on the expected number of air changes per hour, further useful reading can be found in *Woods Practical Guide to Fan Engineering* [1.3].

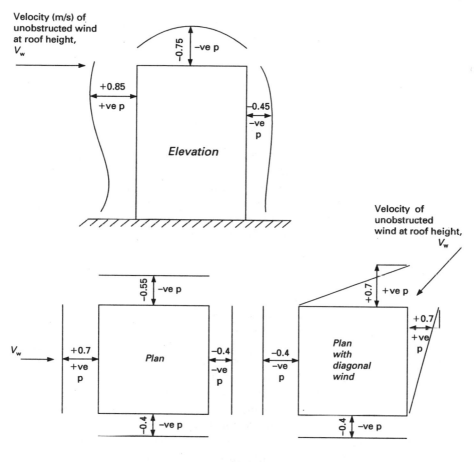

Figure 1.8 Pressure variations on buildings due to the effects of wind velocity

1.6 Need for ventilation of an enclosure

'Natural' ventilation occurs owing to the forces of nature; in some cases it may be regarded as beneficial and in others not. The reasons why 'fresh' air must be introduced to an enclosure may be categorised as follows. There must be sufficient fresh air to:

(1) provide sufficient oxygen (O_2) to sustain life and any combustion process;
(2) remove carbon dioxide (CO_2) and water vapour;
(3) dilute odours;
(4) remove particles causing airborne infection;
(5) promote body cooling;
(6) remove excess heat;
(7) remove possibly noxious industrial fumes.

It is found that the amount of air needed to deal with item (3) is normally greater than the amount of air needed for the other requirements, although there will always be some processes where the needs of the other factors take on a relatively greater importance. For example, the need to remove flammable vapours, which are heavier than air, from a paint spray room, is obviously different from that of the amount of ventilation needed for a lecture theatre. Also, any industrial processes requiring operatives to be kept cool will require a greater volume flow rate of air than that needed for the ventilation of a general office.

With the possible exceptions of items (6) and (7), it may be assumed that if the ventilation rate is sufficient to make body odour unnoticeable, then it follows that most of the other items listed above will be satisfied. This condition is ideally suited to the subjective appraisal of a person entering the enclosure 'fresh' from the outside, when his or her sense of smell will rapidly detect the presence of odour. If however it is required to control the ventilation condition automatically and change the operating mode of the ventilation system, then in the absence of the existence of low-cost olfactory sensors outside the human body, use is usually made of signals obtained from carbon dioxide (CO_2) sensors sited in the room.

As the concentration of CO_2 increases, there is a need to increase the volume flow rate of air. This is now easily possible, brought about by the development of electronic control systems and the use of thyristor and other speed controls on fan motors. The term 'thyristor' is explained in a later section.

According to the 1986 CIBSE Guide [1.4] the maximum allowable concentration of CO_2 for 8-hour exposure for healthy adults in the UK is taken as 0.5% by volume, while in the USA this is taken as 0.25%. The CIBSE Guide gives the minimum ventilation rates required to limit the concentration of CO_2 to 0.5%, over levels of activity ranging from 0.8 litre/s per person for sitting quietly to 5.3–6.4 litre/s per person for very heavy work.

A table of recommended air change rates for various types of building is also published in the 1986 CIBSE Guide, which gives recommendations for the volume flow rate of air/person, as well as the hourly air change rate. Typical air changes/hour are 6–10 for boardrooms, 8–12 for canteens, 10–15 for restaurants, 4–6 for offices and 4–15 for laboratories.

Because statutory regulations must be satisfied, each system must be separately examined and some judgement is required before an air change rate is finally chosen.

1.7 Mechanical ventilation

The operation of natural ventilation, depending as it does on climatological phenomena, implies that the performance will be variable. With the on-set of mass production it is now possible to manufacture a wide range of fans and electric motors, which are readily available at relatively low cost.

The use of fans has almost superseded earlier natural methods of ventilation, as architects no longer recognise the need to design living spaces that are 'lofty', to allow for the natural ventilation process to take place. This reduces the cost of buildings and mechanical ventilation is now used extensively to make the enclosures tolerably comfortable for the occupants. In the case of bathrooms and toilets having no windows, it is mandatory to provide ventilation. The application of mechanical ventilation to an enclosure can be carried out in various ways, and these are classified in the following sections.

1.7.1 Extraction only

Figure 1.9 shows a system in which fans remove air from an enclosure, and it should be noted that it is fundamental to the process of air extraction that planned fixed air inlets must also be provided, in order to ensure satisfactory performance of the scheme.

In the case of a kitchen/dining room complex (KDR), it is common practice to extract air from the kitchen, so removing both heat and cooking smells. By providing suitable fixed openings in the wall between the kitchen and the dining room, air will also flow from the dining room to the kitchen, so ensuring that unwanted kitchen odours are not perceived by the diners. Most enclosures have irregular shapes and, to avoid pockets of stagnant air, it is sometimes necessary to use ductwork for both collection and distribution purposes.

1.7.2 Supply only

Figure 1.10 shows a system in which fans supply air to an enclosure, making it very slightly pressurised, so that air inside the room tends to flow outwards through each of the planned openings; this is found to be helpful in the avoidance of draughts.

If the enclosure happens to be situated in a dusty environment, then the slight pressure within the space will tend to stop the dust coming in through any ill-fitting windows and doors. Window manufacturers refer to the gaps around their products as 'crackage', and nowadays this is expected to be minimal. In dusty conditions the incoming air must be

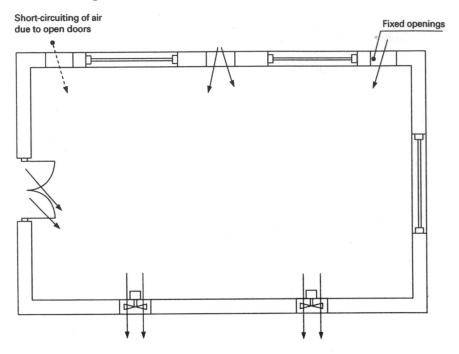

Figure 1.9 Extraction of air using mechanical ventilation

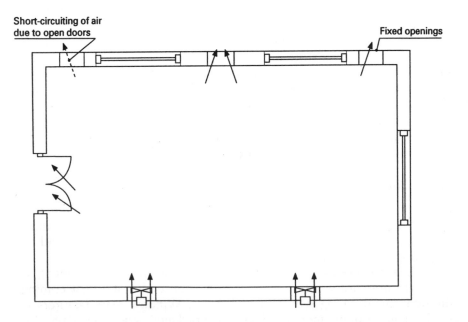

Figure 1.10 Supply of air using mechanical ventilation

filtered and, when this is carried out effectively, the frequency with which the internal decorations need to be renewed will be reduced, producing a saving in maintenance costs. To trap the dust, a viscous filter may be used and, to ensure efficient operation, it can be cleaned continuously and automatically. When the incoming air contains an odour having an organic origin, it is possible to remove it by using a filter containing activated carbon granules.

Whenever stagnant air pockets are to be dealt with by the introduction of fresh and unheated air, the choice of air change rates and the assessed heat input rate are closely linked and, to achieve a solution that will be regarded as being satisfactory, by both the occupants and the client, the consultant must manoeuvre carefully between these two parameters. Clearly, if the space to be ventilated has a heating system which operates indifferently, then provision must be made for the incoming air to be heated, before attempting to ventilate.

1.7.3 Combined extraction and supply

Figure 1.11 shows a system in which air is both supplied to and removed from an enclosure. With this system, the fresh air flow rate can be controlled more closely than in the case of either of the methods detailed in sections 1.7.1 and 1.7.2.

In order to retain the advantage of slight pressurisation of the space as obtained in 1.7.2, it is customary to arrange for the inlet fans to deliver approximately 20% more volume flow rate than the output fans. Also, by controlling the fan speeds automatically, the system can be made to cater for all the ventilation requirements as they vary throughout the time of occupation.

1.8 Thyristor control

The thyristor is a solid-state unidirectional switching device, which operates on the alternating current sine wave in such a way that the wave peaks and parts of the sloping sides of the wave are used to drive the motor.

Using this method of speed control ensures that, even at low speeds, maximum torque is available. Some energy is dissipated in the thyristor, but it is very much easier and cheaper to use than the earlier pole switching devices to be found on older induction motors. This new generation of electronic speed control devices may readily be operated by means of a centralised computer control module, possibly being triggered by a temperature sensor.

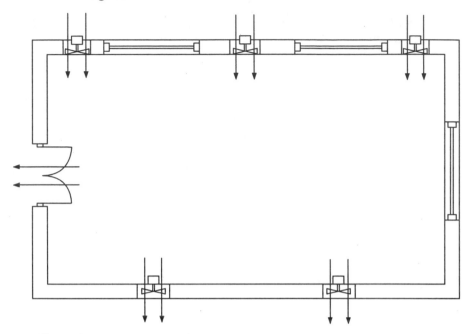

Notes:
1. Inlet volume flow rate is
   20% greater than extract rate.
   This ensures slight pressurisation
   of the enclosure with air
   tending to flow outwards,
   producing no externally
   induced draughts.
2. The positions of the fans
   in the thickness of the walls
   is diagrammatic.

Figure 1.11   Combined extraction and supply of air using mechanical
ventilation

1.9   Ventilation system using combined extraction and supply to a kitchen/dining room complex (KDR)

The use of a ventilation system using combined extraction and supply
makes it possible to provide a separate flow of warmed or cooled air
which can be supplied to, and subsequently extracted from, the dining
area, in the knowledge that it will prevent heat and cooking smells that
emanate from the kitchen from reaching the diners. A suggested traditional duct layout for such a complex is shown in figure 1.12. Note that
the main supply and exhaust ducting are positioned at the centre of the
complex, so avoiding long runs of ducting having inordinately large cross-
sectional areas.

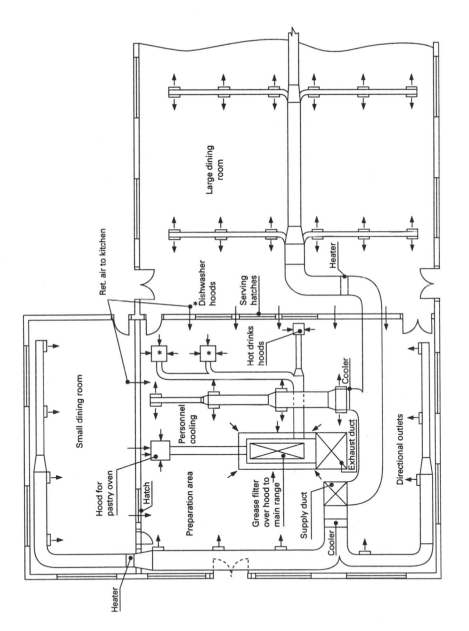

Figure 1.12 Arrangement of ducts for a KDR complex

1.9.1   Notes on sizing a ventilation plant for a KDR

It is always prudent to approach each problem using at least two differ-ent methods of calculation. Provided that the original assumptions are valid in both cases, then it is likely that the results from each method will be comparable and a useful check will then have been made.

Firstly, the kitchen must be under negative pressure so that air will flow into the kitchen from the surrounding areas. In order to achieve this, it is customary to allow for the kitchen supply fans to deliver, say, 85% of fresh air and to allow the remaining 15% to infiltrate from the dining area. The extraction fans should remove 100% of kitchen air.

The capacities of hoods, ducts and fans may be assessed by evaluating the following three items.

Item 1

Extracted air in litres per second is approximately equal to the (number of meals served/h) × either 10, or 15, depending on the type of meal being served [1.4].

Item 2

The summation of the energy consumption rates of the cooking appli-ances is a good indication of the amount of heat available per second to be absorbed by the air in the kitchen. In the UK it is generally assumed that in summer time temperatures in the kitchen should not be greater than 30°C, when the outside air temperature is 25°C.

The necessary hourly air change rate can be calculated from the equation:

$$E = 1200 \ V \times \Delta t$$

where E = energy input (W)

1200 $(J/m^3°C)$ is a constant, allowing for the specific heat capacity of the air at constant pressure and its density

V = volume flow rate (m^3/s)

Δt = temperature difference between kitchen air and incoming ven-tilation air (°C).

It is known that comfort conditions in a kitchen cannot always be ob-tained simply by providing fresh outside air so, at this stage, some allow-ance should be made for the projected volume flow rate to deal with local spot cooling. The food preparation and grill areas will need local spot cooling, and some air conditioning may also be required. Air conditioning implies that the air is delivered either warmed or cooled and having a planned moisture content (see Chapter 2). Currently, it is unusual for full air conditioning to be applied to the whole kitchen area on the grounds of high capital and running costs.

Item 3

Based on the sizes of the canopies (or hoods) over the kitchen ranges, allow for a minimum inlet velocity of 0.35 m/s through the hood openings.

The volume flow rates, calculated from items 1, 2 and 3, may now be compared with yet another estimate based on a table contained in the CIBSE Guide, giving nominal extract rates for kitchen appliances (litre/s) per unit or per m^2 net area of appliance. Typical unit figures are 300 for pastry ovens, 450 for fish fryers and steak grills, 300 for steamers and 150–250 for tea sets [1.4]. According to the Guide, the volume flow rate of extracted air must not be smaller than 17.5 litre/s for each unit of floor area, nor less than 20–30 air changes/h. In fact, in some cases, volume flow rates of extracted air can approach 120 air changes/h.

The differing approaches to the problem suggested above will each produce a slightly different solution, and a final investigation should now be made to check that the sizes of plant specified are not only valid, but realistic. It will then be possible to size the ducts using the constant-pressure drop method. Note that the constant-velocity method of duct sizing is reserved for use when dealing with the conveying of dusts and light particles, in which case the velocity of the air stream should not fall below a settling velocity for the particles being transported.

When specifying ductwork sizes, allowances must be made for providing grease filters, fitted close to the source of production. The ductwork should contain inspection doors that are large enough to permit access for cleaning purposes, and these should be placed at frequent intervals along the duct. In order to allow for the scraping, or pressure hosing, of any painted and galvanised ductwork in the kitchen area, it must be manufactured of material having a thickness of at-least 1 mm. Other likely materials from which the ducts may be manufactured are stainless steel or anodised aluminium. Glass is often used for the canopies. Wherever fat is used for cooking purposes, the effects of a fat fire must always be considered (details of suitable fire precautions are given in Chapter 8).

1.10 Use of a special-purpose false ceiling in a KDR complex

A recently developed alternative method of dealing with the problems associated with the ventilation of kitchens, having a plan area greater than about 40 m^2, is to install a false ceiling, made up of purpose-designed square, stainless steel modules, of which there are four different types.

Type 1

The exhaust module operates as a grease and condensate trap, also removing steam and hot air from the cooking ranges and fat fryers by natural means.

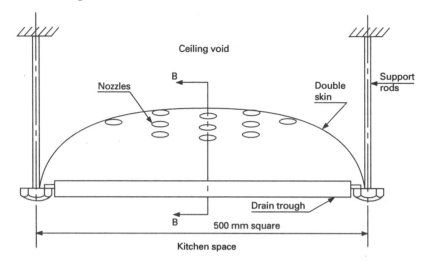

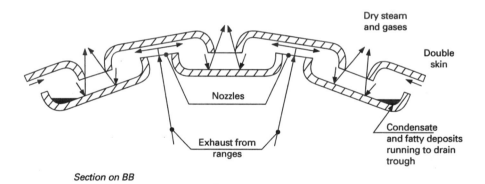

Section on BB

Figure 1.13 Exhaust module (type 1)

Figure 1.13 shows that the underskin of the module has holes in it, pressed into the shape of nozzles pointing upwards; these discharge against the underside of the upper skin. The nozzles serve to increase the velocity of the gases and, owing to the venturi effect (described earlier), the pressure is lowered. As a result of this, the water vapour may exist in a drier state and some of the moisture 'flashes off', to form more water vapour, producing a slight drying effect. The remaining water droplets and other fatty particulates continue in the direction of the main stream of gas and impinge on the underside of the upper skin, falling eventually into the collection troughs pressed into the edges of the lower skin of the module.

For the purposes of cleaning, the whole module may be lifted clear of its supports and immersed in a standard kitchen dishwasher. Spare clean

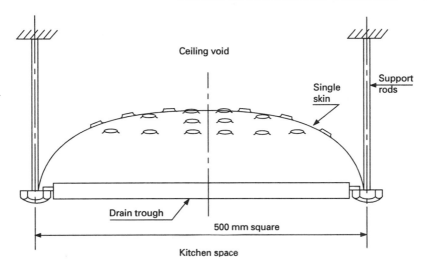

Figure 1.14   Air supply module (type 2)

modules may be inserted, so that the kitchen need never be out of service while cleaning takes place. This results in a considerable saving in maintenance costs, over the traditional KDR exhaust hood ventilation system as previously described.

Type 2

This unit functions as an air supply module, through which fresh or conditioned air is delivered to the perimeter of the kitchen, see figure 1.14.

Type 3

This unit is referred to as a lighting module, through which fresh or conditioned air passes via the casings of the luminaires. This keeps the fittings very slightly pressurised and serves the useful function of excluding steam and fat from the surfaces of the fluorescent tubes, see figure 1.15.

Type 4

This module is made of a single skin of plain steel and is used for infill purposes.

   In practice, type 1 exhaust modules are grouped together over the cooking ranges and fat fryers, and the space above them in the ceiling void is partitioned off from the other modules and then connected to the exhaust ducting. Similarly, the space above the type 2 air supply modules is connected to the air supply ducting. Separate flexible connections are made to each type 3 lighting module and connected to the air supply. Figure 1.16 shows one possible way of using the modular ceiling on the KDR complex previously described.

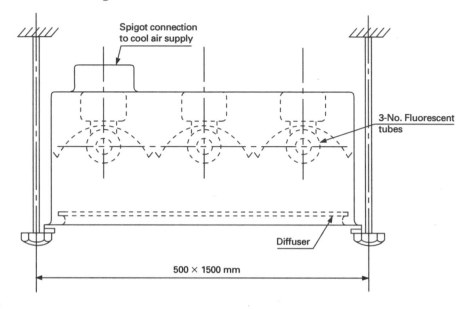

Spigot connection
to cool air supply

3-No. Fluorescent
tubes

Diffuser

500 × 1500 mm

Figure 1.15   Lighting module (type 3)

1.11   Further examples of the need for mechanical ventilation

The comments made earlier with regard to the principle of containment
of kitchen odours apply to the planning of many other systems. Mechanical ventilation is required for those industrial processes producing noxious
gases, which are situated within a general manufacturing area, as in the
case of plating works, steelworks and foundries.

Another obvious case where careful planning of the direction of the
airflow is required is that of the mechanical ventilation of toilet blocks.
As mentioned earlier, such ventilation is the subject of statutory regula-
tion, and the safest way of dealing with the requirements of the regulations is to approach the environmental health officer for initial guidance
and possible help in the interpretation of the local council bye-laws.
Typically, a toilet complex in an office block would consist of a system
of horizontal ducts joining into a common vertical duct having twin fans,
one to act as a standby and having an automatic changeover switch. The
ductwork is connected to the suction side of the fans and the duty fan
operates continuously.

Mechanical ventilation is also required for crematoria, where it is possible that a cremation service and some cremations will be occurring at
the same time. It is normal practice for the cremation furnaces to be pre-
warmed by hot air, and this air, together with the products of combustion, are then exhausted to the chimney using induced draught fans.

Key

1. Extract air modules, surrounded by an airtight enclosure in the ceiling void.
2. Exhaust duct.
3. Exhaust stack.
4. Air supply stack.
5. Air supply duct.
6. Warm air to staff dining room.
7. Warm air to cafeteria.
8. Heater battery.
9. Cooler battery.
10. Cooled air supply to air modules placed where required in false ceiling.
11. Flexible ducting to lighting modules sited in or near exhaust zones.
–○– Positions of lighting fittings (fluorescent tubes).

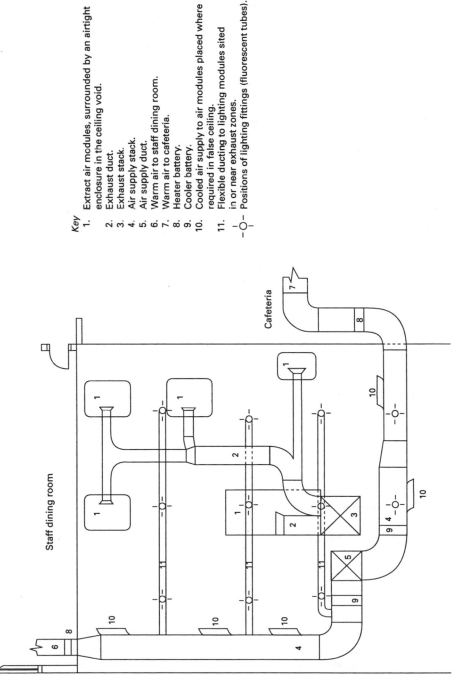

Figure 1.16 Application of the modular ceiling to a KDR complex

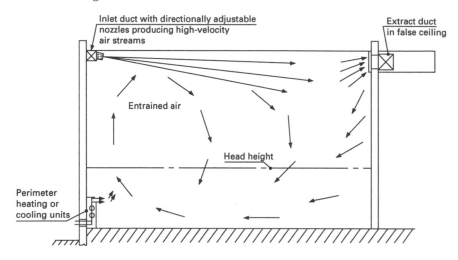

Figure 1.17 Ventilation of a tall enclosure using high-velocity air jets

1.12 Ventilation using high-velocity air jets

It is known that a high-velocity stream of air entrains adjacent air which then travels with it in the direction of movement of the jet.

This principle may be used to ventilate tall enclosures, such as theatres, indoor stadia and law courts. The original air jet and the air entrained with it are then exhausted at high level from the other side of the building. This effect is illustrated in figure 1.17, which also shows the incoming air being heated, or cooled as required, by finned tubes. Provision must be made for access to these low-level heaters, to facilitate the subsequent cleaning of fins and tubes. Figure 1.18 shows the principle of high-velocity ventilation applied to a theatre. If the velocity of a jet of air is too great, it produces unwanted noise which is due to the magnitude of the shear stresses set up by the high-velocity gradients that exist between the moving and the stationary air. This effect becomes a limiting factor on the velocity and possible 'throw' of the jet.

1.13 Distribution of air

Examination of the effects of air distribution in buildings nearly always shows that parts of the enclosure are ill served and are sometimes short-circuited.

This point is illustrated in a simple way in figures 1.9 and 1.10. However, the sketches should make it clear that for good overall coverage of the space, much depends upon the skill with which inlets, outlets and distribution ductwork are positioned and, from the designer's viewpoint, not least of his considerations must be the amount of money available for the work.

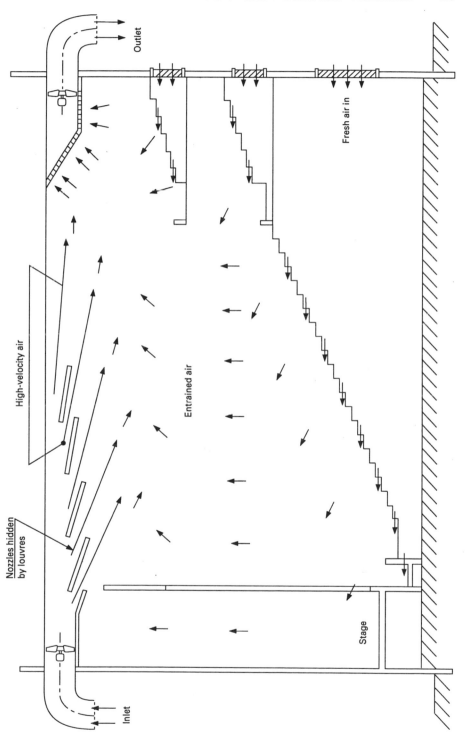

Figure 1.18 Ventilation of a theatre using high-velocity air jets

With regard to the positioning of air outlets, it is often stated that 'no outlets should be sited on a wall of the building which faces the direction of the prevailing wind'. This condition is too stringent if uniform airflow across the ventilated space is to be achieved at all times. In an attempt to overcome the effect of wind pressure, such planned outlets should be fitted with wind cowls, see figure 1.19. Sometimes, self-opening and closing shutters may be used where the noise of operation will not be too obtrusive. Louvres that are virtually waterproof to all but horizontally driving rain may be obtained and are sometimes used for air exits.

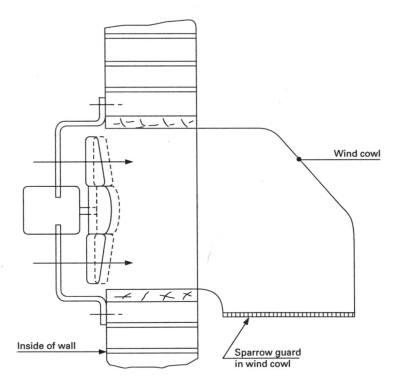

Wind cowl

Inside of wall

Sparrow guard
in wind cowl

Notes

1. When the generation of pressure is important (i.e. when driving into ductwork), place the fan in the broken-line position.
2. When the volume flow rate is important, position the front face of the fan in line with the inner face of the wall.

Figure 1.19 Fan fitted with a wind cowl

1.14   Noise control in ductwork

The noise associated with ductwork may be carried by the air passing through the system and it may also emanate from the panels from which it is constructed.

1.14.1   Material-borne sound

The ductwork used for the distribution of air is frequently made of steel and, in order to limit the first cost, light gauge steel may be specified.

This sometimes results in some sections of the ductwork setting up vibrations which are in phase with, and excited by, the fan and motor, or by some other adjacent machinery. These effects can be avoided by changing the natural frequency of vibration of the offending panel, through the addition of stiffeners. These may take the form of angle stiffeners, riveted, welded or bolted diagonally to the vibrating panel, or of grooves pressed into the panel in the form of a pattern. Both methods increase the second moment of area of the sheet about the plane of bending and make the panel more rigid, so that the subsequent deflection is less and the natural frequency of vibration of the panel is changed. Alternatively, panels having relatively large areas may be stiffened during manufacture by the use of profiling. Profiling is the term used to describe the provision of grooves, which are press-formed (sometimes diagonally) into the panels which make up the ducting; this works in the same way as the other methods previously described.

1.14.2   Fluid-borne sound

Unwanted noise is frequently transmitted by air in the ductwork to all parts of the system, and it may consist of the sound of speech referred to as 'cross talk', coming from a general office, or even a noisy managing director's office. By far the greatest proportion of sounds will come from the operation of the fan and motor, particularly if the former is of the single-stage axial flow type. Some of the unwanted sound may be the result of an abrupt air entry to the fan, the closeness of a bend to the fan, or the effect of slackness in a flexible duct connector, see figure 1.20. All of these faults can be avoided by careful design. The object is to try to make sure that air streams change direction smoothly, so that extreme local air turbulence, giving rise to noise (and incidentally loss in pressure), is avoided.

Unwanted sound energy can be attenuated by lining the inside of the ductwork with a foamed plastic material, placed downstream of the fan and continued for a distance sufficient to make the sound output from the grilles acceptable. Sound absorption occurs whenever a sound pressure wave collides with the elastic foam. The greatest number of collisions

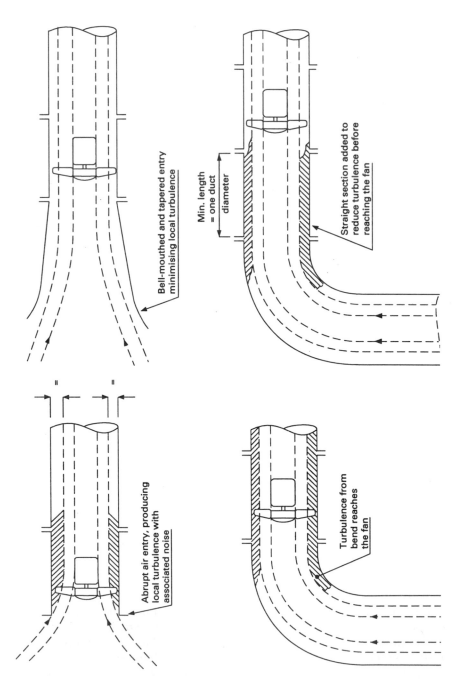

Bell-mouthed and tapered entry minimising local turbulence

Min. length = one duct diameter

Straight section added to reduce turbulence before reaching the fan

Abrupt air entry, producing local turbulence with associated noise

Turbulence from bend reaches the fan

Figure 1.20   Two causes of airborne noise in ductwork (with solutions)

will occur at sharp, right-angled bends, so these should not be neglected in the lining process. It is also likely that at such bends, some of the sound waves will be reflected back towards the source.

Normally, in order to conserve the air pressure generated in the duct by the fan, ductwork designers prefer to use right-angled bends that have a large radius; these are referred to as 'easy sweep' bends. It is singularly unfortunate that the sound attenuation effect of such easy sweep bends is not as good as that of a sharp right-angled bend. The effect of lining the internal surfaces of the ductwork is to increase the resistance set up to the moving air, so that the net duct pressure available becomes less. It follows that if suitable results are to be achieved, the need to attenuate noise in a ductwork system must be recognised and dealt with at the initial design stage. For those interested in carrying out the necessary calculations, information on these and many other problems of noise control related to the built environment is available [1.3, 1.5].

1.15 Effect of short-circuiting in a ventilation system

Some industrial buildings have large door openings through which fork-lift trucks pass, and these openings often present an unplanned inlet or exit for the ventilation air destined for the occupied space. Associated with this, the velocity of the air through the opening is usually high enough to cause personal nuisance to the operatives; the doors may also close noisily.

One solution to this problem is to decrease the pressure drop across the doors by providing more openings:

(a) in the doors (if possible);
(b) in the space surrounding the door frames;
(c) in the walls around the rest of the building, so that air may exit from it with less of a pressure differential.

Note that the effectiveness of any ventilation scheme can only be the result of a series of compromises between the various physical limitations, which are sometimes randomly imposed.

1.16 Clean rooms

An extreme example of good air distribution occurs in the 'clean rooms' used in some high technology industries. For instance, some industrial processes in the electronics and space industries require the manufacture and assembly of component parts to take place in a near dust-free environment.

In an attempt to satisfy this requirement, a bank of very fine filters is arranged to occupy the whole of one end wall of the assembly space.

Fans then move the inlet air across these filters and into the room. An extraction duct is located across the opposite end wall and, in effect, the room becomes one large air duct. Assembly of parts requiring the cleanest of conditions takes place at the end of the room nearer to the filters, while assemblies requiring less exacting conditions are carried out further along the room, towards the extraction duct.

Special precautions are taken with regard to the clothes worn by the operatives, heads are covered, and each person enters and leaves the room through an air lock cubicle, which provides an 'air shower'. In fact, the human body in the process of normal living and locomotion gives off showers of dust and particles of dead skin, and it is this effect that makes the extraction end of the room less 'clean' than the inlet end. For further reading about the relative sizes of different forms of dust particles, see [1.4] (section B3-43).

1.17   Effect of 'open' underfloor heating ducts

Those who attend churches and cathedrals which are still heated by means of large diameter cast iron pipes, some of which are laid beneath gratings in the floor of the central aisle, will be familiar with the characteristic smell that permeates the air. The pedestrian traffic along the central aisles is at times quite dense, and it follows that the heavier dust particles that arise from this movement will fall on to the heated pipes below the gratings. It is from this source that the characteristic odour arises and it would seem that the placing of heating surfaces under gratings in floor ducts in public places, such as foyers, where a high density of traffic is expected, should be avoided.

1.18   Specialised systems of mechanical ventilation

The needs of industry are many and various, and each case is almost certain to require a purpose-designed solution. Clearly, some technical data about the requirements may initially be obtained from the client, but it must be expected that by far the greatest proportion of data will have to be culled from suitable reference books. Much detailed information with regard to the mechanical ventilation of different types of buildings is given in [1.4] (sections B2-8 to 14), and some of the topics dealt with are now listed:

(1) laboratories and fume cupboards, together with precautions needed when dealing with radioactive substances;
(2) animal houses;
(3) pharmaceutical establishments;
(4) printing works;
(5) TV studios.

1.19　Threshold limit values for noxious substances

Many industrial processes use chemicals that are potentially dangerous to the operatives, for instance mineral and toxic dusts, fumes and spray mists, and also gases and vapours.

It is a matter of increasing concern that these substances should, by proper ventilation, be kept to suitably low levels of concentration in the working environment. The Threshold Limits Committee of the American Conference of Governmental Industrial Hygienists produced equations reproduced in the 1970 CIBSE Guide which will determine the Threshold Limit Value (TLV) in mg/m^3 for upwards of 400 chemical substances and mixtures.

The threshold limit value should be regarded only as an approximate guide to an acceptable level of concentration of substances taken over an 8-hour working period and a 40-hour week. The limit may be exceeded for short periods, provided that an equivalent period occurs when the level is lower. But this does not apply to all substances, and it cannot be emphasised too strongly that the results obtained by application of the equations given, should first be verified by an expert industrial chemist before any plant is specified to deal with handling problems. It may well be that the calculated TLVs are still not applicable to safe and healthy conditions for the particular substance being handled, especially if a poisoning effect is cumulative, or the noxious effect is due to, say, relatively short periods of high mixture concentrations. The latter are referred to as 'ceiling limits' which should not be exceeded. The 1970 CIBSE Guide contains an extensive list of TLVs (mg/m^3) for toxic dusts, fumes and mists, for example ammonium chloride fume has a value of 10.0, arsenic 0.5, soluble components of barium 0.5, and manganese 5.0.

1.20　Comfort conditions

Any man-made system must always be examined to see whether or not it comes up to specification. In the case of some ill-prescribed and badly designed ventilation projects, it may be possible to prove that the details of the specification have been met, even though some of the occupants report that they feel 'uncomfortable'. Predictably, this has resulted in the search for an 'Index of Comfort'.

Much research has been carried out and many papers have been written on the subject of comfort and, while enough information has been gathered for 'zones' of comfort to be prescribed, the actual application of such information in normal budget building projects has, in the author's experience to date, proved to be somewhat problematical. Perusal of Bedford's classic work [1.6] will serve to indicate the difficulties to be experienced in assessing what exactly is meant by 'comfort'. The comfort zone for sedentary occupations proposed by the CIBSE Guide is linked to outside air temperature. The following forms of temperature, referred to as temperature 'indices', are used in defining the comfort zone and are now defined.

1.20.1 Air temperature t_a (°C)

This is usually taken to be the mean air temperature obtained using a conventional mercury in glass thermometer. Note that t_{ai} = air temperature inside and t_{ao} = air temperature outside.

1.20.2 Dry resultant temperature t_{res} (°C)

This is the temperature obtained from a thermometer having its bulb placed at the centre of a 100 mm diameter sphere, which has been given a matt black finish to make the reading sensitive to thermal rediation. The value of the air velocity also affects the thermometer reading. This temperature scale was devised by Missenard in 1935. It is widely used in Europe and is often referred to as degrees Missenard (M), to honour its originator. The following relationship applies:

$$t_{res} = \frac{(t_r + 3.17 \times t_{ai} \times v^{1/2})}{(1 + 3.17 \times v^{1/2})} \tag{1.1}$$

where t_{ai} = air temperature inside (°C)

v = air velocity (m/s)

and t_r = mean radiant temperature (°C).

1.20.3 Mean radiant temperature t_r (°C)

This temperature can be obtained from equation (1.1) provided that v, t_{ai} and t_{res} are known.

For still conditions inside an office, or dwelling, a maximum velocity v of 0.1 m/s is a realistic value, and if this is substituted into equation (1.1), t_{res} may be obtained from:

$$t_{res} = 0.5 \times t_r + 0.5 \times t_{ai}$$

Recommended design values for dry resultant temperatures for various buildings are contained in the 1986 CIBSE Guide. Typical examples of t_{res} (°C) are art galleries and museums 20; lecture theatres 18; churches 18; factories (heavy work) 13; living rooms of dwellings 21; bedrooms 18; hospital wards 18; libraries 20; offices 20; and departmental stores 18.

As yet, no mention has been made of the form of dress of the occupants, whether or not they are engaged in vigorous activity, what effect the colour of the decorations has on their perception of warmth, how long they have been subjected to such conditions, or what effect the presence of moisture in the atmosphere exerts on their metabolism.

In the opinion of the author, the search for universal conditions of comfort in the built environment is likely to be unproductive. However, in no way should such a statement deter practising professionals from aspiring to attain the 'comfort state' and still less should it discourage anyone from designing and running experiments to evaluate the interaction of the many parameters involved.

Humphreys [1.7] has summarised the work done by up to 56 separate authors and in his concluding discussion he helpfully points the way for those who wish to gain more knowledge in the difficult field of 'human comfort'. One of his suggestions is that future researchers, when using mathematical models for computer simulation processes, should investigate the effect of time on the human reactions and that the results of such investigations should then be compared with those obtained in the field.

1.21 Importance of air velocity for human comfort

Those who frequent building sites may have noticed the preoccupation of ventilation engineers with the measurement of air velocities, both inside and outside their ducts.

Provided that the temperature of the air does not change appreciably from one measuring point to the next, the continuity equation for the flow of an incompressible fluid may be applied, with small loss in accuracy, to the flow of a compressible fluid (air) in the ducts. Hence:

Volume flow rate = cross-sectional area of duct × air velocity
   [Q (m³/s)]   [a (m²)]   [u (m/s)]

By measuring the velocity of air inside a duct and knowing the cross-sectional area at the point of measurement, the continuity equation $Q = a \times u$ may be used to determine the volume flow rate. This result may then be compared with the design value for that section of the duct and the system dampers may then be altered until subsequent measurements indicate that the design figure has been met. This process is referred to as 'calibrating' the system and is pertinent to the successful commissioning of any arrangement of ductwork.

By measuring the velocity of air in the occupied space at head level, a rough indication of whether or not the occupant will feel comfortable may be obtained. If the air velocity is less than 0.01 m/s at a height of 2 m above the floor, then the air is considered to be stagnant and the conditions are conducive to the occupants having a listless attitude and developing headaches.

If in a temperate climate, with the enclosure at a temperature of 20°C, the air velocity has a value of 0.2 m/s, 2 m above the floor, then these conditions may be regarded as being representative of a 'suitable' state of ventilation, likely to foster feelings of comfort. If the room temperature is higher at say 25°C and the air is heavy with moisture, then air velocities of 0.3 m/s and upwards may also be acceptable.

1.22   Energy balance equation

This equation occurs in various forms depending upon the subject to which it is applied. In the study of thermodynamics it appears as the steady (or non-steady) energy flow equation and is particularly useful in solving problems dealing with the energy balance of engines, turbines, refrigerators and other plant.

When studying the thermal performance of the human body it appears in the following form:

$$M - W = C + E + R + S$$

where M = metabolic rate (watts)
   W = mechanical working rate (watts)
   C = convective heat loss rate (watts)
   E = evaporative heat loss rate (watts)
   R = radiative heat loss rate (watts)
   S = body heat storage rate (watts).

A popular mnemonic among students is $M = SCREW$, although no doubt there are other letter combinations which would also serve. In attempting to make the built environment as habitable as possible, it is clear that ventilation in some form is a necessary requirement. It will also be appreciated that in some cases, in order to try and approach the comfort condition, there is a need for air conditioning to be installed, and this is dealt with in the next chapter.

References

1.1. Bahadori M.N. Paper 6.2 (Solar energy utilisation), in *Proceedings of the First International Conference on Solar Building Technology*, Vol. 2, 1977. RIBA, London.
1.2. Jackman P.J. *A study of the natural ventilation of tall office buildings*, Laboratory Report No. 53, 1969. HVRA, Bracknell, Berkshire.
1.3. Daly B.B. *Woods Practical Guide to Fan Engineering*, 1978 edition. Woods of Colchester Ltd.
1.4. *CIBSE Guide*, Vol. B. *Installation and Equipment Data*, 1986. CIBSE, Delta House, 222 Balham High Road, London SW12 9BS.

1.5. Sharland I. *Woods Practical Guide to Noise Control*, 1979 edition. Woods of Colchester Ltd.

1.6 Bedford T. *The Basic Principles of Ventilation and Heating*, 1978. H.K. Lewis, London.

1.7 Humphreys M.A. *Field studies of thermal comfort compared and applied*. CP 76/75. HMSO, London.

2 Air Conditioning

2.1 Introduction

Air conditioning may be defined as a system or process for controlling the temperature, humidity and cleanliness of air. The term 'humidity' refers to the amount of water vapour which can be 'carried' or 'supported'. It is found that warm air can support more water vapour than cold air and whereas warming air is a simple process, cooling it is more complicated, requiring the use of special-purpose plant. The warming or cooling of air is usually carried out by arranging for it to flow over a collection of tubes, sometimes referred to as a 'nest of tubes', or a heat exchange battery. The area of surface available for any heat exchange may be increased by attaching circular fins to the outside of each tube along its length. The substance producing the heating or cooling is then passed through the inside of the tubes.

2.2 Methods of heat transfer

There are three ways of achieving heat transfer: it may occur as a result of a chemical process, the existence of a temperature gradient, or a change in the state of a substance.

2.2.1 Heat transfer due to a chemical change

Heat transfer occurs when one substance has a great affinity for another. For example, ammonia has a high affinity for water and this means that ammonia gas is readily absorbed into water, giving up its heat of affinity from the chemical reaction. Such reactions are termed 'exothermic', when heat is given out, and 'endothermic', when heat is absorbed.

2.2.2 Heat transfer due to a temperature gradient

This is the form of heat transfer with which the lay person is most familiar. Its magnitude depends on the product of the mass of the substance, its specific heat capacity and the value of the temperature gradient.

2.2.3 Heat transfer due to a change of state

A substance can exist in various forms and generally these are referred to as solid, liquid and vapour phases. When the substance in a heat exchange process does not change phase (that is, solid remains as a solid, liquid remains as a liquid, and vapour remains as a vapour), then energy will be transferred only when a temperature difference occurs. However, when a phase change occurs, energy is transferred at constant temperature. In this case, the amount of energy being exchanged is referred to as either the enthalpy of fusion, when liquid changes to a solid, or the enthalpy of evaporation, when liquid changes to a vapour.

Engineers have found that whenever energy exchanges involve a phase change, smaller heat exchangers are required than if the transfer occurs as a result of the existence of a temperature gradient; consequently, in air conditioning it has become common to use phase change processes for energy exchanges. Different substances change phase under widely differing temperatures and pressures, and the choice of a suitable working substance depends on the conditions needed inside the air conditioning plant. A whole range of substances has been manufactured to serve the needs of the heat transfer industry, and these substances are referred to as refrigerants, the modern versions of which are produced from fluorinated hydrocarbons.

These refrigerants are characterised by rather long names, such as dichlorodifluoromethane and pentachlorofluoromethane. By agreement, each of these fluoro hydrocarbons and many other refrigerants have been classified using the letter 'R' (for refrigerant), followed by a series of numbers, the value and position of which generally serve to denote the chemical make-up of the refrigerant. Details of the rules of the numbering system or code are given in BS4580: 1970. Dichlorodifluoromethane is known as R12 (a popular refrigerant), and pentachlorofluoromethane is known as R111.

2.3 Depletion of the ozone layer

Currently there is considerable concern over the thickness of the ozone layer which is above the poles of the earth. It is believed that these layers are being made thinner by the action of sunlight and free fluorinated hydrocarbon molecules, which may have originated from the spent propellants in aerosol cans and also from the discharge of fluorinated hydrocarbons released from old domestic and industrial refrigeration plants.

The ozone layer serves to prevent most of the ultra-violet light, which is harmful to humans, from reaching the Earth, and if only for this reason, any further depletion of the layer must be avoided. It follows that in the changeover to alternative refrigerants, users of plant containing fluorinated hydrocarbons should never allow such substances to escape, or be discharged freely to the atmosphere. Some companies which manufacture these refrigerants now have a section of their organisation whose function it is to deal with the disposal of these refrigerants.

2.4 Types of refrigerator

There are two main types of refrigerator, both of which use the phase change method of energy exchange, but in each type the change of phase is brought about in a different way. Most refrigerators commonly in use may be classed as (a) the vapour compression type and (b) the absorption type.

2.5 Principle of operation of the vapour compression refrigerator

The operation of this plant may best be understood by first referring to figure 2.1. The cycle can be described as follows:

From 1 to 2

Gaseous refrigerant is drawn into the suction side of the compressor from the low-pressure side of the plant, compressed to a higher pressure with a consequent temperature rise and delivered to the inlet of the condenser, on the high-pressure side.

From 2 to 3

The gas is cooled and condensed, giving off heat to the atmosphere.

From 3 to 4

The relatively high-pressure liquid is passed through the pressure-reducing valve, to the low-pressure side of the system. Because the liquid is now at a lower pressure, its boiling point is lowered, with the result that the liquid boils and turns into a vapour. This stage is sometimes referred to as a 'flashing off' process.

From 4 to 1

The action of 'flashing off' is continued in the evaporator, with the necessary enthalpy of evaporation being taken from the circulating fluid, which becomes chilled. The complete process is then repeated.

2.5.1 The flashing off process

A common example of the result of a 'flashing off' process occurs when a cylinder containing liquid butane under pressure is connected to a gas-burning appliance. The effect of opening the valve on the cylinder is to

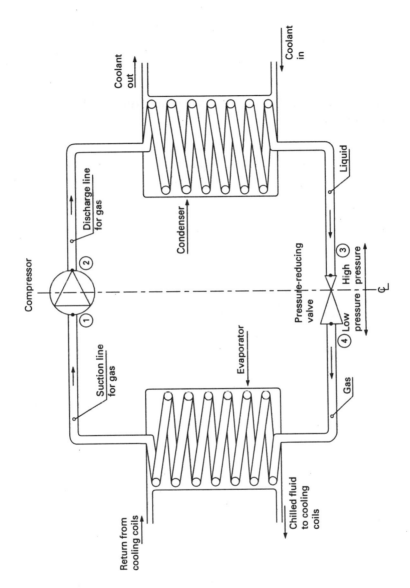

Figure 2.1 Vapour compression refrigerator

allow the pressure inside it to fall slightly and, in order to make good the drop in pressure, some of the liquid butane then turns to gas. In effect the liquid boils, taking the necessary enthalpy of evaporation from the walls of the containing cylinder. After a time, the walls of the cylinder become very cold so that water from the atmosphere condenses on them and, if the volume flow rate of the gas to the appliance is high enough, then this water of condensation may freeze, turning into white frost crystals.

2.6 Principle of operation of the absorption refrigerator using ammonia and water

In this plant, the mechanically driven compressor is replaced with a circulating pump, a heat source and generator and an absorber, see figure 2.2. This is, of course, a more complicated arrangement, but it can be justified because the work done in circulating the refrigerant is much less than that which would be required to drive the compressor of the vapour compression refrigerator previously described. Details of the cycle are as follows:

From 1 to 2

Energy is supplied to the generator (or boiler) by the use of heat, which may be obtained from steam, a flame, an electrical resistance, or solar energy; this separates the ammonia from the solution of water and ammonia, sometimes referred to as an 'ammoniacal solution'.

From 2 to 3

The ammonia vapour is now cooled and condensed to a liquid, giving up its enthalpy of evaporation to the condenser cooling water.

From 3 to 4

The liquid ammonia is then 'flashed off' through the pressure-regulating valve to a lower pressure, which lowers its boiling point. Hence the low-pressure liquid boils, taking the heat needed from the surrounding water in the evaporator and chilling it.

From 4 to 5

Note that the pressure in the evaporator is kept low because of the high affinity that exists between gaseous ammonia and water. The ammonia readily goes into solution, giving up its heat of affinity from the chemical reaction. In this case the chemical reaction is termed 'exothermic' (heat is given out), and this takes place in the absorber.

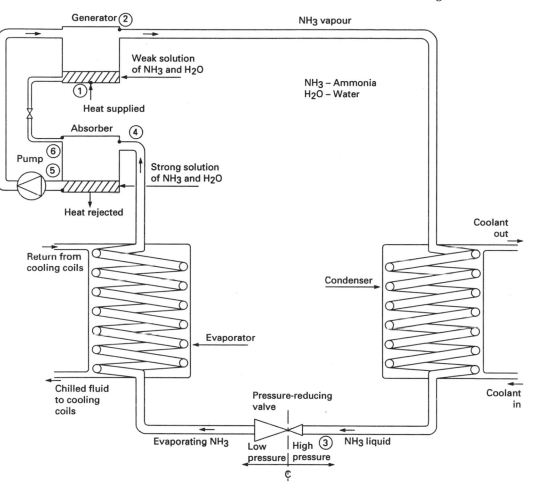

Figure 2.2 Absorption refrigerator

From 5 to 1

The circulating pump draws some of the strong ammoniacal solution from the absorber and delivers it to the generator ready for the ammonia to be boiled off.

From 1 to 6

In order to replenish the liquid in the absorber, the remaining water in the generator is allowed to flow from the relatively high pressure of the generator back to the lower pressure of the absorber via a preset throttling valve. In actual plants some of the heat of the exothermic reaction taking place in the absorber is used to supplement the energy needed in the generator to heat the ammoniacal solution, and this is done using a separate heat exchanger called an economiser (not shown in the figure). Some larger absorption refrigeration plants use a solution of lithium bromide and water instead of ammonia and water.

2.6.1   Use of solar energy as a heat source in the absorption cycle

Devotees of solar energy have been quick to realise that it could be an attractive proposition to use energy from the sun as the heat source for the generator. This is because when the energy output from the sun is at a maximum, the required cooling load is also likely to be a maximum.

When ammonia and water are used, this approach requires a temperature of about 100°C to 110°C at the generator and it follows that if a flat plate solar heat collector is to be used, then the application is best carried out in a tropical climate and the water system should be pressurised to prevent the water from boiling at temperatures just above 100°C. It may be found beneficial to use a flat plate solar heat collector of the Philips type, which consists of evacuated glass tubes in physical contact with the working substance to be heated. The evacuated tubes reduce convection losses within the collector, making it thermally more efficient so that the overall energy loss is much less. Alternatively, a concentrating collector could be used. For those who are interested in the use of solar energy applied to absorption refrigeration machines in the tropics, many papers have been written on the subject, see, for example, [2.1].

2.7   Use of a direct expansion coil

It is common practice to arrange for the evaporator coil to consist of a nest of finned tubes actually fitted into the air conditioning duct. Such an arrangement is referred to as a direct expansion coil, DX for short; its inclusion obviates the need to provide a separate heat exchange battery and circuit for the chilled fluid, which is shown leaving the evaporator and returning to it in figure 2.1. Figure 2.3 shows another variation of plant which would be suitable for a cold room, used for storing foodstuffs. The heat exchanger tubes would in this case have a motor-driven fan mounted behind them, which would serve to circulate the air to be chilled across the tubes and back into the space to be refrigerated.

Mention must be made of the cooling circuit associated with the condenser in figure 2.1. It is common to use cooling water obtained initially from the town's water main and, in order to recycle the water, it too must be cooled, before being reused in the condenser. This process is frequently carried out in cooling towers which, if not maintained and operated properly, can become a source of the *Legionella pneumophilia* bacterium. Readers will know that legionnaires' disease can produce fatal results among operatives and occupants of the building as well as the general public (see section 2.19).

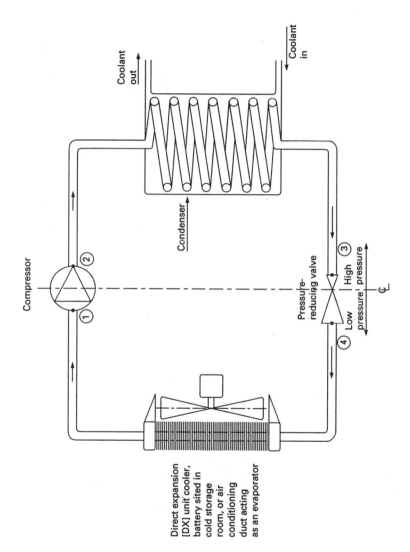

Figure 2.3 Vapour compression refrigerator with DX coil

2.8 Use of water as a refrigerant

Readers who are enthusiasts of paradox will be interested to note that the use of steam to bring about the production of a cool environment is currently gaining ground in the technologies associated with drying food and chemicals (freeze drying). Figure 2.4 shows the layout of the basic system. The water to be chilled at, say, 27°C is sprayed into the evaporator which is kept at a pressure of, say, 0.03 bar, at which pressure the temperature of saturation (boiling point) of the water is 24.1°C. Under these conditions some of the water boils and turns into vapour, taking the enthalpy of evaporation needed from the vessel and its contents, and chilling them in the process. The evaporator is kept at the pressure of 0.03 bar by means of a steam ejector. This device makes use of the property of a high-velocity jet of fluid to entrain with it the surrounding relatively stationary fluid. In this case the steam is passed through a nozzle, which produces a high-velocity jet, entraining the water vapour that leaves the evaporator. The mixture then passes to the condenser via a diffuser, the purpose of which is to reduce the velocity and lessen the possibility of erosion occurring in the condenser.

2.9 Performance of refrigerators

When the refrigeration cycle is used to remove energy from a cold room, keeping the contents at a temperature less than that of the surroundings, the amount of heat extracted is important to the operator; in this case the ratio of heat extracted/work done by the compressor is called the coefficient of performance of the refrigerator (COP_r). Typical values of COP_r range from 1.5 to 4.0 depending on the type of installation.

Examination of the refrigeration cycle will show that the net effect of the operation of the plant is (a) to remove energy from the evaporator, by means of the refrigerant, and (b) to do work on the refrigerant as it passes through the compressor. Both these components of energy are then passed to the condenser and subsequently dissipated in the atmosphere. An alternative method of using the refrigeration cycle is to arrange for the evaporating refrigerant to take energy from a coil which is fitted into, say, the warm air extract duct from a kitchen/dining room complex (KDR). By adding the work of compression to the refrigerant (as before), it is then found that the sum of the two energy components, when passed to the condenser, reaches a temperature which is high enough to provide the input to the coil of a hot water cylinder used for domestic hot water. In this case the amount of heat passed to the coil of the hot water cylinder is important. The plant is said to be operating as a 'heat pump' and the ratio of heat rejected/work done in the compressor is referred to as the coefficient of performance of a heat pump (COP_h).

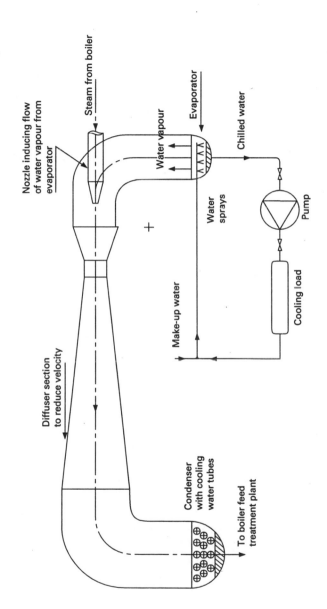

Figure 2.4 Steam jet refrigerator

Clearly, other scenarios are possible and the refrigerant in the evaporator coil could take energy from:

(1) a canal or river;
(2) the atmosphere;
(3) the earth;
(4) the fluid in a series of flat plate solar heat collectors;
(5) any warm outpourings from a kitchen or void in a false ceiling;
(6) any industrial process, be it a foundry or a chemical plant.

After the work of compression has been added to the refrigerant, it could then give energy to:

(a) the cold air entering a warm air heating plant;
(b) the cool return water entering the boiler of a space heating plant.

Past experience with heat pumps has shown that whenever energy is withdrawn from coils of tubing buried in the ground, the route taken by the piping becomes visible after a few months of operation. In winter, the energy taken from the moist soil surrounding the pipes results in the water freezing and expanding to produce 'heave' on the surface. In summer, there is sufficient energy flowing from the surrounding soil to prevent the freezing of the water adjacent to the pipes, and consequently signs of the previous 'heave' disappear. For those who value a well-kept lawn, heat pump evaporator pipes should *not* be buried under greensward. From this point of view it is better to use a battery of finned tubes sited in the open air where it has been found possible (in the south of England), even in the depths of winter, to obtain energy from the surrounding air for use in the home.

It must be mentioned that the flow temperature generally obtainable by means of a heat pump for space heating using radiators (about 50°C) is lower than the normal design temperature of, say, 80°C obtainable from a conventional energy converter or boiler, so that for the heat pump installation it will generally be found necessary to allow for a greater surface area of radiators than would normally be the case.

Over the last 50 years engineers have been fascinated by the possibilities of using the basic refrigerating cycle as a heat pump, but have been disappointed by the lack of cost-effectiveness of their schemes. The author is of the opinion that two interrelated factors will ensure that the heat pump has a viable future. Firstly, the relative cost of energy is gradually rising, and secondly, there is a consequent need for users to be more energy conscious. Both these factors will ensure that clients will be prepared to provide more initial capital, to make heat husbandry a much more common exercise than it has been over the last half century.

2.10   Psychrometric chart and air conditioning

Engineers use psychrometric charts to plan and plot the thermodynamic paths of air through installations which are either existing or proposed. In this way, with the aid of data from the manufacturers of the equipment, a suitable air conditioning system may be prescribed for any given set of conditions.

2.10.1   *Example of the use of the psychrometric chart*

Referring to figure 2.5, air at an outside temperature of, say, 4°C, relative humidity (RH) 80% may be preheated to, say, 18°C, RH 30%, as indicated on the figure of the psychrometric chart by line AB. The air at point B is much drier than it was at A, simply because at the higher temperature it can 'carry' more moisture. For comfort conditions, the air should be delivered at, say, 20°C and 50% RH, and one way of achieving this is to spray warmed or chilled water (depending on the season) into the air, when it will follow the path BC. Final heating will then follow the path CD, producing air at the desired 20°C and 50% RH.

2.11   Reasons for controlling the humidity of air

(1) Dry air encourages the build-up of static electricity so that if, as in a hospital environment, combustible gases are present and sparking occurs, an explosion may result. In modern offices and in computer rooms, the presence of static electricity is unwanted because of the possible destruction of electronic records and equipment; the introduction of air at 50% RH makes such happenings less likely.
(2) Readers will know from personal experience that the mucus membranes present in the nose, throat and lungs are irritated by an excessively dry atmosphere and that high humidity reduces the rate of evaporation of water from the surface of the skin, causing excessive perspiration and so affecting thermal comfort.
(3) It has been found that the life span of the bacteria pneumococcus is less than 10 minutes when the RH is 50%, but of the order of 2 hours when the RH is greater than or less than 50%, see [2.2].
(4) The storage, shelf life and handling of wood, paper, fabric and yarns are greatly affected by humidity.

2.12   Components of a centralised air conditioning plant for a large open plan office

The general arrangements of the components required are as shown in figure 2.6. The calculations and graphical solutions of the designer must take into account the amount of sensible heat gain and evaporative moisture

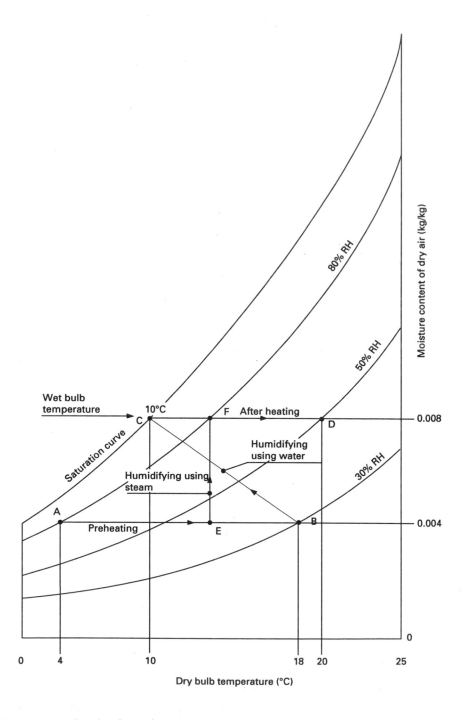

Figure 2.5   Sketch of psychrometric chart

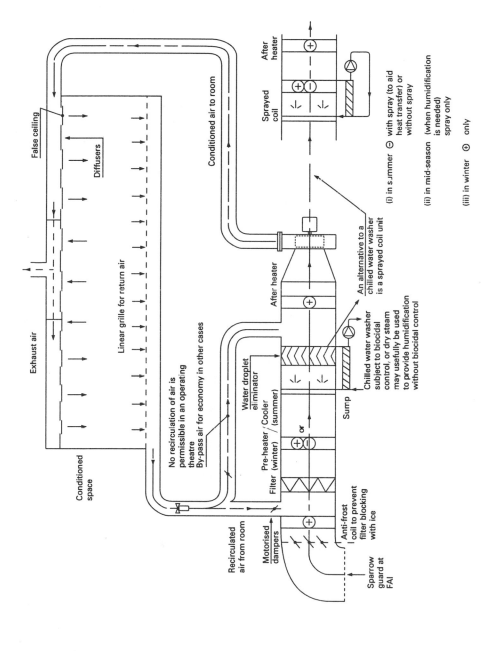

Figure 2.6 Sketch of central refrigeration plant

gain in the spaces being conditioned. When, as is usually the case, the moisture content has to be altered, this can be arranged by using chilled or warmed water sprays, as described in subsection 2.10.1, with or without a direct expansion coil (DX).

Note that the water used in the sprays should either be 'once through', or be subject to biocidal control to prevent the multiplication of *Legionella* bacteria. To facilitate this in health establishments, it is now likely that dry steam injection will be used instead of a water spray system. For installations that do not have steam facilities on site, small steam genera- tors are available, which work on the electrode boiler principle, as explained in Chapter 3. The path of the humidifying process when dry steam is used is indicated by line EF on figure 2.5. Note that less energy is required from both the pre-heater and the after-heater in this case.

The calculations for the variation in thermal loading of the enclosure can be long and rather tedious when done manually, but in recent years, with the advent of computers, the work has been greatly facilitated. The centralised plant illustrated is ideally suited for application to an enclosure such as an open plan office area, where the thermal loading, while varying diurnally, does not vary widely throughout the enclosure at any one time. It can be said that a centralised plant generally provides a constant-volume flow rate of air, but varies the air temperature. Figure 2.7 shows a suggested layout.

Stale exhaust air is drawn into a purpose-made ductwork system and, in buildings which have corridors, these systems have sometimes been used as ready-made return routes for exhaust air. In the event of a fire occurring it is obvious that these corridors will almost immediately be filled with smoke, thus making their use as escape routes impossible. Clearly, exhaust ducting should be provided.

2.13   Use of fan coil units with a centralised system

It is often the case that one floor of an office block is partitioned so that there is a large open plan general office together with satellite offices spaced around, say, two sides of each floor, see figure 2.8. These offices, particularly if they face the sun, will have disproportionately large and different insolation loads from each other and also from the general office, so that it becomes impracticable for a central plant to deal with the different loadings occurring at any one time in all parts of the building. This difficulty can be overcome by the use of fan coil units, illustrated in figure 2.9, which each deal separately with the thermal loading pertaining to the space in which they are fitted.

The fan coil units can be mounted in a false ceiling in each room, or in a false ceiling in an adjacent corridor, or on the wall in each room. Current versions can have pneumatic or direct digital electronic controls which can be removed from each unit and placed in a centralised posi-

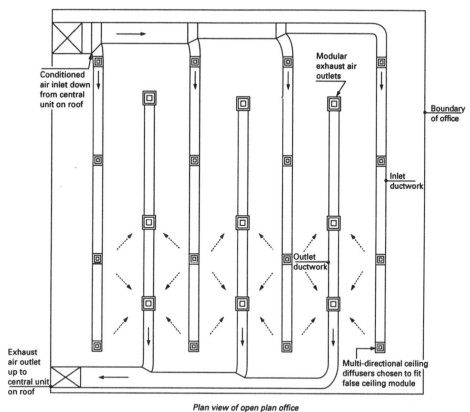

Figure 2.7 Open plan office with air conditioning from a central unit

tion if required, or be linked to a management control system. The units incorporate filters which may be selected to suit the environmental requirements of each room. The conditioned air supplied to each fan coil unit comes from the central plant and is carried in a duct which may conveniently be sited in a false ceiling. The fans are often of the cross-flow type, sometimes referred to as 'drum' or 'squirrel cage' fans, and extend almost the whole length of the outlet grille, so ensuring an even delivery of air to the room. It is important to provide a condensate tray under each cooling coil which should, if possible, drain to waste under the action of gravitational forces, though pumped systems are available.

Depending on the season, the coil of each fan coil unit is supplied with chilled or warmed water and the operation of the fan is linked to the temperature conditions in the room served by the unit. In this way the central plant is able to function under relatively constant conditions of thermal equilibrium, while in practice the enclosure presents a varying thermal load.

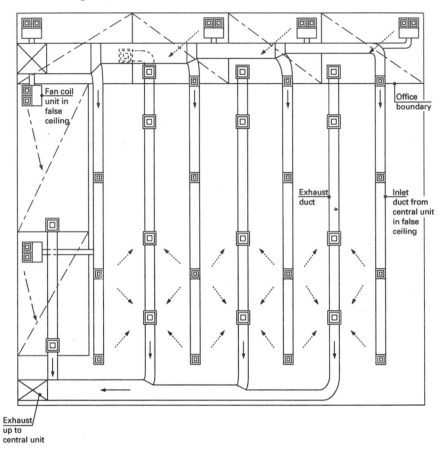

Figure 2.8 Plan of general office, having peripheral offices each served by a fan coil unit dealing with diverse thermal loading

Whenever fan coil units are used, it is essential to make sure that the 'throw' of the air into the room and its volume flow rate are sufficient to deal adequately with the physical size of the room, and that the positioning of the exhaust outlets avoids, as far as possible, short-circuiting of the air.

2.14   Use of a 'split system' to air-condition rooms in a building

When it is necessary to air-condition just a few rooms in a building, it is possible to do this using what is referred to as a 'split' system. It has been found convenient and advantageous to house the basic refrigerator cycle plant in two cabinets, each of which complement one another. When the refrigeration cycle is acting as a cooling unit to the enclosure, the arrangement shown in figure 2.10(a) is used. When the refrigeration

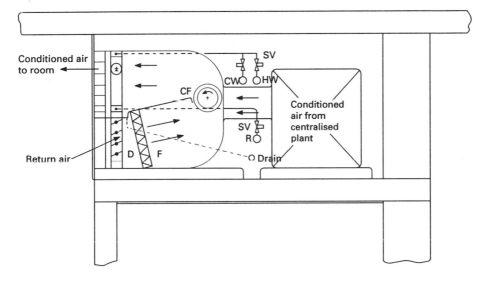

CF Cross-flow fan
R   Common return
HW Hot water
CW Chilled water
F   Filter
D   Dampers
SV Solenoid valve

Figure 2.9   Fan coil unit in a false ceiling

cycle is acting as a heating unit to the enclosure, the arrangement shown in figure 2.10(b) is used, and it is then said to be operating as a heat pump. When wall space is at a premium, the fan coil unit can be placed at high level, or in a false ceiling if available. Figures 2.10(a) and (b) show the two units in juxtaposition, but it is possible to extend the pipework between them, making the two units up to 100 metres apart. The advantages of this system are that: (a) no ductwork is required, and (b) it is obviously easier to run and conceal pipework than ductwork. However, the pipework contains refrigerant, which must be charged when the plant is commissioned.

2.15   Use of the single duct induction system of air conditioning

The application of air conditioning to larger and taller buildings has resulted in the need to deliver relatively greater volume flow rates of air through ductwork and, in order to keep the physical size of the ducts within acceptable limits, it has been found necessary to increase the velocity and pressure of the air being handled by a factor of 2 or more. Also, in order to cut down on the area occupied by the ductwork, it becomes attractive to provide a single air duct, together with cooling/heating water

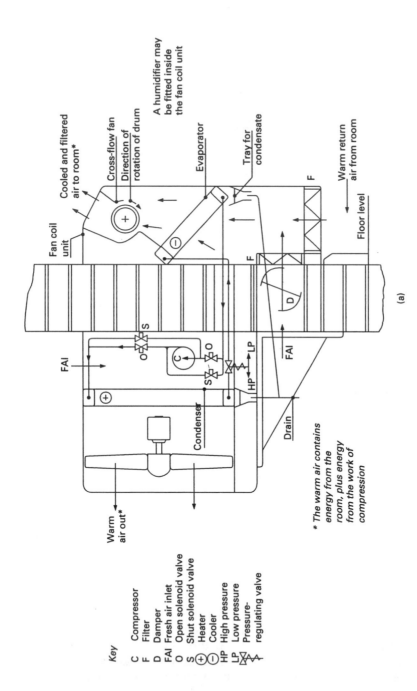

Key

C   Compressor
F   Filter
D   Damper
FAI Fresh air inlet
O   Open solenoid valve
S   Shut solenoid valve
⊕ⓘ Heater Cooler
HP High pressure
LP Low pressure
⧓ Pressure-
regulating valve

Figure 2.10 (a) Heat pump details – cooling mode.

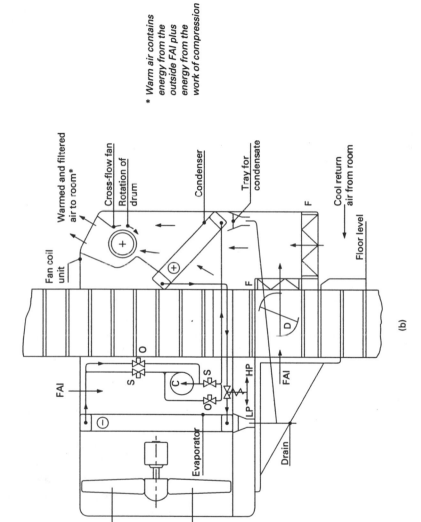

* Warm air contains
energy from the
outside FAI plus
energy from the
work of compression

Warmed and filtered
air to room*

Cross-flow fan
Rotation of
drum

Condenser

Tray for
condensate

Cool return
air from room

Floor level

F

F

Fan coil
unit

FAI

O

S

S

C

O

HP

LP

FAI

D

Drain

Cooled
air out

Evaporator

I

(b)

Figure 2.10 (b) Heat pump details – heating mode

Table 2.1 Typical air velocities in ducts

Type of building	Average velocity (m/s)			
	Main ducts	Branch ducts	Supply grilles	Type of system
Private dwellings	4	3	2	
Public buildings (quiet)	6	5	4	
Public buildings (noisy)	9	7	6	Low velocity
Industrial buildings	11	8	7	
Public buildings	18	15	5	High velocity

Compiled from *Woods Practical Guide to Fan Engineering*, 2nd impression, page 84, table 6.2. © Woods of Colchester Ltd.

flow and return pipes; such a system is referred to as an air/water system. This arrangement is used in the single duct induction system of air conditioning.

Note that a direct result of increasing the velocity of the air in the ducts is to increase the noise produced by the movement of the air through the system. Table 2.1 gives an indication of the air velocities in different sections of the ductwork in different types of building, when using low-velocity and high-velocity air distribution systems. Noise from a fan can be reduced by placing the fan between two purpose-made silencers, consisting of cylindrical sound-absorbing sections of duct, lined with polyurethane foam or fibre glass at least 100 mm thick and contained in an annular cylinder of perforated metal.

The holes in the metal lining, together with the air immediately behind each hole, produce a sound-attenuating effect, sometimes referred to as the 'Helmholtz effect'. The air just behind and around each perforation behaves as a tiny spring, and the amplitude of oscillation of the spring is produced and maintained by alternate pressurisation and rarefaction of the air when a sound wave is directed towards each hole. The vibration of the small air spring produces a minute frictional drag effect and whenever a force is moved through a distance, energy is expended. This energy is absorbed from the sound wave and the sound is reduced. Information on this topic may be found in [2.3].

Further sound attenuation may be achieved where main ducts change to branch ducts, by arranging for the main duct to open out into a chamber lined with sound-absorbing material and referred to as a plenum chamber, octopus box, or rectangular attenuator. Note that if the outlets from these units are offset from the inlet, this method of construction produces further sound attenuation.

In the single duct system referred to at the head of this section, the ductwork after the rectangular attenuator is known as the primary air supply ductwork. Each primary duct then discharges into a terminal rectangular attenuator, or silencer, which is fitted inside the room air induction unit, as shown in figure 2.11. After the air leaves the silencer, it is then discharged through nozzles inclined in an upward direction through the unit and these nozzles increase the velocity of the air at the expense of the pressure in the silencer. The emerging high-velocity air streams are then used to entrain the air surrounding them, so inducing a flow of air from the room across the heater (or cooler) battery, where it is finally delivered through a grille having a relatively large free discharge area, to minimise the noise which may be generated at the exit from the unit.

The heat exchanger may be controlled by a thermostatic valve. The unit does not require a local fan and, if the noise generated by the high-velocity distribution system has been suitably attenuated and the induction unit has been chosen for its low sound emission, then the overall effect for the occupants of the room will be acceptable.

The system is eminently suited to air conditioning the rooms occupying the periphery of buildings and will successfully deal with a depth of up to 4 metres from the external walls. For deeper buildings, the extra ventilation and cooling needed have to be dealt with using a different method.

2.16 Dual duct system of air conditioning

In this system, hot and cold air are circulated in primary ducts and taken to individual blending units which are usually situated in wall cabinets, see figure 2.12. There are many different types of blender with different methods of control, mainly pneumatically operated. The CIBSE Guide 1986, section B3-28, details some of the control functions and associated problems dealing with the effect of variation of pressure in the ducts on the output of the blender.

Because air alone is used as the working substance (an all-air system), treble the volume flow rate of air is required to carry the energy that is needed, compared with the single duct system using air and water. It follows that the extra air supplied must also be exhausted from the building and, because it is now considered to be cost-effective to recover the energy given off from the lighting fittings, office machinery and occupants, this exhaust air may conveniently be used as the vehicle for this purpose. In buildings with a storey height tall enough to accommodate a false ceiling, the space above the ceiling is used to collect the warm exhaust air prior to returning it to the main plant and the heat recovery apparatus.

Under these conditions, the dual duct system provides air conditioning and ventilation for any depth of room, together with a useful element of energy recovery. Considerable operating difficulties have occurred in the past when the control dampers in the blender and the temperature sensor were connected by means of small-diameter plastic tubes, which at some

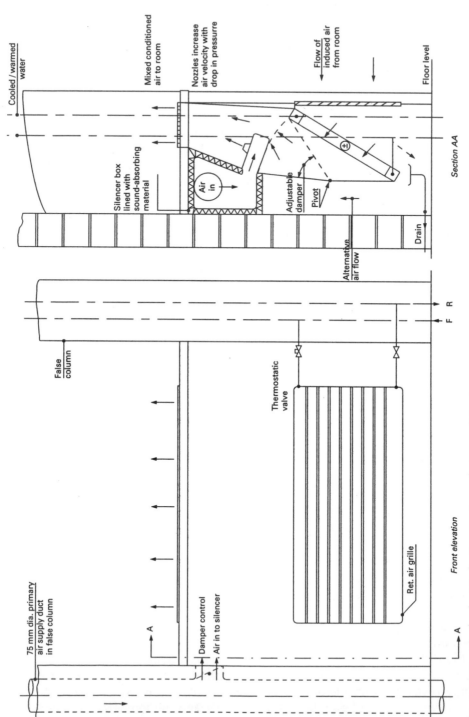

Figure 2.11 Single duct induction system of air conditioning

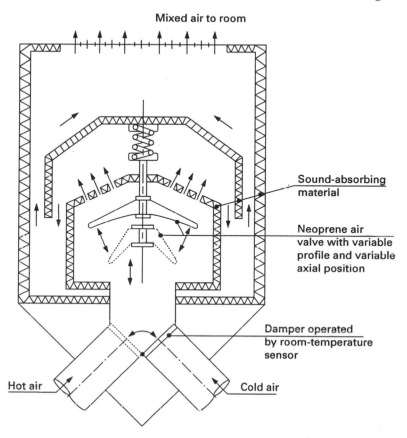

Mixed air to room

Sound-absorbing
material

Neoprene air
valve with variable
profile and variable
axial position

Damper operated
by room-temperature
sensor

Hot air   Cold air

Figure 2.12 Dual duct constant-volume flow rate blender, using a mechanical
method to deal with variations in duct static pressures

point in their length were exposed to the ultra-violet rays of the sun.
These rays degraded the plastic, producing random pressure losses which
made the operation of the control mechanism unreliable.

2.17 Use of constant-volume, fan-assisted terminal units in high-velocity systems

These acoustically lined units take a supply of cool, conditioned, high-
velocity air from a central plant (primary air) and mix it with warm air,
usually taken from the ceiling void (secondary air), to produce a constant-volume flow rate, which is discharged to the space being served,
see figure 2.13. If additional heat is required, this can be provided from
within the unit, by using hot water heating coils and, conversely, if heat
needs to be removed, then cooling coils may be used. The amount of
primary air needed is controlled automatically by an electronic controller

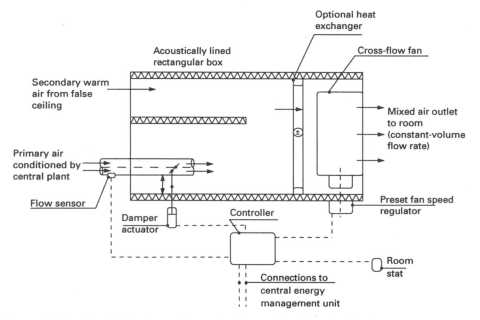

Figure 2.13   Constant-volume flow rate, fan-assisted terminal unit

which can operate an electric or pneumatically controlled damper fitted to the primary air duct. The option is open for each controller to be part of a central energy management system.

2.18   Use of variable air volume (VAV) terminal units and ceiling diffuser strips in high-velocity systems

In this case a different kind of terminal unit is often used which gives the air conditioning engineer one more dimension of control over the use of energy. The enclosure is kept at the desired condition by varying not only the temperature of the discharged air, but also its volume flow rate, see figure 2.14. Control of the unit may be accomplished using a central management system. Because the unit is used in a high-velocity air system, it is lined acoustically.

The unit operates in the following way:

(1) The flow sensor generates a signal, the magnitude of which is compared by the controller with a set point signal representing the desired volume flow rate, which can be variable between preset maximum and minimum values.
(2) If a difference is detected between the signals, the controller generates an output proportional to the difference, and this output signal is then used to change the position of the damper in the primary air supply duct, so achieving the desired volume flow rate to the zone.

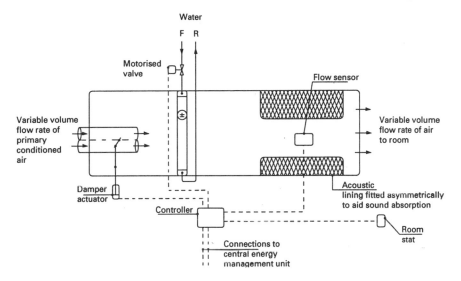

Figure 2.14 Variable-volume flow rate terminal unit

(3) The room thermostat may call for some reheat and this can be accomplished by arranging for the controller to produce a proportional signal, which will operate a motorised valve on the reheat coil.

The unit thus delivers a selected volume flow rate to the zone, even though the static pressure of the primary conditioned air is varying. It also produces a variation in energy input, depending on the local need.

An alternative type of VAV unit can be found in buildings having false ceilings which include linear (or strip) air diffusers. By fitting a room temperature sensor, connected via relays and motors to a movable plate contained in the diffuser, the area of diffuser open to the room may be changed and the volume flow rate altered accordingly. When a central air conditioning plant is supplying a number of zones each requiring different thermal inputs, it follows that, during the 24 hour cycle, local variations will affect the volume flow rate and pressure in the main supply duct. To overcome this, a pressure sensor in the main duct can be used to change the speed of the main fan (or fans), so saving a considerable amount of electrical and thermal energy.

For those who wish to investigate the problems associated with the design and operation of air conditioning installations and equipment, reference to the CIBSE Guide 1986, section B3, will help them realise the size of the task involved, together with some of the pitfalls inherent in the application of air conditioning techniques to the built environment. Readers should also consult *Heating and Air Conditioning of Buildings*, 7th edition, by P.L. Martin and D.R. Oughton, which, traditionally and deservedly, has been the Bible for service engineers since 1936 when it

was first published and when the first authors were Oscar Faber and
J.R. Kell.

2.19   Effects of *Legionella* bacteria on the built environment

The second report of the Committee of Enquiry into the outbreak of Legionnaires' disease in Stafford in April 1985 was presented to Parliament
in 1987. The report specifically excluded domestic premises from its recommendations because of the lack of evidence of any *Legionella* outbreaks therein. The report does suggest, however, that with the new and
improved method of identifying Legionnaires' disease using monoclonal
antibodies, some association may subsequently be demonstrated.

The types of buildings and systems which are categorised as being at
risk are 'older' hotels and health care premises. But following recent
outbreaks, it would be an irresponsible building manager who insisted
that 'nothing need be done' to make a building safe for both the occupiers and the general public. The main types of system at risk, taken in
order, are as follows:

(1) conventional calorifiers, poorly maintained, connected in parallel and
     not having temperature-averaging pumped circulatory systems;
(2) showers that are used infrequently;
(3) newly tested pipework and systems that have contained water for long
     periods before being commissioned;
(4) badly sited 'fresh' air intakes, possibly drawing in aerosols from an
     adjacent cooling tower;
(5) low domestic hot water outlet temperatures (less than 40°C).

The percentages of *Legionella* detected in various locations in enclosures [2.4] are: calorifiers 38; showers 28; other hot sources 7.4; cooling
towers 11.3; and all other sources 5.3.

The Stafford report [2.5] draws attention to several Codes of Practice
dealing with *Legionella* in the built environment. Other useful information may be obtained from CIBSE publications TM13 (1991): 'Minimising the risk of Legionnaires' disease' and GN3 (1993): 'Legionellosis
– an interpretation of the requirements of the HSC's Approved Code
of Practice'.

By following up these references, building managers may choose to
follow whichever Code of Practice is deemed to suit their establishment
best and may then start to implement the relevant recommendations. The
author proposes to deal with HMSO Code of Practice entitled *The Control of Legionella in Health Care Premises* (January 1990), and will para-
phrase the major recommendations in order that the reader may become
acquainted with the kind of problems that will be faced by surveyors,
designers of services, builders, commissioning engineers and maintenance

personnel, when they are dealing with the likely effects of the multiplication of the *Legionella* bacteria in the built environment.

High concentrations of the bacteria can produce the disease among the
following persons:

(1) those members of the population aged 50 years and over;
(2) male members of the population (who are three times more susceptible than females);
(3) those having existing lung infections, possibly brought on by excessive
smoking;
(4) those whose natural defences have been weakened because of dis-
eases;
(5) those whose natural defence mechanisms have been weakened by
taking immuno suppressant drugs;
(6) those who work in or around an air-conditioned enclosure, where
the services are not designed and maintained to control *Legionella*
bacteria.

The recommendations contained in the code of practice are aimed at
managers of health care premises, but clearly much of what is suggested
is highly relevant to the design and operation of existing and proposed
buildings outside this sector.

2.19.1   Epidemiology of the bacteria

The bacteria have been found in hot water systems, particularly in showers
and calorifiers and in cooling water systems serving air conditioning plant;
also in whirlpool spas where the water is recirculated and in wall-mounted
humidifiers, which use tap water. The bacteria are transmitted by small
water droplets being carried in the air (aerosols), which can be lodged in
lung tissue by inhalation. Tests carried out in the laboratory indicate that
the multiplication rate of the bacteria is greatest at the temperature of
human blood, that is 37°C. Multiplication stops at 46°C. The bacteria
will survive for only minutes at 60°C and are killed instantaneously at
70°C. Below temperatures of 20°C, the multiplication rate is insignificant,
but the bacteria lie dormant. The risk of infection is influenced by:

(a) the bacteria count;
(b) the rate of respiration; and
(c) the time of exposure.

Hence building managers must aim to control those factors that are within
their purview.

For example, the bacteria count can be controlled by chlorination of
all non-potable supplies, by the prevention of the formation of stagnant

water in storage tanks and calorifiers and any redundant pipework, by temperature control, and by water treatment. Control of exposure time of a person to the bacteria is possible to a very great extent. The normal exposure to aerosols when taking a shower is about 5 minutes; but for the air intake to a building drawing in contaminated aerosols from a poorly maintained cooling tower, the exposure time for the occupants can be as long as their period of occupancy. It is obvious that careful siting, or resiting of the air intake relative to any cooling tower outlet, is of paramount importance.

2.19.2 Avoiding build-up of Legionella *bacteria at the design stage*

Sizing of cold water storage tanks should be such that no more than 24 hours of storage capacity is provided. The tanks should preferably be connected in series, in such a way that each tank can be isolated for cleaning purposes, without interrupting the supply to the building. Figure 2.15 shows the valving arrangements. The drain valves and vents are used to empty the pipes when they are not in use. The internal connections to each tank should be designed so that the stratification of the water within the tank will be minimised. If storage capacity is to be provided for future extensions, it should not be currently operable.

2.19.3 Avoiding build-up of Legionella *bacteria in an existing building*

It is likely that the tanks will be connected in parallel and, when this is the case, a check should be made on the rate of fall in level of the free surface in the tanks taken over a typical diurnal cycle. If too much capacity has been provided, it should then be possible to lower the level of the free surface in each tank, so cutting down the storage capacity. Figure 2.16 illustrates the suggested method of connection. All storage tanks, especially those containing potable water, must be insulated and have a fixed cover. In addition the overflows must be screened and the covers should be finished with a reflective surface which will help to keep the tanks cool in summer.

2.19.4 Treatment of hot and cold distribution pipes and 'dead legs'

All distribution pipes should be insulated. This will ensure that, in summer, within 2 minutes of opening a tap, the temperature of the water will be less than 20°C. If cold distribution pipes pass close to hot water pipes, then suitable insulation or rerouting is desirable. In the past, hot water dead legs of a length proportional to their diameter have been allowed. However, in order to avoid the build-up of *Legionella* bacteria in those sections of pipe that constitute a dead leg, it is now recommended that

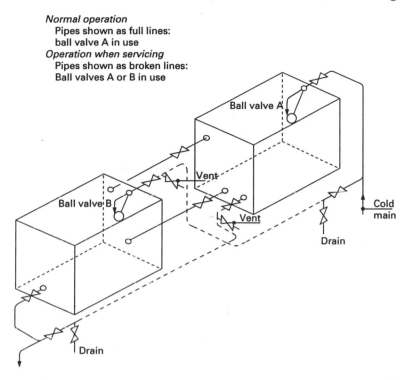

Figure 2.15 Series connection of cold water storage (CWS) tanks to increase the volume flow rate through each tank

for health premises, a flow and return pipe should be provided up to 300 mm from the discharge point, see figure 2.17. Designers and others should also note the points which follow:

(1) The return pipe can be smaller than the flow pipe, as it only needs to be large enough to supply a mass flow rate of water sufficient to make good the thermal emission from the two insulated pipes.
(2) The water will arrive at the discharge point from two directions through two pipes, so that the mass flow rate available for a given head will be increased.
(3) The pipes should be laid to a gradient of about 1 in 100, rising in the direction of flow and falling in the return direction. The gradient will assist air venting and subsequently be useful when draining and disinfecting processes are carried out. Because the domestic hot water is 'fresh' (unlike the hot water in a space heating system), it will contain its full complement of dissolved oxygen and other gases, which will be continuously liberated in the pipes. These gases will collect at any high point and progressively block the flow of water. The use of automatic air vents will alleviate this effect, if they are fitted at the high points of the system.

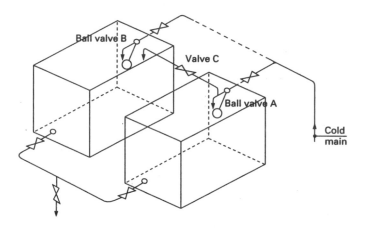

Normal operation
  Pipes shown as full lines:
  valves A and C open, valve B closed
Operation when servicing
  Pipes shown as broken lines:
  valves A and C closed, valve B open

Ball valve B

Valve C

Ball valve A

Cold
main

Figure 2.16 Suggested parallel connections to CWS tanks

(4) In some cases it may not be possible to arrange for thermosiphonic flow of the hot water to occur and, in these cases, a small circulator fitted to the flow and return pipes will overcome the problem.

2.19.5 Installation and treatment of calorifiers

Calorifiers, as with cold water storage tanks, should be connected and operated in series, to ensure a greater mass flow rate of water through each vessel. The following points should be noted:

(1) All calorifiers should have adequate access for cleaning purposes.
(2) If new vertical calorifiers are to be fitted, then the use of those having ends that are concave should be avoided, so that sediment will not build up in the corners of the lower end, see figure 2.17(a).
(3) Drain valves should be installed at the lowest point of the vessel and be of such a size that the discharge rate is sufficient to draw out any sediment that has settled in the bottom of the calorifier. A good guide is to size the drain valve such that the vessel may be emptied over a period of between 0.5 to 1 hour. In health premises, it is recommended that the removal of accumulated sludge should be carried out quarterly, by opening and closing the drain valve during the process, so that a non-uniform flow regime is set up inside the vessel, to help in dislodging any remaining pockets of sludge.

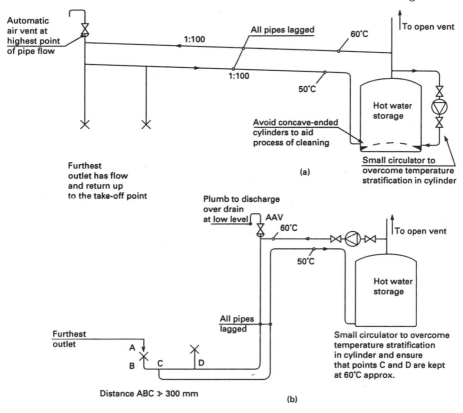

Figure 2.17 (a) Pipework suitable for thermosiphonic flow to furthest outlet.
(b) Pipework requiring a circulator

2.19.6 *Effect of vertical temperature gradients in a calorifier*

Vertical calorifiers have a vertical temperature gradient of about 40°C between a maximum temperature of, say, 60°C and a minimum temperature of 20°C, and it follows that, about half way up the vessel, the temperature can be expected to be of the order of 40°C.

From the information already given, it will be realised that the lower half of the calorifier presents a suitable environment for the multiplication of *Legionella* bacteria. By reducing the vertical temperature gradient, this situation can be avoided and this is readily accomplished by fitting a small circulating pump, drawing water from near the top of the vessel and returning it to a position near to the bottom. When this is done, the temperature gradient is reduced to about 3° or 4°C; also the volume of hot water stored is increased and the mean temperature at which the vessel is operating is raised to approximately 55°C, which temperature is not conducive to the multiplication of *Legionella* bacteria. The

Code of Practice suggests that the mean circulating temperature should not be less than 50°C.

2.20 Installation and operation of pumps fitted to rising mains containing hot water

Sometimes these pumps are fitted with a non-return valve to discourage the movement of water through the pump in the reverse direction when it is stopped. In general, because of the inherent unreliability of these valves, the use of a non-return valve is not sufficient to guarantee that the fluid will not flow in the wrong direction. Note that Water Authorities usually fit two non-return valves in series when they are intending to prevent the occurrence of reverse flow.

In some organisations, it is common practice to fit duplicate pumps with change-over valves, so that in the event of either electrical or mechanical breakdown, the standby pump may be brought into service and the duty pump be repaired, without interruption to the domestic hot water supply. For most installations, however, this extra expense is not justified and it is usually considered sufficient to have a spare motor and pump fixed on an adjacent wall so that a relatively swift substitution can be made. Commonly the duplicate pump is often fitted on a by-pass and, if the pump is not used regularly, the water in the by-pass will become stagnant at a temperature that is favourable to the multiplication of the *Legionella* bacteria.

If it is considered essential to have an uninterrupted flow of domestic hot water available at all times and duplicate pumps are deemed to be justified, then it is sensible to arrange for a double pole time switch to deactivate the duty pump and activate the standby pump, possibly on a bi-weekly basis. The author suggests, however, that it might be better in the long term to operate the standby pump for, say, one day per half week before returning to the duty pump for the remaining two or three days. In this way, one can try to ensure that both pumps do not wear out at the same time.

2.21 Effects of night shutdown

According to the Code of Practice, the activation of a domestic hot water circulating pump, discharging into a ring main after a 12-hour shutdown, is allowable, provided no water is drawn off until the return temperature from the system reaches 50°C. In addition, the temperature of water issuing from an outlet furthest from the system should be between 50°C and 60°C, within one minute of full flow.

2.22 Precautions to be taken at the building commissioning stage

These will consist of a process of disinfection using chlorine, and details of the methods used are given in clause 2032 of the Code of Practice. Before employing these methods, Contractors are advised to contact the local Water Authorities and the local Council in order that they may become acquainted with the rules and regulations regarding the avoidance of back siphonage into the public water supply and also discharge into foul sewers of treated water.

Those responsible for chlorine dosing should note that chlorine destroys the resin beds of the Zeolite base exchange process used in water softening, so clearly these beds must be isolated before any chlorine dosing is begun.

2.23 Use of evaporative cooling towers with air conditioning plant

There are several variations of this type of cooler and three such variations are shown in figures 2.18(a), (b) and (c). Over the years, these coolers have proved to be the most cost-effective solution to providing a supply of cooling water for use in the condensers of refrigerating plant. The units also take up less space and use less power than other methods of cooling.

The cooling capabilities rely on a change of phase of some of the cooling water from liquid to vapour. This change takes the enthalpy of evaporation from the remaining droplets of water, cooling them in the process. These cool droplets then gravitate to a sump, from which the cooled water supply is then circulated through the condenser of the central refrigerating plant. The loss of water by evaporation is made up from a ball valve in the sump. Note that any organic cells and inorganic dissolved solids present in the make-up water are not themselves subject to the process of evaporation so that, as time passes, the total dissolved solids (TDS) contained in the cooling water increase, as well as the amount of organic cells.

It is thought that living cells, such as algae and amoeba, may provide a source of nutrients for the *Legionella* bacteria. Likewise, the use of rubber, leather, boss white jointing compound, hemp and some mastics are known to be nutrients; it is therefore suggested that designers and building managers should check the lists of preferred materials, as issued by the local Water Authorities for use in building systems, before continuing to use traditional materials on site. It should be noted that ferric oxide sludge is known to be a nutrient 'par excellence' for the *Legionella* bacteria. Hence the removal of this type of sludge is essential.

The nutrient cells referred to above may be killed using antimicrobial chemicals, while the concentration of organic substances (TDS) can be removed by an automatic process of periodic replenishment of the contents

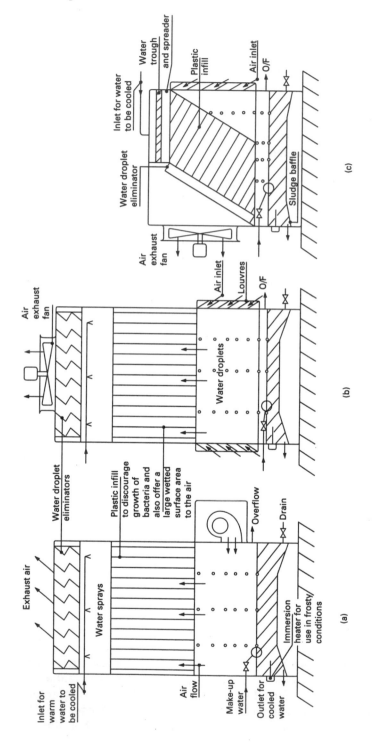

Figure 2.18 (a) Forced draught cooling tower. (b) Induced draught cooling tower. (c) Cross-flow cooling tower

of the sump of the cooling tower using fresh water. This is referred to as a 'bleeding' process. The Code of Practice gives detailed recommendations about the frequency and manner in which such cleaning processes shall be carried out, so that the TDS are never greater than 2000 parts per million.

Readers should know that *Legionella* bacteria have been shown to become resistant to a given biocide over a certain time. However, by changing to a different biocide when this occurs, the bacteria are again killed, until a new resistance is built up, when a return to the original biocide again becomes desirable. This process is termed 'duo chemical' biocide treatment.

This particular problem has been solved by a company [2.6] that currently uses a combination of fine filtering and the continuous application of ultra-violet light, to kill 99.9% of the bacteria (see figure 2.19). The function of the 20 micron filter is to remove dirt particles, dissolved solids and nutrient algae, behind which the *Legionella* bacteria (size 3 microns) may be lurking. Part of the filtered water is then irradiated with ultra-violet light and, over a 24-hour period, the complete contents of the cooling tower sump will have been circulated at least 14 times, making the irradiation virtually continuous. The system also incorporates an automatic backwash for the filter and an automatic 'bleed' period, controlled by sensors that detect the change in electrical conductivity of the water to keep the dissolved solids to a reasonable level.

Other refinements, such as automatic dosing of the water with scale inhibitor, together with electromagnetic scale control, are also offered. A big advantage of the system is that no expensive toxic chemicals are needed, which also means that there can be no problem in disposing of the effluent when draining down.

The outlet from any cooling tower will contain saturated air and fine water droplets (aerosols), which if the cleaning processes mentioned above and detailed fully in the Code of Practice have been complied with, will be harmless to humans. The possibility that some 'drift' from a poorly maintained cooling tower may enter an air intake sited close to the cooling unit could be likely, and it is obviously good practice to ensure that this situation cannot arise by careful placing of the cooling tower outlet and the fresh air inlet. It should be noted that, in the case of some cooling towers it was possible to install water droplet eliminators upside down, so that their performance characteristics were not what the designer intended.

The cooling tower should be placed so that members of the public and users of adjacent buildings are not subjected to the contents of the 'drift' coming from it. However, this suggestion may in many cases prove impossible to implement.

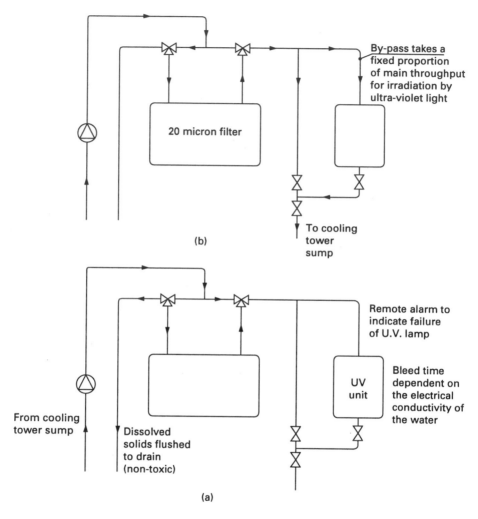

Figure 2.19   System for controlling bacteria using ultra-violet light: (a) back
flushing (bleed) mode; (b) normal operating mode

2.24   Use of other forms of cooling with air conditioning plant

Where new plant is being designed, alternative methods to those of
evaporative cooling may have to be considered.

2.24.1   Replacement option 1

Air-cooled condensing units are available, which take the refrigerant to
be condensed from the central plant and circulate it through a coil over

which air is blown. This is known as the DX option. If the central plant is an existing one, using this method will result in the sacrifice of some refrigerating effect.

2.24.2 Replacement option 2

Water from the condenser of the central refrigeration plant may be circulated through a coil, over which air is blown; these coolers are referred to as 'air blast coolers'.

As no change of phase occurs in either replacement option 1 or 2, both options take up to two to three times more space and use 20–25% more power than a conventional cooling tower. When contemplating carrying out changes to an existing system, there are many factors to consider and clauses 4518–4520 of the Code of Practice will be found to be very helpful in outlining the points that must be considered before any course of action is decided.

References

2.1. Chinnappa J.C.V. The utilisation of solar energy for space cooling, in *Proceedings of the First International Conference on Solar Building Technology*, Vol. 2, 1977. RIBA, London.
2.2. Spirax Sarco Ltd. *The Use of Steam Injection Humidifiers*, October 1989. Technical literature, from Spirax Sarco, Cheltenham.
2.3. Sharland I. *Woods Guide to Noise Control*, 1979 edition. Woods 'Acoustics', a division of Woods of Colchester Ltd.
2.4. Harper D. *Legionnaires' Disease in Buildings*. CIBSE regional meeting at the Moat House, Brentwood, 3 April 1990.
2.5. *The Second Report of the Committee of Enquiry into the Outbreak of Legionnaires' Disease in Stafford (England) in April 1985* (Chairman Sir John Badenoch).
2.6. System Uvex Ltd, Cranbourne Industrial Estate, Potters Bar. Trade literature, 1990.

3 Collection and Storage of Energy

3.1 Geothermal energy

In some parts of the earth's interior, energy is produced from the decay of the radioactive isotopes of uranium, thorium and potassium. It is known that energy at a temperature suitable for space heating is to be found in underground reservoirs of hot water (aquifers) and also in hot dry rocks (HDR).

In 1991, in the UK, the Southampton City Council, in conjunction with the private sector, carried out a district heating project monitored by the Department of Energy (DOE). In the European Union, 'Atlases of Sub Surface Temperatures' and of 'Geothermal Resources' are being compiled. HDR research is currently progressing near Gothenburg in Sweden, Japan, Los Alamos in the USA, and at Rosemanowes in Cornwall, the latter using personnel from the Camborne School of Mines.

In the UK, the research has been investigated, monitored and reported, by the DOE, who have concluded that to make the production of electrical energy viable, the rock temperature should be in excess of 200°C and that this would require boreholes to be drilled to depths of up to 6 km in very hard rocks. It is thought that fissures in the rock can be produced using either explosives or hydraulic fracturing techniques. Cold water would then be pumped continuously into the injection borehole in the expectation that hot water would emerge via the production borehole at the surface. It would then be passed through heat exchangers and returned to the underground reservoir. The energy so obtained may be used directly to supply a district heating scheme, or if it is hot enough, to produce steam, for use in a turbine powering an electrical generator.

The cost of developing such systems is high and their location is influenced by subterranean geography. It must also be noted that at the time of writing, in spite of pumping in cold water over a protracted period, the exit pipe at the Rosemanowes plant has only produced mud, instead of the hot water expected. It follows that the use of such systems cannot, at the moment, be regarded as an acceptable option for the provision of space heating, even in an area close to the source.

3.2 Passive collection of solar energy

This is brought about by designing a building such that whenever insolation falls upon it, the energy received is made available to the interior primarily by natural means (often using the stack effect), without the use of pumps or fans (see figures 1.2 and 1.3 in Chapter 1). In the UK, the Department of Energy has carried out a research and development programme dealing with design and field studies, over a range of 40 domestic and other buildings. It also carried out work on computer modelling, backed up by test cell studies.

3.3 Active collection of solar energy

This phrase is used when purpose-made collectors are utilised and these may be focusing or non-focusing. The most popular type, which belongs to the latter category, is the flat plate solar heat collector (fpshc) which is normally covered with one or two glass or plastic covers, see figure 3.1.

The author has approximately 10 m² of such collectors mounted on his roof (near Chelmsford, Essex, UK), which provide warm (and sometimes very hot) water whenever insolation falls upon them. The energy collected is transmitted by thermosiphonic flow to two horizontal hot water storage cylinders, one direct and the other indirect, both of which are sited in the loft space. The hot water outlet is then connected in place of the original cold feed to a third cylinder sited in the linen cupboard. An electronic logic device is incorporated in the system to prevent reverse flow occurring from the horizontal cylinders to the collectors when the sun no longer shines (see figure 3.2). There are times when the two horizontal cylinders contain water that is merely tepid and under such conditions any *Legionella* bacteria which are present will flourish. However, this problem is overcome by ensuring that the water in the third and final cylinder is always heated to a temperature in excess of 70°C (see section 2.19.1).

When fpshc are operated at a mean temperature that is not much greater than the atmospheric temperature, the energy losses from the collectors are very low and the thermal efficiency can be as high as 60%, which is very good news for a man-made appliance. Figure 3.3 shows the typical thermal performance lines for an fpshc, having two glass covers, and gives an indication of some possible thermal efficiencies when it is operating at both high and low temperatures.

In researching the thermal performance of fpshc in the UK (see [3.1] and [3.2]), the author found it necessary to use a temperature index called the 't value' (t_v°C). This is the number of degrees Celsius that the collector is operating above the ambient temperature, and when using figure 3.3 it will be seen that it does provide an indication of the magnitude of the losses to be expected for different values of insolation.

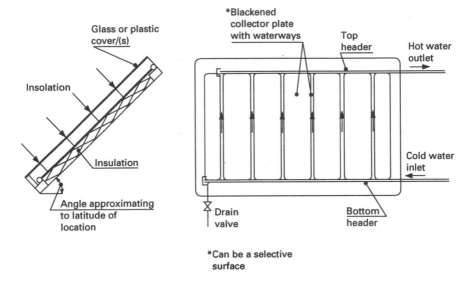

Figure 3.1   Flat plate solar heat collector

The *t* value (t_v°C) may be defined as follows:

$$t_v = (t_f + t_r)/2 - t_a$$

where t_f = temperature of water flowing from the collector (°C)
      t_r = temperature of water returning to the collector (°C)
      t_a = the ambient temperature (°C).

Further inspection of figure 3.3 will show that when *t* values are low, the thermal efficiency of the collector (output/input) is greater than when the *t* value is high. It follows that to obtain the highest thermal efficiencies, fpshc should be designed to operate for most of the time at low *t* values, so taking advantage of their thermal characteristics, and when operating in the UK, fortuitously making the most of the climate.

Because of this characteristic, fhspc may be used to good effect to warm the water contained in a swimming pool. Commonly such collectors have no glass or any other cover, as they operate most of the time only just above atmospheric temperature; reference [3.3] gives a method for determining the size of collector needed to carry out such a task. In hotter climates it is possible to operate the fpshc, albeit with some loss of efficiency, at temperatures that are frequently well above that of the atmosphere, and under these conditions the collector will provide a good supply of domestic hot water.

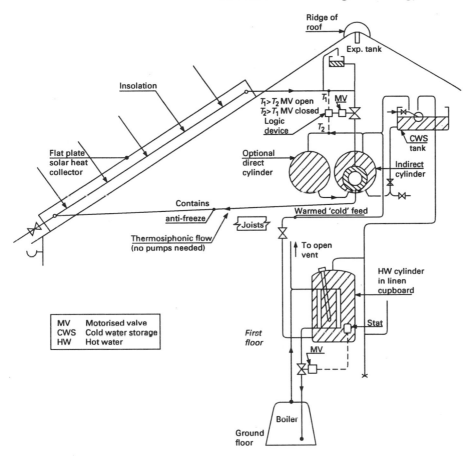

Figure 3.2 Collection of solar energy for domestic purposes

3.4 Wind and wave energy, conventional hydroelectric dam systems and the use of tidal flow

The energy obtained from the first three of these sources also originates from the sun and involves the change of kinetic or potential energy, first into mechanical work and then into electrical energy, which may then be used for space heating.

Tidal flow relies on gravitational forces and, with the possible exception of geothermal energy, all the sources of energy mentioned so far may be described as being renewable.

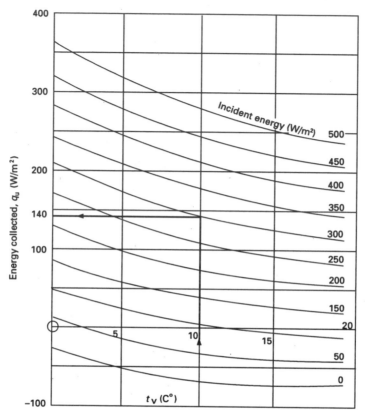

Figure 3.3 Thermal performance curves for an fpshc having two glass covers

3.5 Availability of thermal energy processes

Thermal energy is available from many processes, not all of which are utilised in the built environment; of the seven processes discussed in this section, only those listed as items 3, 6 and 7 are currently viable for space heating purposes.

Item 1: Nuclear fission

This process is used mainly in the production of electrical energy, and space heating projects in the built environment do not normally utilise thermal energy from this source.

Item 2: Nuclear fusion

Nuclear fusion is the process believed to take place in the centre of the sun, which, predictably, produces solar energy. This cannot yet be replicated continuously by man on earth, but obviously scientists are greatly intrigued by the phenomenon and developments are awaited.

Item 3: Phase change of substances

The phase change of water from vapour (steam) to liquid is commonly used in the industrial sector, and the phase change of refrigerants is also used in refrigerators and heat pumps.

Item 4: Chemical changes

These have been used successfully to store energy in non-cyclic processes only.

Item 5: Mechanical work

Mechanical work is not used for space heating purposes. Manufacturers that test large engines, convert the mechanical work produced by means of the dynamometers into heat, or use the engines directly to drive electrical generators, which produce energy for subsequent dissipation by resistive means. Note that all forms of energy are mutually interchangeable, but there is a fixed rate of exchange and it is found that, according to the Second Law of Thermodynamics, more energy has to be expended to obtain mechanical work than to obtain heat; thus it is not a sensible endeavour to use mechanical work to obtain thermal energy.

Item 6: Resistive effect of electrical current

This relies on the heating effect produced when an electrical current is passed through an electrical conductor. Its use pre-supposes that one has access to an electrical distribution system whereby no storage facilities will be needed and no provision for a flue is necessary. The power is instantly available at the touch of a switch and it is easy to appreciate how potentially attractive electricity is as a source of energy. It may be categorised as a 'secondary' source of energy, the implication being that in order to be obtained, it must first be produced from a 'primary' source of energy. It can of course be produced by using 'primary' fuels such as coal, natural gas and oil, by harnessing the enormous amounts of energy available in nuclear fuels, or as previously discussed it may be produced using gravitational effects. Details of the generation, transmission, distribution and emission of electrical power are given in Chapter 6.

Item 7: Combustion of matter

The combustion of matter is the oldest and most common way of obtaining thermal energy. The equipment required to ensure the safe burning of such fuels as coal, gas and oil is described in Chapter 5. The characteristics of the fuels contained in the following list are each dealt with separately:

(1) biofuels such as straw, wood and gas;
(2) municipal waste;
(3) natural gas;
(4) coal and its man-made by-products;
(5) oil and its various distillates.

3.6   Bio-fuels

These can be in solid, liquid, or gaseous form; they are derived from organic sources and details of some of these fuels are as follows.

3.6.1   Straw

For straw the energy of combustion is approximately 12 MJ/kg. In the UK there are plentiful supplies of straw available for burning. At Woburn Abbey, straw bales are shredded and burnt in purpose-made boilers, the energy so produced then being used to provide space heating and domestic hot water. The process has been automated but the assembling of the bales at the place of burning requires considerable work input. It also remains to be seen whether such appliances can be readily adapted to comply with the provisions of the Environmental Protection Act 1990.

3.6.2   Wood

The energy of combustion of wood is 15.5 to 24.5 MJ/kg. In the UK, the rising cost of conventional fuels, coupled with the availability of storm-produced timber, has revived interest in the use of wood as a fuel. The Department of Energy has a research and development programme, covering the growth, harvesting and utilisation of wood, produced using short rotation forestry, on sites in Northern Ireland, Scotland and England. In rural areas, wood is sometimes used as a fuel in domestic premises, but it is not in common use for commercial or industrial buildings.

3.6.3   Biogas

The energy of combustion of biogas is 18 MJ/m³. This gas can be obtained from landfill sites where decomposition of organic waste is taking place. The decomposition occurs in three consecutive stages, but each stage need not occur at the same time on every part of the site, owing to the variable nature of the waste.

Stage 1

Aerobic bacteria which thrive on oxygen contained in pockets of air, formed when the waste was first laid down, produce carbon dioxide, water and heat. The bacteria also partially degrade the residual organic matter.

Stage 2

When the oxygen has been exhausted, anaerobic bacteria which thrive on carbohydrates start to produce organic acids, hydrogen, nitrogen and carbon dioxide; this process reaches a production peak between 11 to 40 days after placement of the refuse.

Stage 3

Between 180 to 500 days after placement, a different type of anaerobic bacteria appear; these thrive on the organic acids and produce methane, carbon dioxide and water as by-products. Generally, it can be expected that between one to two years after the tip was laid, the steady production of methane and carbon dioxide will have been initiated [3.4].

Computer modelling [3.5] suggests that it is economic to use the first three years in extracting the gas at maximum output and arranging for the rest of extraction to be completed within five years. The payback period is expected to be approximately three years, when dealing with a tip containing one million tonne of refuse. The study also shows that it is economical to supply many users with gas when they are situated at a distance of no more than 2 to 3 km from the tip, but for a distance up to 10 km there should preferably be only one user.

In order to extract the gases from the site, 'wells' are sunk with a pitch of 50 to 100 metres. The wells are then connected to a grid system of plastic pipes and headers which lead into a filter, before being connected to the suction side of a compressor. In some sites it is necessary to control the horizontal movement of gas from the edges (referred to as 'migration'), to prevent it becoming an environmental nuisance. In these cases, horizontal trenches are dug around the site perimeter and filled with gravel. These are then vented to the atmosphere so that the gases can be burnt off.

In those landfill sites where gas harvesting does not take place and where the gases are allowed to escape into the atmosphere, there is obviously a potentially unacceptable environmental hazard. It follows that, taking up the option to harvest the gas would seem to make good sense, giving as it does, the bonus of converting a liability into an asset. In the event, the exercise has proved to be economically viable and in the UK to date there are at least 16 companies with the technology to produce the gas from landfill sites. Commonly, the gas has been used as a fuel for kilns in brickworks, or as a fuel in boilers. Provided that the corrosive constituents of the gas are removed, it may also be used in gas turbines or gas engines, which may power electrical generators. In the UK this method of using biogas is growing in popularity, as it is easier to 'export' electricity over a distance than it is gas. Reference [3.6] gives information on the rate of growth of the different methods of utilisation taken over the years 1986 to 1989.

3.6.4 *The storage of gas*

Whenever gas for combustion is manufactured locally, it will be found necessary to find a means of storing it, in order to ensure continuity of supply to the consumers under all possible demand conditions. When town gas was used for space heating, it was usually manufactured by the local gas company and such storage facilities took the form of an inverted steel cylinder, closed at the top and mounted with the open end immersed in water, commonly known as a 'gasometer'.

With the introduction of natural gas, delivered under pressure from a national grid, the gas mains themselves provide a large storage volume, so that storage cylinders are no longer required in large numbers. However, storage cylinders will be required whenever biogas is harvested.

3.7 Municipal waste

The average *net* energy of combustion of municipal waste is 8 MJ/kg. During the decade from 1950 to 1960, this source of energy was used by some local councils in the UK to provide space heating and domestic hot water for the buildings sited adjacent to the point of burning. It should be realised that the cost of energy at that time was much lower than it is today and, during this period, the incentives to utilise municipal waste as a source of energy were correspondingly reduced. In the UK, the Borough Council of East Ham (now Newham) believed that it was more cost-effective to use the domestic waste to carry out landfill projects so that the council would have the option of developing such sites in future years. Accordingly, it (and many other councils), discontinued the use of its municipal incinerators, in the belief that they were labour-intensive, space-consuming and generally uneconomic. At that time, possible effects of air pollution were hardly ever discussed.

In Nottingham, the City Council realised that they had a shortage of local landfill sites that would be suitable for refuse disposal and in conjunction with the National Coal Board, it developed sites in the city centre, incorporating district heating and separate power generation, using energy obtained from its refuse incinerator. This scheme was taken over by Nottinghamshire County Council in 1974 and continues to provide a convenient base load for the incinerator which handles approximately 120 000 tonne/year of refuse from Nottingham City and surrounding areas. Electrostatic filters are used to separate the dust from the flue gases coming from the furnaces.

The City of Sheffield Cleansing Services Department also uses refuse as a source of energy for space heating and domestic hot water. Annually, approximately 130 000 tonne of refuse are incinerated supplying a total heat load of 30 MW from this source. Two 6 MW gas-fired boilers are provided for 'topping-up' purposes, for possible use during peak load conditions. An impressive list of buildings is supplied with energy in this way, including the Town Hall and many other civic buildings, together with more than 4300 domestic properties.

In 1993 in the UK there were only two other complexes using mass burn incineration with energy recovery and sale; these were Coventry and Edmonton. The latter has the largest unit, burning 360 000 tonne/year of municipal refuse. In 1994, the South East London Combined Heat and Power plant at Deptford commenced operations. It was designed to burn 420 000 tonne/year of screened municipal waste, while producing flue gas emissions that complied with the latest environmental standards. The plant produces 32 MW of power and enough hot water to heat 7500 homes and four schools in Southwark.

Taken together, the total quantity of refuse dealt with by all five schemes is approximately 1.2×10^6 tonne/year. This must be compared with an estimated 30×10^6 tonne/year of refuse for the UK as a whole [3.7].

In the rest of Europe and North America, the practice of mass burn incineration with energy recovery is very much more common than it is in the UK, and the latest developments are concentrating on selective screening of the refuse intake, resulting in more closely controlled operating conditions, together with enhanced treatment of the flue gases, involving the use of line scrubbers and extra and more effective filtration. From an environmental viewpoint, clearly such initiatives should be encouraged, particularly as the latest European Directive for Municipal Waste Incineration (New Plant 89/369 EEC) prescribes limits for the emission and composition of the combustion gases, together with limits for combustion conditions and monitoring requirements.

3.7.1 Refuse derived fuel (RDF)

The energy of combustion of refuse derived fuel is 15.9 MJ/kg. The handling plant and boilers required for mass burn incineration are large and specialised. It is possible to produce a refuse derived fuel (RDF), taking the form of pellets, which can be burnt in conventional smaller boilers that have been modified to suit the characteristics of the new fuel. The rate of production of RDF is, at the time of writing, greater than the demand and it is true to say that much work is being channelled into developing the different combustion technologies needed, to make the fuel an acceptable alternative (or addition) to conventional fuels. Provided the resulting products of combustion are seen to comply with the EEC Directive (New Plant 89/369) and provided it is still cost-effective, its continuing use may be expected.

3.8 Natural gas

The energy of combustion of natural gas is 38.6 MJ/m^3. In the UK, natural gas contains approximately 93% of methane and is made up of a blend of gases taken from deposits under the North Sea. The supply company has a statutory obligation to control the magnitude of the energy of combustion of the gas, within closely defined limits, in order to ensure that the consumer is at all times paying for a guaranteed amount of energy. In the UK, natural gas is currently available; it requires no storage on site, is continuously available and there is no residue after burning. Special tariffs may sometimes be negotiated if the building is classified as being 'large commercial' or 'large industrial'. The volume of gas consumed is metered at the intake position, and the connections to the meter for commercial or industrial premises are arranged to have a by-pass so that maintenance to the meter can be effected without the supply being discontinued, see figure 3.4.

Gas meters are usually sited in an enclosure which must be permanently ventilated, at both high and low levels. The objects are: (1) to ensure that a possible gas leak does not have a chance to fill up the enclosure, with a potentially explosive mixture of gas and oxygen, as this could be activated by a spark; and (2) to encourage the thermosiphonic flow of air within the chamber, so helping to keep the materials inside the gas meter at an equable temperature. Nowadays a filter is fitted in the main and each appliance has an adjustable constant-pressure governor, so that diurnal variations in the mains gas pressure do not affect the performance of any of the equipment.

3.9 Coal

The energy of combustion of coal varies from 21 to 31 MJ/kg with a mean value of 26 MJ/kg being used for the rating of equipment. Coal, being a natural product which is mined in different areas, possesses many different qualities. It is classified initially by a numbered coding system, given to it by British Coal, the value of which may range from rank 100 to 900. Rank 100 is referred to as a high ranking coal; it contains the most carbon and is the least volatile. Lower ranking coals contain relatively less carbon, but are more volatile. Table 3.1 gives some of the properties of the coal associated with its rank. Within each band of rank numbers, the coals are given a discrete number.

All coals from ranks 100 to 900 are further graded by passing them

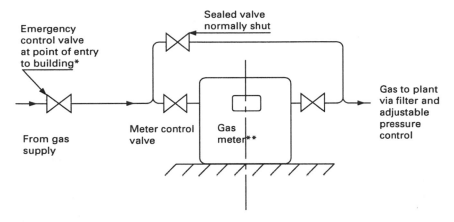

* For further information,
  see British Gas Publication
  IM/21 *Guidance Notes on the
  Gas Safety (Installation and Use)
  Regulations 1984.*
** Meters with electrical outputs
   are available for connection
   to energy management systems,
   see British Gas Publication
   IM/23 *Guidance Notes on the
   Connection of Electrical
   Equipment to Gas Meters.*

Figure 3.4   Industrial gas meter with by-pass

Table 3.1   Properties of coal associated with its rank

British Coal coding rank	Net energy of combustion (MJ/kg)	Volatile content	Use	Types
100–200	29.75	Low	Gravity-fed boilers	Anthracite Dry steam coals Coking steam coals
200–400	29.3	Medium	Mechanical stokers coals and pulverised fuel burners	Bituminous
400–900	26	High		

Table 3.2 Classification and screen size of coals

Name	Diameter of hole (mm)	
	Upper limit	*Lower limit*
Large cobbles	>150	75
Cobbles	100–150	50–100
Trebles/large nuts	63–100	38–63
Doubles/nuts	38–63	25–38
Singles	25–38	13–18

Reproduced from CIBSE Guide C, *Reference data* (1986), by permission of the Chartered Insitution of Building Services Engineers.

through screens having circular holes of particular sizes. The emerging coal is then described according to the screen size as illustrated in table 3.2.

There are many derivatives of coal and these include coke and coalite which, in the UK, are mainly used in the domestic market. A by-product of coal was town gas, which was formerly produced in the UK for use both in the home and in industry.

3.9.1 The storage of solid fuel (coal)

Physical handling of the fuel is nowadays considered to be unacceptable, inconvenient and uneconomic, and it follows that any installation must be designed to be, as nearly as possible, automatic in operation. The fuel must generally be off-loaded from a tipper lorry, directly into the coal bunker. The bunker must be so positioned that the coal can be delivered automatically to the holding hoppers, mounted over each boiler, so that each automatic stoker will then feed the fuel on to each grate. At the end of the burning process, the remaining clinker and ash must also be removed (if possible automatically), cooled and delivered to a holding hopper, ready for batch delivery, to an ash disposal vehicle.

The fuel and ash can be moved through the plant pneumatically (a noisy process), or by means of bucket elevators, belt conveyors, vibratory conveyors or rotating helical screws operating in ducts, or by means of the force of gravity. A good deal of specialised knowledge and forward planning are required, together with considerable capital expenditure, to bring such projects to a successful conclusion.

Architects, engineers and builders, who may be faced with first designing and then constructing such a scheme, will find a large amount of useful information is contained in the publications of the British Coal Technical Service, and also the CIBSE Guide, section B13-4. British Coal's publication *Industrial Solid Fuel Plant* details the precautions to be taken with regard to the possible occurrence of spontaneous combustion in storage silos and in stockpiles, as well as much more information on the design parameters dealing with the smooth functioning of the complete plant.

Table 3.3 Classification and properties of oils

	Distillate oil		Residual oils		
	Grade C2 (Kerosine)	Grade D (Gas oil)	Grade E (Light fuel oil)	Grade F (Medium fuel oil)	Grade G (Heavy fuel oil)
Kinematic viscosity (cSt [centistokes] mm²/s) at 38°C	1–2	6			
Kinematic viscosity (cSt) at 82°C			12.5	30	70
Energy of combustion (MJ/kg)	43.6	42.7	41.0	40.5	40.0
Minimum storage temperature (°C)			7	20	32
Minimum handling temperature (°C)			7	27	38

3.10 Oil, its residuals and distillates

In the UK, oil is produced in various grades to specifications prescribed by the relevant British Standards. There are six recognised grades of fuel and some of the properties of each of five of the most common are shown in table 3.3.

3.10.1 Distillate fuel graded C1 (paraffin)

This grade of oil is not shown in table 3.3. It is mainly used for the space heating of single rooms in dwelling houses, greenhouses, some garages and light industrial buildings, using free standing portable appliances. Normally no flue is provided, so that the products of combustion are allowed to remain in the space being heated, which means that all the energy of combustion of the fuel is available for heating purposes.

When any fuel is used in this way, fixed ventilation must be provided to sustain life. The products of combustion of paraffin contain a high proportion of water vapour, which will condense on any cold surface and mould will appear in a badly ventilated room. As a rough guide, one part by volume of paraffin vapour will produce the same volume of water vapour, and the same is true of natural gas. The humid conditions so produced are sometimes regarded by gardeners as being beneficial to plant life.

3.10.2   Distillate fuel graded C2 (kerosene)

This is the least viscous of the oils shown in table 3.3. In the built environment, it is mainly used for space heating in domestic premises, using vaporising, or pressure jet burners, fitted to a flued boiler. In the UK, the oil can be stored out of doors in an uninsulated tank and will present no flow problems in the coldest of winters, as its freezing point is −40°C. This grade of oil is also used as an aviation fuel and in the event of an oil crisis its price is likely to change more than those of other fuels.

3.10.3   Distillate fuel graded class D (gas oil)

This is referred to as diesel oil, after Dr Rudolf Diesel, who developed the early air blast oil injection engines. It can be used for space heating, or as a fuel in modern diesel engines. In the UK it carries a tax when used in vehicles using the public highway, but currently no tax is levied when it is used for space heating or for agricultural purposes. When this oil is supplied for use as a heating oil, it is coloured with a red dye, so that Customs and Excise Officers may detect when such oil is being used for transport purposes on the public highway. At the time of going to press, value added tax is now being collected from all fuels used for space heating and cooking in the UK.

   Gas oil for heating is used in domestic, light commercial and industrial premises, and in pressure jet type burners, fitted to flued boilers. It can be stored in an uninsulated and sheltered tank, sited outside the property. Crystals of naphtha start to form when the temperature is within the band −4°C to −7°C, making the oil appear cloudy – this is referred to as the 'cloud point'. These crystals block the small holes in the oil filter, causing the flow of oil to the burner to become restricted and eventually to stop. An almost instant solution is to place a 60 W bulb under the bowl of the filter.

3.10.4   Residual or blended fuel graded class E (light fuel oil)

This is used in commercial and light industrial premises for space heating purposes. It is generally introduced to a flued combustion chamber using pressure jet burners, at which point it is heated to approximately 60°C. The oil may be stored in an uninsulated tank, sited inside the building, where its temperature should not be allowed to fall below 7°C, as pumping the oil will then become difficult. In the UK this grade of oil is not used in such large quantities as it once was. The blending process is known to be critical and over a protracted period, it is possible for it to

Table 3.4 Storage and handling conditions for different grades of oil

Class of oil	Suitable storage conditions in the UK	Special handling details
C2	External tank uninsulated	None
D	External tank in sheltered position	None
E	(a) External tank insulated, oil temperature must be greater than 7°C (b) Internal tank uninsulated (c) Tank in chamber underground uninsulated, temperature 10°C all year round	Heat oil at burner to approximately 60°C
F	Internal tank insulated	Trace heat oil pipes from tank, heat oil at burner
G	Internal tank insulated	Trace heat oil pipes from tank, heat oil at burner, heat oil in tank

stratify in the storage tank, forming crystals of naptha.

Table 3.4 gives a summary of the storage possibilities for each of the commonly used heating oils in the UK, together with details of any special handling requirements. Details of a typical underground oil storage tank suitable for use with class D or E oil are shown in figure 3.5. Before drawing up a specification for such a tank, BS 799 *Oil burning equipment* should be consulted. It must be noted that a more cost-effective way to store oil underground is to introduce it into an externally tanked and oil-proofed concrete storage chamber; these can be provided by specialist companies. Plastic storage tanks made from polyethylene, which has been stabilised against the action of ultra-violet light, are now available and one manufacturer gives a 10 year guarantee with this product when sited above ground.

When oil is being introduced into any container, the displaced air must be allowed to escape to the atmosphere by means of a vent pipe. In order to avoid subjecting the underground tank to excessive hydraulic pressure in the event of overfilling, this pipe should terminate no more than one metre above ground level. Note that if the vent pipe is full of oil and is as high as the bottom of the tank is deep, the pressure at the lowest part of the tank will be twice the normal maximum working pressure and the tank will be under extra pressure, until the oil in the vent pipe has fallen to a point that is level with the top of the tank.

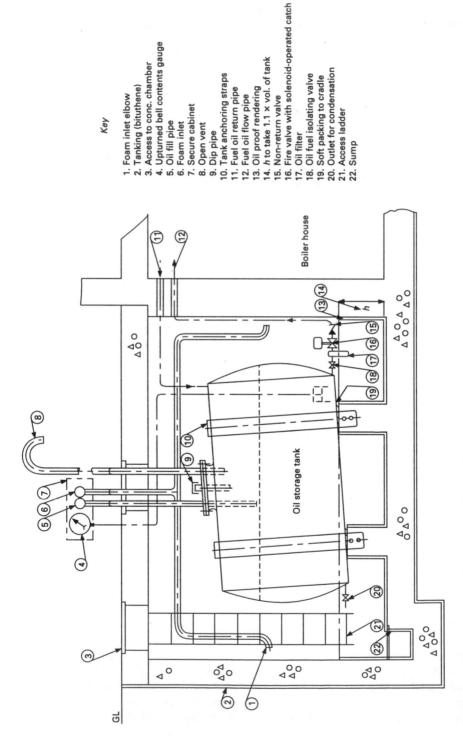

Key

1. Foam inlet elbow
2. Tanking (bituthene)
3. Access to conc. chamber
4. Upturned bell contents gauge
5. Oil fill pipe
6. Foam inlet
7. Secure cabinet
8. Open vent
9. Dip pipe
10. Tank anchoring straps
11. Fuel oil return pipe
12. Fuel oil flow pipe
13. Oil proof rendering
14. h to take 1.1 × vol. of tank
15. Non-return valve
16. Fire valve with solenoid-operated catch
17. Oil filter
18. Oil fuel isolating valve
19. Soft packing to cradle
20. Outlet for condensation
21. Access ladder
22. Sump

Boiler house

Oil storage tank

GL

Figure 3.5   Underground oil storage tank and chamber

When oil is being delivered, there should be a reliable means of determining the moment when the flow of oil should be stopped. An elegantly simple solution to this problem is to fit a whistle to the vent pipe inside the tank. The frequency of sound emitted by the whistle will change as the level of oil in the tank approaches the top, and the whistle will cease to sound as soon as the oil level rises above it, so giving the operative sufficient time to close the delivery valve.

There are many different ways of determining the amount of oil contained in a tank. Some readers will be familiar with the hypsometer method, whereby a connection, with a low-level cock, is made at the top and bottom of the tank and the two connections are then joined using a transparent plastic tube placed inside a guard. This method is obviously vulnerable to mechanical damage, with the subsequent loss of the contents of the tank followed by the extra cost of any environmental disturbance due to the spillage. However, this drawback may be overcome by arranging for the cock to remain in the closed position, except when it is necessary to obtain an indication of the depth of oil in the tank. When class D oil (gas oil) is being used, the inside of the plastic tube becomes stained and in order for the user to be able to detect the oil level with any certainty, the tube should be renewed at approximately two-yearly intervals.

Another way of obtaining an indication of the quantity of oil in a tank is to use the so-called 'cat and mouse' method. It this case, a float resting on the surface of the oil is linked by means of a Bowden cable running over a pulley to an indicator, which moves up or down a vertical scale, indicating the volumetric contents of the tank. This system, though inherently simple, is known to be unreliable and problems can arise from the following causes:

(1) mechanical damage may occur to the float when the tank is being filled;
(2) the slider may be prevented from moving by friction or even by the careless placing of rubbish;
(3) the rate at which the float and the indicator moves is imperceptible over a short period, so that instant remote testing can be inaccurate.

A reliable way of determining the contents of the tank is to use the 'upturned bell' method, where the pressure of the air contained in the bell is measured, using a pressure gauge having its scale calibrated in litres, see figure 3.5. More elegantly, a series of electrodes of increasing length, which can be mounted in a circle, may be fitted inside the tank, such that when connected electrically, they give an indication in a series of finite steps of the variation in depth of the oil.

All tanks containing heating oil have a fire valve fitted in the outlet pipe, and in the event of the occurrence of an unscheduled fire, this valve cuts off the flow of oil to the burners. In the past, the mass on the lever arm of the valve was held in the up (open) position, using a Bowden cable (or catenary wire) running over pulleys and connecting in series a number of fusible links, each one of which was mounted above a burner. A fusible link consists of two metal plates (usually brass) which have been soft soldered together using a low melting point solder. This system has sometimes been found to be unsatisfactory, owing to the supporting wire becoming less flexible and kinking, so that subsequent operation of the arm of the valve proved to be unpredictable.

A more reliable method is to arrange to support the mass on the arm of the valve, by means of a solenoid-operated catch. The electrical current operating the solenoid may then be passed through a circuit containing any number of fusible links. If, owing to excessive local heating, the solder on any link melts, the current will be interrupted, de-energising the solenoid and releasing the catch supporting the mass. The subsequent motion of the arm will then close the valve. Note also that bi-metallic links may be used in place of fusible links.

Whenever oil-fired heating boilers are installed on the roof of a building, it is common practice to provide a small-capacity oil service supply tank sited adjacent to the burners, and the size of this tank must be decided in conjunction with the local fire prevention officer. One of his job functions is to limit the size of any future conflagration and, in pursuance of this, it is likely that he will supply details of the type of fire valve required, together with information on the possible provision of pipes designed to carry foam directly to the source of a fire. Chapter 8 gives more details.

3.10.5   Residual or blended fuel graded class F (medium fuel oil)

This is used in larger sized industrial premises, where the space heating and process heating loads are high. When the oil becomes more dense and more viscous, it becomes difficult to pump from the storage tanks to the place of burning and, in order to make the operation possible, heat must be supplied to the circulation pipes. Accordingly, in an attempt to make this grade of oil more attractive to a potential customer, the oil companies have made the cost of class F oil lower than those of the grades previously mentioned.

3.10.6   Residual or blended fuel graded class G (heavy fuel oil)

This is the cheapest of all the grades and it also has the lowest energy of combustion of 40 MJ/kg. It has a relative density (to that of water) of 0.97 which is greater than all the other oils, so that a greater mass of oil is obtained per litre delivered. The use of this oil is usually reserved for large hospitals and large industrial complexes, where the high rate of oil consumption can justify the extra energy required to handle this highly viscous fuel. The oil has to be heated in an insulated tank and in the supply pipes, and this energy is often provided by circulating steam, or hot water, through heating coils immersed in the tank. Alternatively, electrically operated immersion heaters may be fitted to a large-diameter draw-off tube located inside the tank. A practical drawback to the use of class G oil becomes apparent whenever maintenance work is done that necessitates breaking into the oil pipes or tanks. In the experience of the author, when this occurs the subsequent spread of black filth can be extensive. Operators of such plant will know that the subsequent dirty state of the various surfaces in the boiler house and the poor state of morale of the workforce do not figure in the cost of the oil. Hence, it is not unknown for cost-conscious managers and accountants to be able to demonstrate that the use of such oil is indeed cost-effective, albeit extremely messy in the event of spillage.

3.10.7 Two other gaseous distillates of oil: butane and propane

These may conveniently be stored in liquefied form, under pressure, in cylinders which are then inappropriately referred to as 'gas bottles'. Liquid butane under pressure in such a cylinder has a boiling point close to 0°C, so that whenever the cylinder outlet valve is opened, in an ambient temperature just greater than 0°C, the liquid contained in the pressurised cylinder starts to boil and the butane gas given off may then be burnt in the appliance to which it is connected. The cost of these gases is slightly greater than the oil from which they are derived, which difference reflects the cost of supplying the bottles and of compressing the gas. Butane gas has an energy of combustion of 113 MJ/m^3 or 46 MJ/kg.

Propane has a boiling point of −45°C and can therefore be used under any conditions where the ambient temperature is greater than −45°C. It has an energy of combustion of 86 MJ/m^3 or 46.5 MJ/kg. These gases are widely used as a portable energy source by the building industry ('the ubiquitous red cylinder'). In the absence of a supply of natural gas, propane is frequently used in catering and domestic premises for cooking and space heating purposes. In these cases, pressurised cylindrical storage containers are installed on site and these may be hired from the supply company and sited in an open space, to assist dispersal in the event of a gas leak occurring. Both gases are heavier than air, so that any unventilated enclosure will fill up with gas from the bottom. It follows that in any land-based project involving the use of butane or propane, it is essential that adequate fixed ventilation shall be provided at low level and no leakage of unburnt gas from the system can be tolerated. When such a system is installed in a boat, such low-level ventilation is obviously not possible. In the event of a gas leak and in the absence of a flameproof electrically operated gas pump, to discharge the gas overboard, a baling-out operation of the gas will be needed.

3.11 Choice of fuel

If a client does not have a preference for a given fuel and a choice of fuel is available, the decision as to which of these fuels may be most suitable depends on many factors, and a suggested list of questions that may usefully be posed follows:

(1) Is the proposed development near to an oil refinery, a natural gas pipeline, or a coalmine and, if the plant is a large one, can preferential tariffs be negotiated?

(2) Could a dual, or triple, fuel energy converter be used,
(a) to take full advantage of any price changes in alternative fuels?
or (b) to minimise the effect of future industrial disputes in any sector, which may result in the cessation of supplies?

(3) If the fuel chosen requires storage, is there a convenient place for this on site?

(4) If solid fuel is chosen, what arrangements should be made for its storage and handling and the eventual disposal of the ash?

(5) Should the process be manual or automatic?

(6) If the process is to be automatic, what are the likely architectural requirements?

(7) Is skilled labour likely to be available from the client's employees, for the purposes of controlling and maintaining the plant?

(8) If the building is an existing one, are the flueing arrangements satisfactory for the proposed fuel?

(9) If the building is proposed, what effect does the third (1981) edition of the *Clean Air Acts Memorandum on Chimney Heights* (HMSO, London) have on the height of any proposed chimney for the fuel chosen? This memorandum specifies the increase in chimney height needed to limit local pollution due to sulphur dioxide (SO_2), depending on the type of locality in which the plant is situated. Readers should note that natural gas contains less sulphur than coal and oil, so that SO_2 emissions are inherently less.

(10) Will the percentage emission of flue gases from the chosen fuel satisfy the current EC directives?

For readers who require technical data to enable them to size a flue and specify the material from which it is to be constructed, reference should be made to the 1986 CIBSE Guide, section B13. Note also that any recommendations subsequently made must also comply with those made by the local Public Health Department and, in the near future, also those of the appropriate European Council.

References

3.1. Hassan G. *The thermal performance of flat plate solar heat collectors.* PhD Thesis, 1969. Faculty of Engineering, University of London.

3.2. Hassan G. The testing of flat plate solar heat collectors, in *Proceedings of the First International Conference on Solar Building Technology*, Vol. 1, 1977. RIBA, London.

3.3. Hassan G. A design procedure suitable for calculating the size of a flat plate solar heat collector needed to warm an outdoor swimming pool in the UK, in *Proceedings of the Summer Conference of the Institution of Heating and Ventilating Engineers*, 1970.

3.4. Schumacher M.M. *Landfill Methane Recovery*, 1983. Noyes Data Corporation, Park Ridge, New Jersey.

3.5. Ader G. *Methane from Landfill.* Ader Associates. [British Library, Boston Spa 16.4.82 3816:465]

3.6. Richards K.M. *Landfill Gas Exploitation in the UK – An Update*, June 1988. ETSU, Harwell.

3.7. Stronach N. J. *Energy from Municipal Solid Waste: The Future?*, 1988. ETSU-L-24, Harwell.

4 Release and Absorption of Energy

4.1 Equipment required to burn natural gas

A typical pipework layout, with controls, for a gas-fired atmospheric burner attached to a domestic boiler is shown in figure 4.1. An atmospheric burner is one in which the air for combustion is naturally aspirated and the pressure inside the boiler combustion chamber is just below atmospheric. The inclusion of a draught diverter in the flueway just above the boiler ensures that any change in the draught pressure which may occur will not materially alter the pressure inside the boiler combustion chamber, or (in the case of an increase in pressure) extinguish the main flame and/or the pilot light.

The lighting sequence for the burner shown in figure 4.1 is as follows:

(1) Close the burner cock, open the main gas valve and the pilot cock.
(2) Press and hold the manual override button on SV1 to open the valve, allowing gas to flow to the pilot jet.
(3) Continue to hold the override button and ignite the pilot either by means of a lighted taper or by pressing the spring loaded 'ignite button' (this button applies a sudden force to a quartz crystal, which then produces a high voltage across the tips of two electrodes, so generating a spark).
(4) Continue holding the override button until the heat from the pilot light has generated sufficient voltage across the ends of the thermocouple to hold SV1 and SV2 open electrically, at which time it may then be released. If the pilot light is extinguished, return to step (2).
(5) Open the burner gas cock, when the main burner will light.

The reader should note that if the pilot light is extinguished for any reason, both SV1 and SV2 will be de-energised, so preventing the flow of gas to the main and pilot burners. Subsequent normal operation of the main burner may be controlled by switching SV2 on or off, by means of a suitably placed room thermostat.

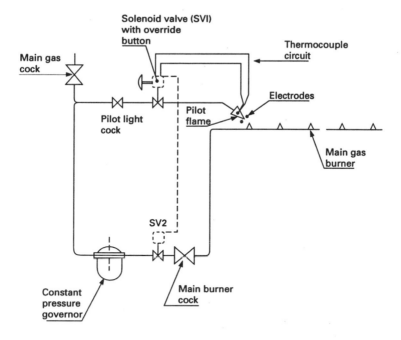

Figure 4.1   Controls and connections for a gas-fired atmospheric burner

The diameter of the gas pipe supplying the main burner for use in industrial or commercial premises is of necessity larger than that used on a domestic boiler, so that the diameter of the solenoid valve in such a pipe would also be large and the valve correspondingly more expensive. This problem may be overcome if the apparatus for the control of such boilers is arranged as shown in figure 4.2. It will also be seen that a flame detector is fitted, which is sensitive to ultra-violet light (UVL) because the light emitted from a natural gas flame contains a large proportion of UVL.

When the frost thermostat is not calling for heat, it will be in the closed position and when this is the case, closure of any one or all of the SV valves 2, 3, 4 and 5 will produce a build-up of pressure to occur on the top side of the diaphragm of the on/off gas controller, causing the gas to the main burner to be shut off. Conversely, gas will be admitted to the main burner whenever the pressure on top of the diaphragm is less than that below it, and this can be caused by the frost thermostat opening and/or by all of the SV valves 2, 3, 4 and 5 being open at the same time.

An alternative type of burner is the pressurised burner in which the air required for combustion is provided by a forced draught fan. In this case the combustion chamber operates at a pressure that is slightly above atmospheric pressure and no draught diverter is required to be fitted into the flueway. These burners are noisy and it follows that they should only be sited where the generation of such noise can be easily attenuated or

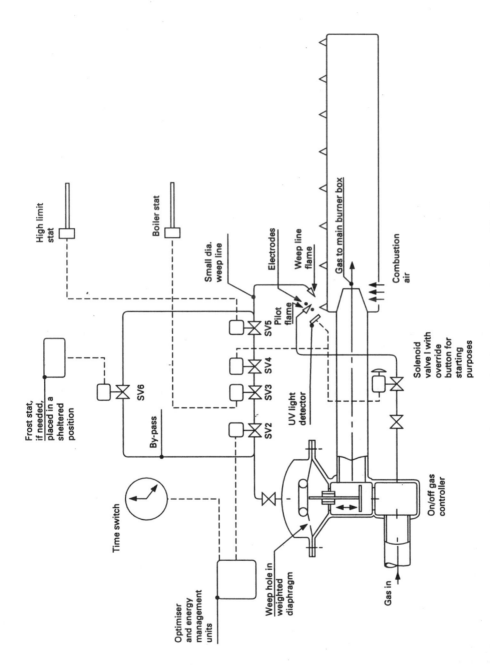

Figure 4.2 Controls and connections for a gas-fired commercial/industrial boiler

tolerated. They are usually fitted to large boilers and, in order to avoid the production of high thermal stresses in the heat exchange surfaces on start up and shut down, the control equipment is designed to limit the flow rate of the gas to the burner, to one-third of full output for a predetermined period before proceeding to maximum flow rate. On shut-down, this sequence occurs in reverse.

In conditions where the gas pressure is too low to supply the amount of energy required for the connected load, it is sometimes possible (with the permission of British Gas) to install a gas pressure booster in the form of a compressor, the suction side of which draws gas from the main, raising it to the required pressure. Generally the pressure of the gas in the main, though varying, is greater than the pressure required at the appliance and it is the purpose of a gas governor to deliver gas to the burner at a lower, but constant pressure. One form of governor consists of a flexible diaphragm, the centre of which carries a valve. The diaphragm can be loaded with lead weights, the value of which can be varied to suit the local pressure conditions in the main and at the appliance.

It should be noted that any organisation supplying gas to a group of buildings must ensure that the pressure of the gas in the mains is at all times above atmospheric, so that there is never a possibility of air entering the gas supply pipe to form a mixture, which, if by chance it were ignited, might explode, resulting in possible loss of life and possible widespread damage.

4.2 Equipment required to burn coal

This may broadly be classified into equipment which produces overfeed combustion and that which produces underfeed combustion. The latter is most likely to succeed in burning the coal with the minimum production of smoke, because the primary air, reaching the burning zone, does so through fuel which is as yet unburnt, so ensuring that it has its normal complement of oxygen necessary for the combustion process.

In overfeed combustion the primary air will already have passed through the burning zone, giving up most of its oxygen before coming into contact with unburnt fuel. In both types of stoker it is important to arrange an adequate supply of secondary air to provide enough oxygen to complete the combustion process. Any unburnt carbon is deposited as soot and the lighter particles are carried away in the flue gases with visible results which are no longer environmentally acceptable. When carbon is partially burnt to carbon monoxide, 10 MJ/kg of energy are released: however, when the same amount of carbon is completely burnt to carbon dioxide, 34 MJ/kg of energy become available. Clearly, it is wasteful to have incomplete combustion occurring anywhere in the combustion chamber, and in practice arrangements are made to provide excess air to that theoretically required in order to complete the burning process.

It should be noted, however, that air contains approximately 79% by volume of nitrogen which, being an inert gas, does not contribute to the combustion process. It absorbs energy from the combustion chamber and other gases, until it reaches the temperature of its surroundings and is ultimately discharged to the flue. For this reason, the provision of too much excess air can be wasteful because of the presence of large volumes of nitrogen.

In practice, the percentage by volume of carbon dioxide in the flue gases affords a useful guide to the amount of excess air that has been provided, and 'target' values of carbon dioxide vary from 10% to 13% depending on the type of fuel being used. With bituminous coal, the percentage of carbon dioxide in the flue gases becomes less, as excess air is increased beyond the amount required for theoretically correct combustion. Useful information on the determination of the magnitude of heat losses to the flue using the Siegert formula may be found in BS 845.

Research in progress at the Coal Research Establishment at Cheltenham, UK, aims at reducing the emission of oxides of nitrogen in flue gases obtained from pulverised fuel burners, by supplying air for combustion in varying amounts at separate stages of combustion. As a result of the research, low nitrogen oxide burners are being fitted to twelve of the largest power stations in the UK using pulverised fuel. For further information on the various types of coal burners, the reader is referred to references [4.1] and [4.2].

4.2.1 Pot burner

Nowadays, coal and its derivatives are nearly always burnt using automatic stoking equipment, and the arrangement shown in figure 4.3 depicts a pot type burner fuelled directly from the storage bunker by means of two helical screw conveyors. Where the boiler is sited directly underneath the fuel store, it may be more convenient to use only one conveyor and to introduce the fuel to the system by means of a gravity fed hopper, and this arrangement is shown by broken lines in the figure.

The screw is liable to incur damage if it receives a loose metal object and, to avoid such an occurrence, the drive mechanism may include a clutch, or a 'shear pin', through which all the power must be transmitted. In determining the size of the shear pin, a certain amount of trial and error is sometimes required, particularly when the machinery has just been commissioned and is not as free running as might be wished. During this period, it is wise to ensure that spare shear pins (possibly of varying sizes) are available, in order that the stoking process will continue without too many unscheduled interruptions.

Reference to figure 4.3 will show that combustion takes place in the body of the coal as it is impelled upwards. The sides of the refractory container (retort) are kept cool by arranging for the primary air needed

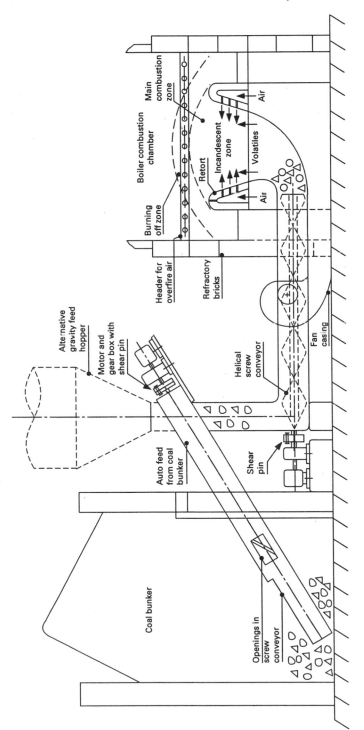

Figure 4.3 Fire pot and fuel bed with automatic feed from coal storage bunker

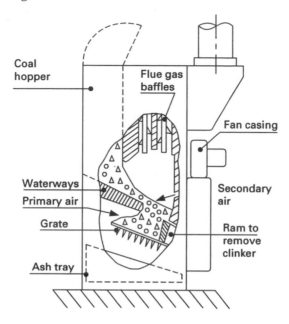

Figure 4.4   The magazine fed burner

for combustion to pass through peripheral slots in its circumference, so cooling it and at the same time preheating the air. This type of burner functions satisfactorily on doubles or singles, having rank numbers 200, 201 and 600 to 900, and it may be fitted to boilers of up to 1.75 MW. The ash, in the form of large pieces of fused clinker, is removed manually. In some boilers, the process has been automated by providing a reciprocating grate, placed to one side of the pot burner, through which the broken clinker can pass into a collector, where a helical screw then removes it to a waiting bin for disposal.

4.2.2 Magazine fed burner

This system is shown in figure 4.4 and relies on the coal self-feeding from a loaded hopper on to a grate. The rate of combustion is controlled by the operation of a forced draught fan, actuated by a thermostat. The resulting ash and clinker are then ejected by means of a ram, which travels across the grate, moving the spent fuel into the ash container. Systems of this type are to be found in both domestic and commercial situations, having outputs ranging from 26 to 600 kW. They must be operated with coal having a low volatile content so that there is little likelihood of the fire spreading upwards into the unburnt fuel in the magazine (or hopper). The preferred fuel is anthracite.

4.2.3   Coking stoker

This type of stoker is shown in figure 4.5. The coal is contained in a hopper and falls, under the action of gravity, into a divider having a reciprocating ram (or pusher), which forces the coal into the hottest part of the furnace and on to the top coking plate. At this stage some of the volatiles are given off and burnt, so starting the process of turning the coal into coke. During the next forward stroke of the ram, fresh fuel displaces the previous charge, moving part of it on to a bottom coking plate, where it continues to give off volatiles that burn on the underside of the fuel bed. The remaining fuel is pushed into the established fire bed where it forms a 'wedge' of coal, which then burns on the top and bottom sides of the wedge.

Subsequent movement of the whole burning mass further into the combustion chamber is accomplished by moving all the firebars together in this direction. The motion is carried out using cams placed so that continued rotation then retracts groups of single firebars until all the bars are again fully retracted. The burning fuel, which is now changing into ash, is eventually pushed over the end of the grate on subsequent forward strokes of the ram. An induced draught fan is required to sustain a high rate of combustion. This type of stoker is usually fitted to shell boilers and small watertube boilers having ratings up to 4.5 MW, and the type of fuel used can be doubles, singles or smalls.

4.2.4   Chain grate stoker

Figure 4.6 shows the stoker, which relies on a gravity fed hopper discharging fuel on to a slowly moving endless grate, which eventually discharges the spent fuel on to the ash removal conveyor. The depth of the firebed is controlled by the vertical position of the entry gate and, in common with the coking stoker, the fuel entering the combustion chamber does so underneath a refractory arch, which radiates energy on to the coal. At this point the volatiles are given off and burnt, and the coal starts to change into coke, burning from the top of the grate. A forced draught fan is provided and the speed of the grate can be infinitely variable, or be driven by a fixed-speed motor operated by a time switch. The preferred coal is singles and washed or untreated smalls having a rank that is not strongly caking. The chain grate stoker may be fitted to shell and watertube boilers up to 80 MW.

4.2.5   Travelling grate stoker

This stoker operates on the same principle as the chain grate stoker but is used for larger thermal loads. The difference between the two stokers lies in the way in which the movement of the grate is obtained. The grate bars of the travelling grate stoker are moved by having their ends fixed to

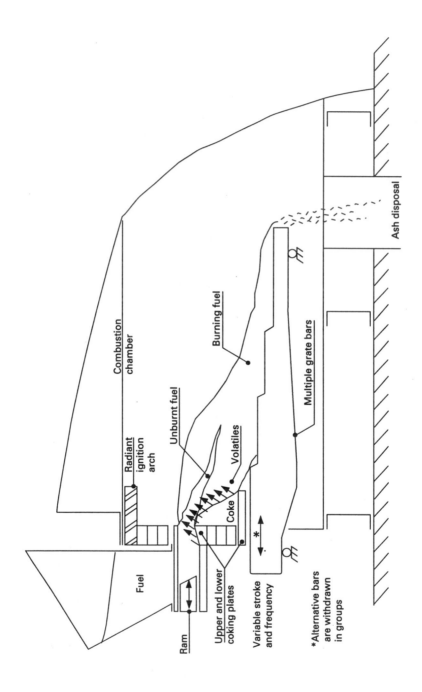

Combustion chamber

Radiant ignition arch

Unburnt fuel

Burning fuel

Volatiles

Coke

Multiple grate bars

Fuel

Ram

Upper and lower coking plates

Variable stroke and frequency

*Alternative bars are withdrawn in groups

Ash disposal

Figure 4.5   The coking stoker

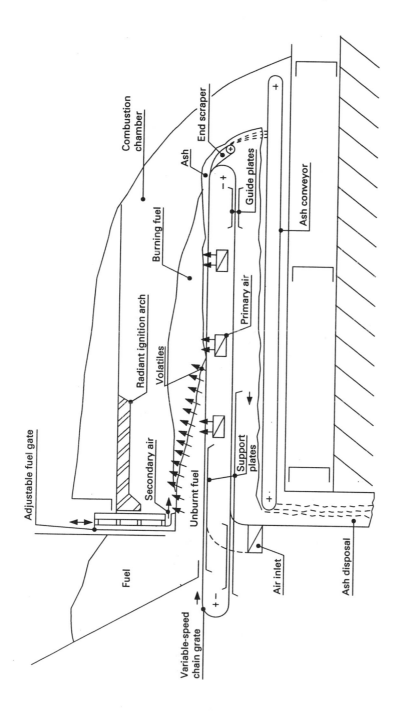

Figure 4.6 The chain grate stoker

two endless chains, one on each side of the boiler. Different types of grate bars are needed for different fuels, and it is important that the manufacturer is informed of the choice of fuel at a sufficiently early stage in the design. The preferred fuel is smalls.

4.2.6 Sprinkler stoker

Early versions of this system were developed to mimic the way in which manual stoking of the highest standards was intended to be performed. The perfect stoker, using his physical dexterity, would always be expected to deliver the fuel evenly along the full length of the grate, thus ensuring that all the combustion space was used. This ideal was approached by arranging for the fuel to fall under the action of gravity, from a hopper on to a rotating drum, when it was given sufficient impetus to be thrown along the whole length of the fixed grate.

Modern versions of this system rely on the coal being pulverised and introduced into the combustion space pneumatically, either through a vertical tube positioned over the centre of the combustion chamber, or through a horizontal tube fitted at the front of the boiler, see figure 4.7. This latter system can be retrofitted to an existing boiler, and all boilers fired in this way have secondary air for combustion introduced at the front and the larger plant also makes use of an additional secondary air supply provided at the rear of the combustion space.

In this system the mass of coal being burnt at any one time is smaller than in other types of boilers, so the response to changes in load are very much improved, making the burner in this respect on a par with oil and gas. In both cases, grit arresters are fitted and the reclaimed partially burnt fuel is then reintroduced into the combustion chamber. Sprinkler stokers are used on boilers having an output ranging from 600 kW to 8.8 MW and the fuel used is bituminous singles. The ash that is formed can be removed either automatically or manually, the task requiring a few minutes every 6 to 8 hours.

4.2.7 Use of a fluidised bed as a coal burner

Fluidised beds were initially developed to function as conveyors of granular and powdered materials. The material, originally at rest in a sloping, enclosed longitudinal conveyor, was blown upwards by air, which was introduced from the underside of the conveyor along its length. The material became airborne and the stream of air then transported the particles in the direction required.

When the fluidised bed is utilised as a combustor, it consists typically of a bed of sand 150 to 250 mm deep, which, when air is introduced from the underside of the container, causes the bed to rise to approximately twice its original depth. By pre-warming the space above the bed to a temperature between 500°C and 600°C using gas or oil burners, and

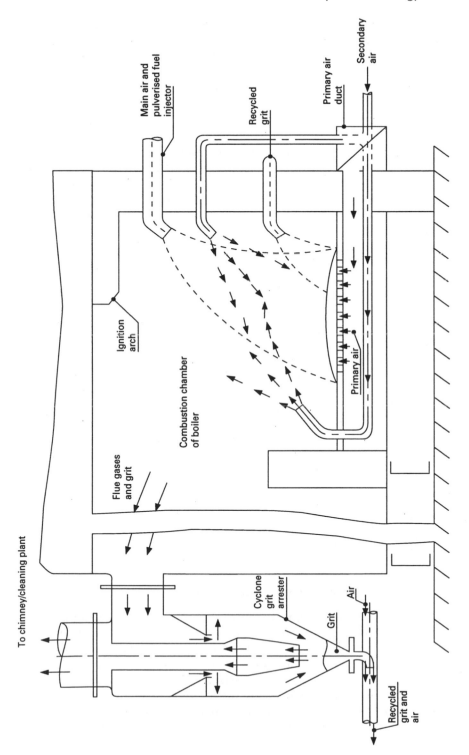

Figure 4.7 The fixed grate front sprinkler firing system

then introducing coal, the operating temperature of the bed rises to between 800°C and 950°C. The distribution of temperature in the depth of
the bed and above it is uniform, making heat transfer conditions excellent, and coils are fitted in these positions.

Very large boilers can be fitted with fluidised beds 1 metre in depth
and requiring a vertical space above the bed of between 3 and 4 metres
to complete combustion. These are referred to as 'deep beds' and they
will burn low grade fuels. It is also possible to include some limestone in
the material of the bed, which, when it combines with the sulphur in the
fuel, reduces the acidic nature of the flue gases. The rate at which energy
can be transferred from a fluidised bed is very high and this results in
boiler sizes being smaller than those of the more conventional types.
Figure 4.8 illustrates a fluidised bed having heat exchange surfaces above
and below it.

At the time of writing, research was being carried out by the British
Coal Research Establishment into the use of fluidised combustors employed as gasifiers and energy converters, used in conjunction with a gas
turbine and compressor generating set and a steam turbine generating
set. The thermodynamic cycle used is referred to as the 'topping cycle'
and it operates using a pressurised fluidised bed combustor (PFBC), which
may be coupled to an atmospheric pressure-circulating fluidised bed combustor to burn the char.

The name of the cycle comes from the act of removing fuel gas from
the pressurised gasifier at the 'top' of the cycle, before introducing it via
a cleaner to the combustor of a gas turbine. When dealing with a generating plant using pulverised fuel having a capacity of 350 MW, it is expected that the use of the topping cycle will produce a 20% reduction in
the cost of generating electricity, a 20% reduction in carbon dioxide production, low emissions of sulphur and nitrogen oxides, and a low capital
cost [4.3].

4.3 Equipment required to burn oil

The equipment used to burn oil depends on the grade of oil being burnt
and it is categorised below.

4.3.1 Oil-fired burners: the pressure jet type

For the range of boilers from 10 kW to 2.5 MW the use of the pressure
jet type of burner is common. This unit uses a gear pump to force oil
through a suitably sized orifice into the combustion chamber. The oil is
split up into tiny droplets, each of which presents a relatively large surface area to the surrounding air, so assisting combustion. The air necessary for combustion is introduced by means of a cross-flow fan, and the
motor driving this fan also drives the oil gear pump. Each unit has an
electronic controller which is designed to 'fail safe' under all possible

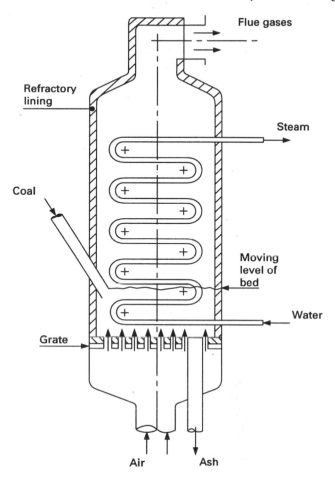

Figure 4.8 Fluidised bed combustion chamber

operating conditions. A typical arrangement of the fan and controls for a domestic burner are shown in figure 4.9, and the starting sequence is as follows:

(1) The electrodes spark for about 20 s before the motor driving the fan and gear pump is energised.
(2) The oil is ignited.
(3) The electrodes continue to spark for a further 20 s in case initial 'flame-out' occurs; the presence or absence of a flame is detected using a photo-electric cell and if the flame is not detected after the second period of 20 s, the electrodes will spark again to ignite the oil and the system will spark intermittently over a period of 10 s in an attempt to establish the flame before finally proceeding to 'lock-out'.

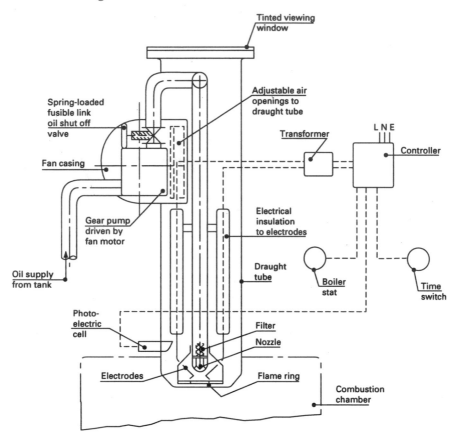

Figure 4.9 Domestic pressure jet oil-fired burner

(4) The controller then indicates by means of a light that this condition has been reached and will not allow another automatic restart until a period of 1 to 2 minutes has elapsed, so giving time for unburnt fuel to be cleared by natural draught from the combustion chamber; the restart button, when pressed, then allows the starting sequence to be repeated.

When commissioning an oil burner, adjustments have to be made to regulate the quantity of air for combustion, so that the oxygen content of the flue gases is approximately between 6.5% and 7.5% by volume and the carbon dioxide content is between 10% and 12% by volume. Such limits ensure that all the carbon in the fuel is burnt completely to carbon dioxide, so causing the minimum of soot formation. In the absence of a flue gas testing kit, the air register can be set temporarily so that the observed colour of the oil flame approaches white from a bright orange. When the flame is dark orange, combustion is incomplete and soot will be deposited on the surrounding surfaces.

In the case of new equipment, it can be assumed that the oil burner will have been assembled in the factory so that only the air register will require adjustment in order to ensure the best combustion conditions. If however, the various settings have been altered, the subsequent setting up of the burner so that it will ignite reliably can be frustrating and time-consuming. Some of the variables affecting this process are:

(1) grade of oil;
(2) oil pressure at the orifice;
(3) presence of air and/or water in the oil supply pipe;
(4) size, type and cleanliness of the orifice and its filter;
(5) position of the flame ring relative to the orifice, and the position of the orifice relative to the end of the air draught tube;
(6) concentricity of the burner and flame tube in the air draught tube;
(7) magnitude of the flue draught and the setting of the air register;
(8) position and cleanliness of the electrodes.

4.3.2   Rotary cup burners

For the range of burners from 150 kW to 5 MW, two kinds of rotary cup units are available.

Type (a)

Air under pressure is directed on to a vaned rotor on the centre-line of which a frustum of a cone is mounted, see figure 4.10. Oil is introduced through the hollow shaft of the rotor and moved by centrifugal force to the large diameter of the cone, where it is mixed with the air emerging from an annulus, formed by the inside of the draught tube and the outside of the cone. This ring of air, having passed through the vaned rotor, serves to split the stream of oil into tiny droplets, which are readily burnt in the combustion chamber. The air required to drive the rotor represents approximately 20% of that needed for combustion; the remaining 80% is supplied by means of a separate fan. It should be noted that if the volume flow rate of the air falls below the design value, the rate of spin of the rotor will fall and a 'blow back' may occur, as a result of which flames from the burning oil will emerge from the front of the boiler.

Type (b)

In this version the spinning cup is motor driven and the air supplied for mixing comes from a fan powered by the same motor. With this arrangement the operation of the unit is more reliable and has a lower fire risk. Larger burners are usually ignited from a gas flame which itself is ignited by means of a spark from electrodes.

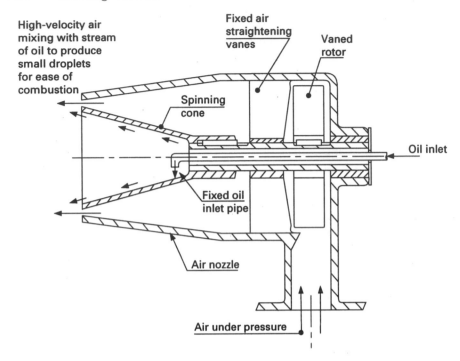

Figure 4.10 Rotary cup burner type (a)

4.4 Equipment required to absorb energy

Thermal energy absorbing appliances, colloquially known as boilers, almost invariably use water as the working substance. Those boilers used in the built environment are made of cast iron or steel, and the various types can be divided broadly into five categories.

4.4.1 Cast iron sectional boiler (up to 1 MW)

This type of boiler, shown in figure 4.11, is constructed of a number of separate sections made of nickel cast iron. The addition of nickel to the iron casting process makes the material of the sections tougher, and this means that they are more able to withstand any sudden mechanical shocks, which could occur during or after assembly. Each section when cast and machined can be connected to other sections to produce a boiler suitable for a given output. Traditionally when coke (a by-product of coal) was used as a fuel, the maximum number of sections was limited to about 13 – this being the maximum length of boiler that it was thought could be stoked efficiently by hand.

Whenever a conversion from manually stoked solid fuel to gas or oil was carried out on an existing cast iron sectional boiler, it was found

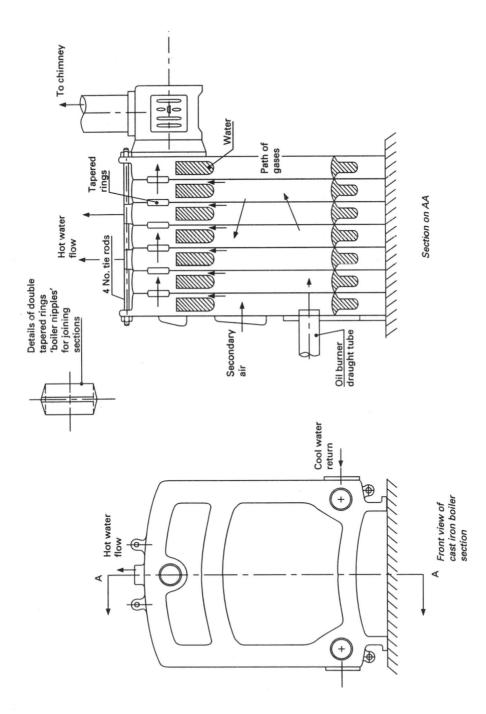

Figure 4.11 Seven-section cast iron oil-fired boiler

that the temperature inside the boiler changed very quickly after initiation of the flame and sometimes leaks developed between the sections, caused by thermal stresses producing expansion and contraction owing to the normal on/off cycling of the burner. This fault was more likely to occur in boilers having a number of sections greater than ten, so that it was not considered good practice to convert boilers having more than ten sections to an 'automatic' fuel.

Earlier, cast iron solid fuel boilers were mounted on a layer of insulating bricks, surrounded by engineering bricks so that the large amount of energy being radiated downwards from the underside of the grate could be reflected back into the combustion space, so avoiding damage to the floor. Cast iron boilers manufactured for use with an automatic solid fuel burner, or oil and gas burners, now have waterways which completely surround the furnace so that generally it is no longer necessary to insulate the plinth on which the boiler stands.

This type of boiler is generally suitable for use where the head of water above it is no greater than 40 metres. In an open vented system (one in which at some point the pipework is open to the atmosphere), this is equivalent to a building height of about 13 storeys. However, some boilers can be specially designed to withstand heads of water greater than 40 m.

It is interesting to note that if the boiler is sited on the roof of a building (rather than, say, in the basement), this will cause the internal water pressure in the boiler to be lower, so that the possible safe operating temperature of the water will also be lower. This means that the amount of energy carried by 1 kg of water is less. It follows that for a given thermal output, the mass flow rate of the circulating water must be greater, resulting in the expenditure of more energy in pumping. However, the act of siting the boiler on the roof will save the significant cost of a chimney being constructed for the full height of the building.

When determining the pressure at the base of a column of fluid, the following equation applies:

$$P = h \times \rho \times g$$

where P = pressure in Pascal (Pa) or (N/m²)
  h = head of fluid in m
  ρ = density of the fluid (kg/m³)
  g = acceleration due to the earth's gravitational field (gravity), 9.81 (m/s²).

4.4.2 Steel shell boiler (100 kW to 3.5 MW)

The development of this type of boiler was helped considerably by the advent of reliable welding and testing techniques, both of which may now be done automatically, making it possible to fabricate a complete

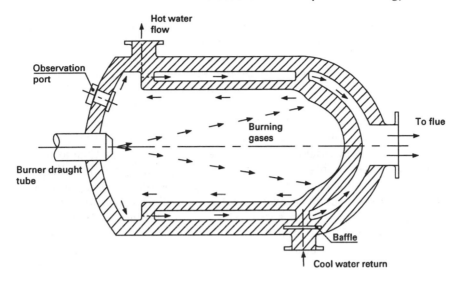

Figure 4.12   Steel shell boiler

unit of a size to suit the required output, see figure 4.12. Mild steel can be used because rusting will only occur on the water side when oxygen is available. In this case the source of oxygen is the water and because the same water is circulated continuously, any oxygen present initially will, after a finite time, be converted to ferric oxide and no further rusting will then take place. In hard water regions, if fresh water is introduced, scale will be deposited on the heating surfaces of the boiler and more rusting will occur. It is for these reasons that it is not good practice to fit a domestic hot water tap to the hot water output side of a boiler.

Whenever any fuel containing sulphur is used, it is possible for corrosion to occur on the combustion side of the mild steel shell and this is called 'thermal corrosion'. This will occur if the minimum temperature of the water inside the boiler falls below about 60°C. If as a result of local cooling the flue gas temperature comes within the range 140°C to 115°C, then the dew point of the sulphur compounds may be reached and sulphuric acid will be condensed on the surface of the steel, resulting in the occurrence of thermal corrosion in the locality of the cooled area. This type of corrosion can occur when the burner is either on or off. It can be kept to a minimum by setting the boiler controls so that the lowest boiler temperature is normally above 60°C.

The temperature control of most domestic heating systems relies upon the adjustment of the boiler thermostat, so that under normal operating conditions, the boiler will be operating for long periods at water temperatures very much lower than 60°C. For systems operating in this way, a life expectancy for any mild steel boiler of anything over 5 or 6 years would be a bonus and it is for this reason that some manufacturers of domestic boilers construct the combustion chambers from cast iron or stainless steel. For householders having a mild steel boiler who want to make it last a little longer, it is necessary for them to resist the temptation to control the temperature of the emitters by adjusting the boiler thermostat to such a low temperature that conditions conducive to the onset of thermal corrosion apply. One solution to this problem is to set the boiler thermostat to 80°C and to control the temperature of the emitters using a by-pass and a motorised mixing valve controlled by a compensator unit. The compensatory control system is discussed in Chapter 6.

Steel boilers are lighter than the equivalent sized cast iron boilers and, unlike boilers using cast iron sections, will not normally fracture if the water contained in them freezes. Steel shell boilers can also withstand a greater head of water than sectional cast iron boilers. The thermal capacity of steel boilers is less than cast iron sectional boilers, so that response times are shorter and on/off losses (cycling losses) are correspondingly reduced. They can be fuelled automatically by solid fuel, oil or gas.

4.4.3 Electrode boiler (up to 3.1 MW)

The electrode boiler is a special-purpose energy converter and absorber based on the shell boiler. Unlike other boilers, it does not require a flue or a burner as it relies on the heating effect of an alternating electric current when it is passed through water. Readers should note that if direct current was used, electrolysis of the water would occur and hydrogen and oxygen bubbles would be produced. The use of alternating current prevents such formations occurring. The addition of sodium carbonate or sodium sulphite to water makes it more resistive to an electric current, which increases the power dissipated (the I^2R effect), resulting in more heat being released. The boiler can be used with water or steam as the working substance, see figures 4.13(a) and (b), but different controlling methods are employed depending on which phase of the water substance is being used.

Control method when water is used as the working substance

The magnitude of the flow of current between the electrodes is proportional to their submerged surface area and as the heat released is proportional to the square of the current (the resistance being largely constant), the thermal output can be controlled by automatically varying the vertical position of the porcelain insulators. When the boiler is cold, the insulating cover is at the lowest point, so that an initially low starting current occurs (a useful characteristic for direct on-line switching). The insulating shield is then raised until it is in the position indicated by broken lines in figure 4.13(a) (the highest point), which is the position for normal operation.

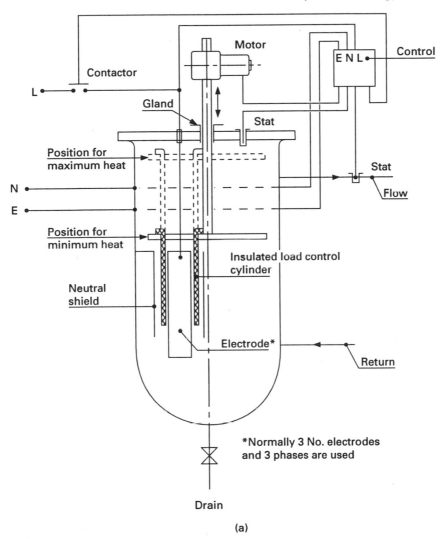

Figure 4.13 (a) Electrode boiler using water as the working substance.

Control method when steam is used as the working substance

In this case, when the boiler is cold it contains air and water up to the level A, as shown in figure 4.13(b). By raising the water level to B as shown, all of the surface area of the electrodes is brought into contact with the water, maximum current flows and maximum heat output is obtained. The steam generated mixes with the air, and the free surface of the water gradually falls, so reducing the value of the current, this action

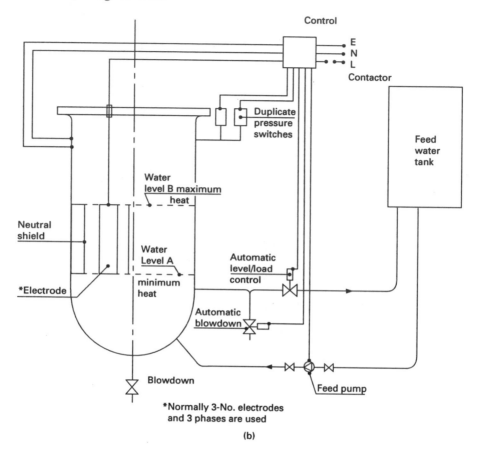

Control

E
N
L

Contactor

Duplicate
pressure
switches

Feed
water
tank

Water
level B maximum
heat

Neutral
shield

Water
Level A

Automatic
level/load
control

minimum
heat

*Electrode

Automatic
blowdown

Blowdown

Feed pump

*Normally 3-No. electrodes
and 3 phases are used

(b)

Figure 4.13   (b) Electrode boiler using steam as the working substance

making the boiler inherently safe. By changing the mass of water inside
the shell of the boiler, the pressure and mass flow rate of the steam can
be controlled.

Electrode boilers producing steam are to be found in centralised air
conditioning units (mainly in health care premises) and also in the process industries. Many are used even when peak rate electricity tariffs are
in force and they provide steam locally, at the point of production, for a
finite period. This can sometimes be shown to be more economical than
the use of a central steam generating plant, which may have relatively
high thermal transmission losses over an extended period.

In order to ensure that the heat exchange surfaces of the boiler do not
become encrusted with scale, making them less efficient, a water treatment plant must be available to remove the hardness salts in the feed
water. The blowdown valves shown in figures 4.13(a) and (b) are opened
from time to time, in order to remove an accumulation of dissolved solids,
which are precipitated by the water softening and heating processes.

4.4.4 *Fire tube boiler (1 to 10 MW)*

For larger thermal outputs, a steel shell boiler is provided with banks of fire tubes containing hot gases on the inside and water on the outside. They are produced in horizontal or vertical form and are constructed so that the flue gases traverse the length of the boiler either two or three times and are then referred to as two or three pass boilers. Figure 4.14 shows a three-pass horizontal shell type 'wet back' fire tube boiler. The term 'wet back' refers to the cross-over duct for the combustion gases at the rear of the boiler being surrounded by water. The term 'dry back' is used to describe the combustion cross-over duct when it contains tubes having either water or steam inside them. When steam is contained in the tubes, it becomes superheated.

The fuel used can be solid, liquid or gaseous and, with the expenditure of some extra capital, it is sometimes possible to change over to an alternative fuel or fuels, if any one becomes cheaper or more readily available. This and other types of boiler are produced by some manufacturers as a 'packaged' unit, which implies that the boiler has been factory tested and comes complete with an integrated electronic control system. The 'Marske' fire tube boiler, having an output of approximately 18 MW, has been developed in order to take advantage of recent developments in fluidised bed techniques.

In an attempt to attract new orders and to keep ahead of their competitors, boiler manufacturers are continuously engaging in development work and currently the main thrust of this is in the field of solid-state electronic controls, using more and more elegant purpose-designed 'chips'. The use of such high technology is a bonus when all systems are 'go', but the author suggests that simple DIY back-up systems should also be incorporated so that in the event of the automatic control system malfunctioning, maintenance personnel can ensure continued operation, be it only on a temporary basis.

4.4.5 *Water tube boiler (500 kW to 10 MW and upwards)*

As the name suggests, this type of boiler consists of tubes containing water and/or steam on the inside, with burning gases in banks that may have any convenient configuration to suit the shape of the boiler and the type of firing required. The output may be hot water under pressure and/or superheated steam (the latter being steam that is not in contact with water and has reached a temperature above the saturation temperature corresponding to the local pressure). These boilers are reserved for the largest thermal loads and, because the diameters of the water tubes are small, it is possible to operate the boiler at much higher pressures than would be possible with a steel shell boiler. When steam is used as the working substance, the higher pressures and temperatures that are produced enable the steam to be used to drive a turbine or a reciprocating

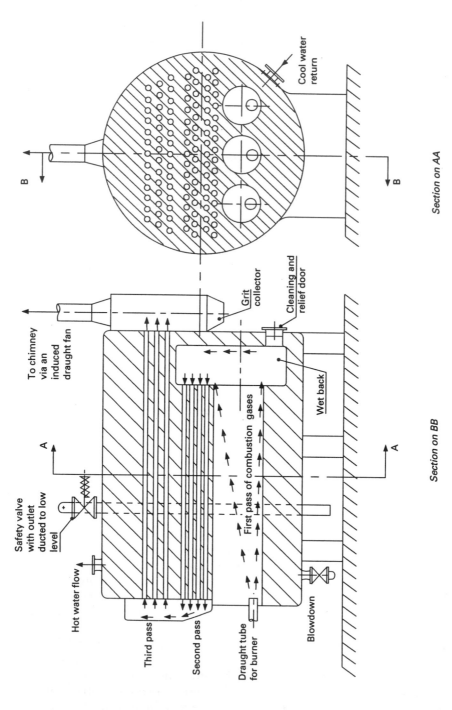

Figure 4.14   A three-pass horizontal shell type wet back fire tube boiler

engine, either of which may be coupled to an electrical generator. Some of the smaller water tube boilers are of the packaged variety, but larger boilers are the result of individual design and are often purpose built to the requirements of the client.

4.5 Flueing arrangements and emissions

All boilers, except electrode boilers, require a system that will continuously remove the products of combustion from the combustion chamber, so that the burning process can continue. In the past it was considered acceptable, in the case of installations having several boilers, to collect the gases from each boiler and direct them into a common duct, running at an angle of just a few degrees above the horizontal and joining into the base of a common vertical chimney. This is no longer considered acceptable, as under low load conditions, the efflux velocity of the flue gases from the top of the chimney will be too low. In addition, if the boilers employ forced draught fans, flue gases from the duty boilers will find their way into the boiler house, via the common flueway connections. Better practice is to ensure that each boiler is connected to an individual stack, normally contained in a common envelope, and figure 4.15 illustrates four possible types of construction.

The recommended velocity of efflux of gases from the top of the chimney for natural draught boilers is 6 m/s. This velocity is sufficient to prevent:

(a) down-washing of flue gases on the outside of the chimney, on the leaward side;
(b) passage of cold outside air down the inside of the chimney, so cooling the upper section.

Frequently the diameter at the top of the chimney is reduced by using a nozzle, in order to ensure that the efflux velocity is increased at the point of exit.

It is essential that chimneys are insulated sufficiently to make sure that the temperature of the gases does not fall within the range 115°C to 140°C, when the dew point (condensation point) of sulphur dioxide is reached and sticky acid smuts form. These adhere loosely to the lining of the chimney, until a subsequent increase in the volume flow rate of the flue gases dislodges them, producing an obvious effect on the surroundings. Note that the use of draught stabilisers which reduce the magnitude of the draught in the chimney by allowing relatively cool air into the system, may also have the effect of helping to cool the gases down to the dew point temperature.

The production of acid rain which is precipitated upon the lakes and forests of surrounding areas is currently a matter of much concern. It is suggested that it is a simplistic solution to raise the height of a chimney

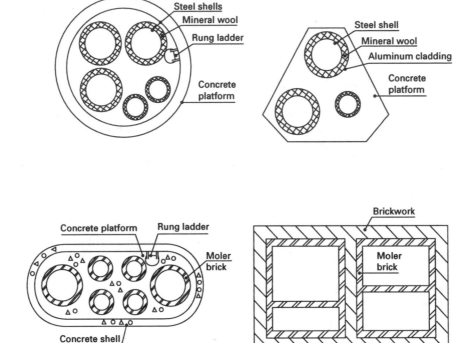

Figure 4.15 Arrangements for multiple flues

so that, in the event, acid rain falls not on the surrounding areas but on distant parts. Current technology has been developed to remove the sulphur compounds from the flue gases of power stations, by desulphurisation at a cost of approximately £250 million per power station. The by-products of the washing processes are sulphuric acid and gypsum, which are both useful commodities, but may eventually be produced in excessive quantities.

Other compounds that contribute to acid rain come from the oxides of nitrogen and are designated as NO_x. They are formed from the nitrogen present in both the fuel and the air used for combustion. When natural gas is burnt, the main constituent of the NO_x compounds formed in the combustion chamber is nitric oxide (NO); this quickly combines with the oxygen in the atmosphere to form nitrogen dioxide (NO_2), which then reacts with other compounds to form nitric acid (HNO_3). From 1992 onwards, a great deal of confidential experimental work was carried out on combustion and flame chemistry, so that the manufacturers of combustion equipment would be able to produce 'low NO_x' combustion chambers. A European limit for NO_x compounds of 250 mg/kWh for boilers of 70 kW and below has been proposed. The relevant Technical Committees are TC 57 and TC 109. Research work was also being done by BSRIA on the characteristics of NO_x and carbon monoxide (CO) emissions from boilers. In 1992, tests had shown that the levels of CO are highest for those boilers having the lowest NO_x values, and a report is available from BSRIA, see also [4.4]).

As already mentioned, the sulphur content of natural gas is negligible when compared with that of coal (1.0 to 1.9% by weight) and oil (0.2 to 4.5% by weight), and a switch to the use of natural gas in power stations (provided the price is realistic) would be an obvious step in the search to reduce sulphur compounds in the flue gases (and the upper atmosphere). For many years, coal has been the fuel from which the greatest proportion of the electrical energy produced in the UK has been obtained. In 1991, about 75% of the electrical energy generated by National Power and Powergen was derived from coal [4.5] but this proportion has subsequently diminished. British Coal was obviously keen to maintain its share of the market, and several interesting new techniques for desulphurisation have been tried and subsequently implemented. In large conventional boilers and those using fluidised beds, burning pulverised coal, lime is added at the point of burning in order to make the flue gases less acid.

4.5.1 Use of the condensing boiler

An interesting development in the search for greater thermal efficiency from boilers burning natural gas is to insert one or two non-corrosive heat exchangers (depending on the size of the boiler) into the gas stream at the exit from the boiler. This has the effect of cooling the gases from about 300°C to 40°C, a temperature which is below both the acid and water dew points. Gas boilers that have these extra heat exchangers fitted are referred to as 'condensing boilers' and the resulting effects of the acidic products so formed are acceptable, provided a suitable non-corrosive drain system is fitted. Because the flue gas is at a very much lower temperature, the stack effect is much reduced and the gases have to be removed by a fan.

At this point, it is apposite to introduce a note of caution about the way in which condensing boilers might be employed: this is concerned with the likely operating temperatures of the heat exchanger in the flueway of the condensing boiler. In order for the heat exchanger to function, there must always be a temperature difference between the fluids on either side of the heat transfer surface. If the leaving flue gas temperature is 40°C and the temperature difference across the surface at this point is 5 K, it follows that the temperature of the water returning to the boiler from the heating load must not be greater than 35°C in order to take full advantage of the conditions available. When this is so, it is claimed by the manufacturers that the thermal efficiency of the boiler will be approximately 94%. Lower return temperatures from the plant may produce even higher efficiencies.

If the heating system supplies embedded coils and panels, it is likely that return water temperatures of 40°C will occur. However, if the system supplies conventional radiators, sized to operate (when at full output) with a flow temperature of 70°C and a return temperature of 60°C, it follows that, in this case, a proportion of the extra condensing heat exchanger surface provided in the boiler will not be contributing fully and the thermal efficiency of the boiler will be enhanced, but by a smaller amount than might be expected. Hence in order to take full advantage of the condensing heat exchanger surface, the temperature of the return must be around 35°C.

4.5.2   Use of nuclear energy to limit noxious emissions

It is tempting to propose that more power stations should be nuclear powered so that the emissions of sulphur compounds, nitrogen oxides and carbon dioxide will be further reduced – the latter being a contributor to the greenhouse effect. Opponents of the use of thermo-nuclear power will affirm that possible accidental emissions of a radioactive nature together with the continuing accumulation of radioactive waste militate against such a choice. It is vital that the radioactive waste so produced is satisfactorily disposed of deep underground in granite bearing rocks or deep under the sea bed.

4.5.3   Use of the balanced flue

This device originated in the gas industry and was the result of attempts to ensure stable (that is, balanced) pressure conditions inside the combustion space of gas-fired boilers. The likelihood of 'flame out' occurring, owing to the incidence of random down-draughts, would clearly be reduced if the pressures of the combustion air and the flue gases were at all times equal. Figure 4.16 shows a balanced flue fitted to a gas appliance and it will be appreciated that because both inlet and outlet orifices are positioned at the same place, there is unlikely to be any large difference in pressure between the two gas streams. However, an obvious exception to this will arise if the common inlet/outlet is positioned close to a windy corner of a building, where violent gusts of wind sometimes occur.

The use of a balanced flue on any gas-fired or oil-fired boiler having up to 60 kW heat input provides much more flexibility in the positioning of such appliances, but installers must be careful to take note of the recommendations made by the manufacturers and British Gas, of those places where it would be inadvisable to fit such a flue, see [4.6]. The document gives details of the Gas Regulations, together with information on ventilation and combustion, as well as a list of minimum distances

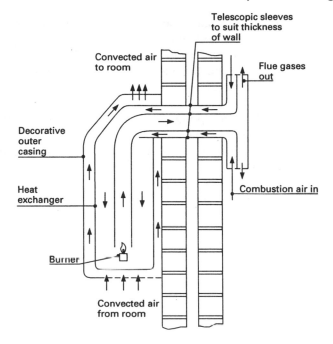

Figure 4.16 Room sealed convector with balanced flue

permissible for the fitting of a balanced flue, from open windows, gutters and plastic soil pipes and the corners of a building, together with information on the method of sleeving and of fireproofing the short stub duct when it has to pass through combustible materials. The discharge of undiluted combustion gases on to a public walkway is not allowed, and any gas-fired water heater fitted to a shower room, bathroom or garage must be of the room sealed type.

In those cases where a suitable outside wall is available to site the flue outlet, but where it is at an angle to, and no more than 2 metres from the appliance, it is possible to fit an extended angled flueway. In such cases it is necessary to use an extractor fan to discharge the combustion gases, and this arrangement is referred to as 'fan assisted'. Note also that where an existing conventional flue is available that is not large enough to deal with the flue gases from a proposed new appliance, it is usually possible to install an induced draught fan to overcome the difficulty.

4.5.4 Use of the SE-duct

The problem of providing flueways to a multiplicity of appliances burning natural gas, as in a large block of flats, may conveniently be dealt with by providing a single vertical flueway, usually made of refractory insulated concrete. If the gas appliances are 'room sealed', that is the air for combustion and the flue gases are kept separate from the rooms in which the appliances are fixed, then a SE-duct (the SE is pronounced 'see') may be built in to the structure. The prefix 'SE' stands for the former South East (gas) Region where the idea was first developed. Figure 4.17 shows three ways in which air may be introduced at the bottom of the SE-duct.

Note that all balanced flue appliances are room sealed, but that not all room sealed appliances have balanced flues. A room sealed unit may have an inlet for combustion air that is drawing air from a second space, while discharging the flue gases to a third space.

The size of the SE-duct must be sufficient to allow for the products of combustion to be diluted to such an extent that when all the units are in operation, the carbon dioxide content in the duct at the uppermost appliance is no more than 2% by volume. In the SE-duct system it is likely that quite different pressures will exist at the air inlet at the bottom of the building and at the flue outlet on the roof, owing to the variable effects of the wind and also the stack effect. If it is considerd likely that the local wind conditions might be too extreme to use a SE-duct, or if it is not possible (perhaps in a retrofit) to provide for an air inlet at low level, then the U-duct may be used.

4.5.5   Use of the U-duct

In the case of a U-duct, both the air for combustion and the flue gases are respectively taken in and exhausted through the same terminal sited on the roof, see figure 4.18(a). The system behaves as a balanced flue, ensuring the differential pressure between the inlet and outlet of all appliances remains constant. This system does, however, require a greater flue area than the SE-duct.

4.5.6   Use of the shunt-duct

When the appliances used are not of the room sealed type, a 'shunt-duct' may be used, see figure 4.18(b), and this particular arrangement requires less space inside the building than the other forms of common duct. The manufacture and supply of complete flue units for all the aforementioned types of duct, together with flues suitable for use in timber framed buildings, are undertaken by specialist companies.

4.5.7   Use of flue gas dilution techniques

In installations where a gas appliance has to be fitted at low level and where, in addition, it is considered not to be practical to install a flue perhaps extending for the full height of the building, the problem of disposing of the products of combustion can be overcome by arranging for their dilution in the ratio of 14:1. In this way the carbon dioxide content

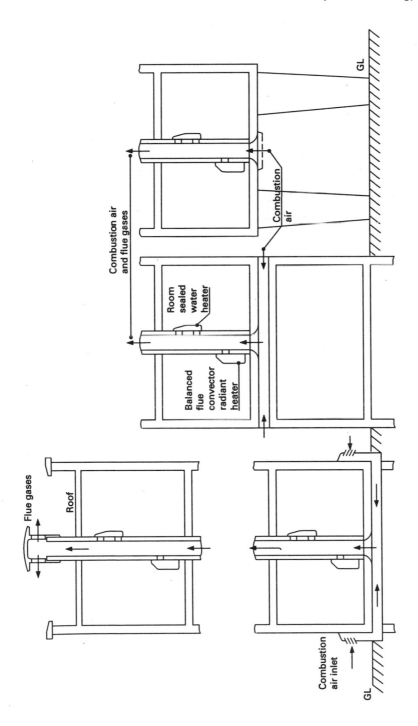

Figure 4.17 SE-duct applied to multi-storey flats

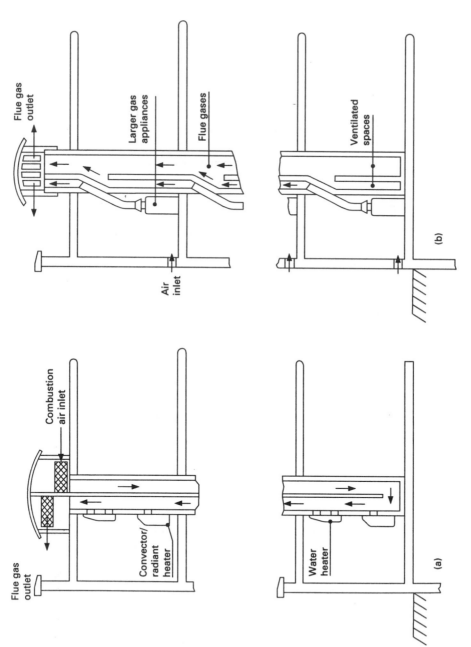

Figure 4.18 (a) U-duct and (b) shunt duct applied to multi-storey flats

of the gases is reduced to less than 1% by volume, when they may safely be discharged at low level to the surroundings, and their presence will not be apparent.

One way of accomplishing this is to use a bifurcated fan (sometimes called a split fan), in which one duct containing outside air and another containing flue gases are connected together into the eye of the fan, where they are then mixed and propelled through a third duct to the atmosphere. The use of a bifurcated fan makes it possible to fix the motor driving the fan in the cooler ambient air, where the hot and potentially corrosive gases cannot harm it. Some local authorities stipulate that a standby fan shall be fitted, which will automatically be energised if the duty fan stops turning. If the standby fan also fails, the appliance must then be shut down by the control system. Another way of carrying out the necessary dilution is to fit an axial flow fan in a duct, which is jointly connected to the outside air and to the flue outlet from the unit.

4.5.8 Supply of air for combustion and ventilation

It is important that all burners are provided with a continous supply of fresh air in order that the oxidation process may be successfully completed. When a sleeping family competes with a burner for a supply of oxygen, the family will lose and be asphyxiated. It is therefore of paramount importance that the designer/installer of any heating system involving the use of a burner ensures that the supply of air for combustion and ventilation is sufficient and will be unhindered at all times.

Flueless appliances should only be used in an enclosure that has at least one opening window, together with multiple fixed air openings, both at high and at low level. The aggregate free area of the grilles is to be as specified by the makers of the heating appliance.

For intermediate sized burners having a conventional flue, the determination of a suitable free area of grille suitable for combustion and ventilation has in the past been obtained by some designers by multiplying the area of the flue by 3. However, for such an important consideration it is always best to abide by the recommendations given in approved document J (1990 edition) of the Building Regulations 1985, and also as given by the manufacturers of the appliance, which should be checked against the details given in BS 5440 *Code of Practice for flues and air supply for gas appliances of rated input not exceeding 60 kW. Part 2: 1976 Air supply*. Another obvious and fruitful source of information is British Gas.

For larger burners (45 kW to 6.0 MW) having a conventional flue, the CIBSE Guide, section B13, gives many design details to enable sufficient air both for combustion and ventilation to be 'built in' to any system.

References

4.1. National Coal Board/British Coal. *Details of Industrial Solid Fuel Plant.* Technical literature, 1965–95. Past publications of British Coal may be obtained from the Publications Section, College of Fuel Technology, and British Coal, Hobart House, Grosvenor Place, London SW1X 7AE.
4.2. *CIBSE Guide*, Vol. B, 1986, sections B13 and 45. CIBSE, Delta House, 222 Balham High Road, London SW12 9BS.
4.3. British Coal Research Establishment. *The British Coal Topping Cycle.* Stoke Orchard, Cheltenham, Gloucestershire GL52 4RZ.
4.4. Teekoram A. Boilers take the knocks. *CIBSE Journal*, December 1992.
4.5. National Power and Powergen. *The First Share Prospectus*, 1991.
4.6. British Gas. *Gas in Housing – A Technical Guide*, 1994.

5 Distribution, Emission and Control of Thermal Energy

5.1 Early distribution of energy

The first application of central heating, where the source of heat is located some distance away from the area that is to be heated, is credited to the Lacedaemonians of Greece [5.1], who in 350 B.C. constructed a temple in Ephesus, in which hot gases, derived from burning lignacite, were passed through conduits built into the floor.

The greatest colonisers around the time of 80 B.C. were the Romans and as they advanced further northwards, the use of fire to warm themselves and their buildings became essential. G. Sergius Orata is reputed to have invented the 'hypocaust', a term used to describe a central heating system in which hot gases obtained from a fire, using the stack effect, passed underneath the floor and through flues built into the walls. The gases, which may be referred to as the 'working substance', were not returned to the furnace, and such a process may be classified as being 'non-cyclic' and 'indirect'.

The Romans also built communal bath houses with living quarters adjacent, providing an environment in which weary travellers and others could clean up and relax. The water for the baths was heated in a wood-burning boiler constructed of lead, which was supported on iron bars. The flue gases were allowed to escape to the atmosphere, carrying with them a relatively large amount of heat that had not been utilised.

5.1.1 Early energy distribution using steam

It is claimed [5.2] that steam heating was an English innovation, first proposed by William Cook in 1745. The same reference records that in 1784, James Watt heated his study indirectly by means of steam passing through a box made of tinned plate, while Matthew Boulton also used steam, contained within pipes, to heat his living room and bathroom indirectly. At the time of these developments, coal was relatively cheap and after the steam had been used it was released to the atmosphere, making the process non-cyclic.

5.1.2   Early energy distribution using water

The idea of replacing a directly acting working substance with one that acts indirectly and is recirculated within the system (a cyclic process) must be regarded as a major advance in the technique of warming the built environment. The first successful use of hot water in this way was by John Evelyn in 1675, followed in 1916 by Sir Martin Triewald who applied the technique to heat a greenhouse [5.2].

In 1816 the Marquis de Chabannes came to England where he knew he could rely on a plentiful supply of cheap coal to use in his ventures into 'indirect heating'. He installed his system in two multi-storey town houses in London and this is said [5.2] to be the first time that a heating system operating at atmospheric pressure and using water as the working substance had been utilised in this way. Heat was introduced to the rooms by means of 'calorifères', each of which consisted of a water tank through which open-ended pipes passed. Air from the room was warmed as it moved by natural circulation through the pipes and the tank was itself encased in an 'ornamental' casing.

5.1.3   Use of the sealed pipe coil and the stopped end steam tube

In 1930 Angier March Perkins heated large buildings using high-pressure hot water obtained from steam boilers. As steam pressures were increased, steam generators required thicker plates and stronger joints to resist the larger forces produced. Good heat transfer required large surface areas so that the boilers became more expensive and potentially more dangerous.

Perkins introduced a sealed pipe coil, having small-diameter pipes containing water, into the combustion chamber and in this way he was able to increase the surface area available for heat transfer, with the possibility of reaching water pressures of the order of 27 MPa before the pipes burst. One-sixth of the length of the pipe coil was placed in the furnace and the remaining length was routed through the space to be heated. The water circulated around the pipe coils owing to the difference in density between rising and falling columns of water, that is by thermosiphonic flow – sometimes referred to in the trade as 'gravity flow'.

In 1865 Loftus Perkins (son of Angier Perkins) took out a patent which dealt with 'the stopped end steam tube', which he later successfully applied to the space heating of buildings and the heating of bakers' ovens.

5.2   Current methods of energy distribution and emission using water

In order to distinguish between the three types of space heating system in which hot water may be used as a working substance, it has become customary to categorise them in terms of the pressure at which each operates. Rather loosely they are described as low-, medium- and high-pressure hot water systems; table 5.1 shows the commonly accepted tem-

Table 5.1 Categorisation of space heating systems using hot water as the working fluid

Type and use	Heating load (MW)	Flow temp. (°C)	Absolute pressure	Temperature difference (°C)
Low pressure	0–2	0–90	100 kPa (atmos. press.)	10–20
Domestic Commercial Office blocks Small factories				
Medium pressure	2–3	120–133	200–300 kPa	25–30
Large factories				
High pressure	>3	150–180	500 kPa–1 MPa	45–65
Very large industrial complexes Airports				

perature and pressure differences between the three main types. The table also gives an indication of the size and type of central heating load likely to be served by each system.

5.2.1 Pipework arrangements for low-pressure hot water space heating systems

The use of low-pressure hot water as the working substance for space heating in the built environment is currently the most popular option for the majority of developments. It is rare that the pipework system in any particular building will conform strictly to the recognised patterns of connections, and in practice it is not necessary that they should. It must be mentioned that without the help of drawings (preferably isometric), such patterns are not easily recognisable when walking through the building.

Figures 5.1(a), (b) and (c) depict some of the conventional arrangements of pipework for use in low-pressure hot water systems. It is possible to use 'mixed' systems of pipework, and most designers tend to place the heat emitters where they are required and then arrange the pipework and connections to produce the most efficient and cost-effective arrangement, which may, or may not, be consistent with the subsequent aesthetic strictures of the architect.

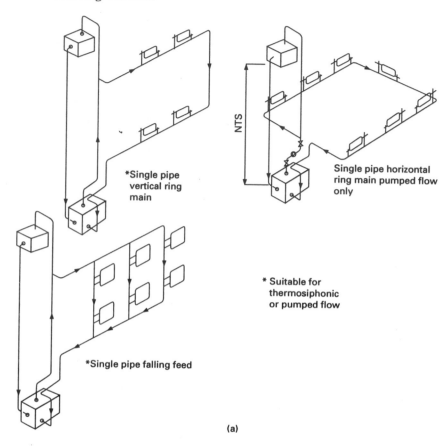

*Single pipe
vertical ring
main

Single pipe horizontal
ring main pumped flow
only

NTS

* Suitable for
thermosiphonic
or pumped flow

*Single pipe falling feed

(a)

Figure 5.1   (a) Pipework for low-pressure hot water heating using single pipes
for thermosiphonic or pumped flow

5.2.2 Single pipe system

This is eminently suited to small schemes where the size of the single
supply pipe is no greater than 28 mm. When the diameter of the pipe
becomes greater than this, the response time becomes too long for the
emitters to adjust to changing thermal requirements as they occur. This
state of affairs may go unnoticed if the system is operated continuously,
as in an old people's home, or where round-the-clock nursing is being
practised, but it is not recommended for those establishments requiring
the plant to function on a diurnal basis.

When commissioning a single pipe system, it will be found that when
all the valves across each emitter are open, the mean temperature of the
second emitter will be lower than that of the first and so on, with the last
unit having the lowest temperature of all. It will also be found that by
almost closing the lockshield valve on the first, second and third heaters,

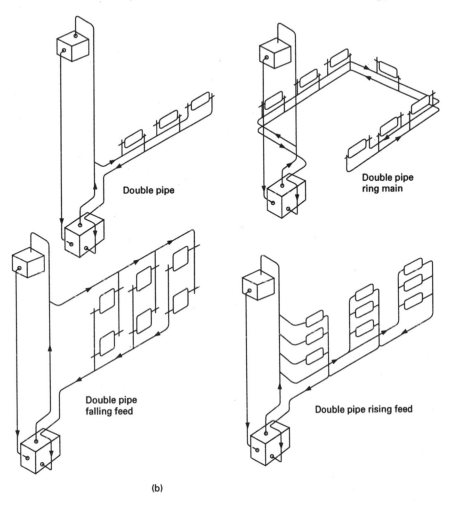

(b)

Figure 5.1 (b) Pipework for low-pressure hot water heating using double pipes for thermosiphonic or pumped flow.

the mean temperatures of all the emitters will tend to converge. It is tempting to conclude that by assiduously continuing the process (referred to as 'balancing' or 'calibrating' the system), the mean temperature of each emitter could become the same. In practice, however, this situation is hardly ever encountered and the main reason for this is as follows.

If the system has a pump and the emitter takes the form of a convector (convectors are found to operate indifferently without pumped flow), and if one assumes that most of the energy being emitted is convected, the emission of heat will be proportional to the temperature difference between the surrounding air and the mean temperature of the water flowing through the unit raised to the power of approximately 1.3. By almost

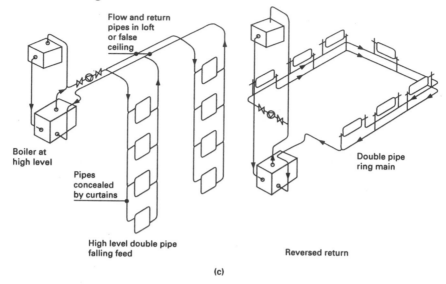

Figure 5.1   (c) Pipework for low-pressure hot water heating using double pipes for pumped flow only

closing the lockshield valve on the convector, it is the mass flow rate of the water that is altered by an unspecified amount; this then changes the heat emission rate indirectly. Over the whole system only a small convergence of mean temperature can be expected, making it unrewarding to spend time calibrating the system. This difficulty with valves occurs with all systems, but the effect is more noticeable with single pipe flow.

5.2.3   Double pipe system

This system is widely used. A flow and return pipe is provided for each emitter and because there are two pipes, instead of only one, as in the single pipe system, they can be smaller in diameter and a greater thermal load can be carried.

Inspection of that part of figure 5.1(b) which shows a double pipe ring main reveals that the last emitter in the circuit requires the water to travel the greatest distance to and from the boiler, while the water entering the first emitter travels the shortest distance. As a result of this, the pressure drop from the boiler to the first heater is less than that from the boiler to the last heater, resulting in a larger volume flow rate of water (and a higher temperature) where the pressure drop is least, and a progressively smaller volume flow rate (and a lower temperature) where the pressure drop is greatest. This chain of events makes it necessary to calibrate the system when it is commissioned and in fact the task is easier than when calibrating a single pipe flow system; however, because of difficulties with the valve flow characteristics, it will not be possible to obtain perfect equalisation of the emitter temperatures.

Figure 5.1(c) shows a 'reverse return' system which has been designed with the object of making the calibration process easier. Inspection of the drawing shows that the distance of travel from the boiler to any emitter has a constant value and the temperature at each emitter should be the same. Some calibration is still necessary, however, because of the limitations imposed in the choosing of suitable standard pipe sizes and the system will not function without a pump.

Figure 5.1(c) also shows a high-level, double pipe, falling feed system and this can be attractive to those who dislike surface pipework. Provided that the emitters are placed underneath the windows, most of the vertical pipework can be concealed by the curtains. Curtains should be of such a length that the bottom hem finishes about 25 mm above the top of the heater, so that when they are drawn, the emission from the unit is not masked in any way. The horizontal flow and return pipes may usually be concealed in the loft or in a false ceiling, and the boiler can be conveniently mounted at high level in the building. The use of a pump is recommended.

5.2.4 Thermosiphonic flow

Because the layout of pipework suitable for thermosiphonic flow can still be found in older buildings, such systems have been included in figures 5.2(a) and (b). In these cases it was normal to provide a basement boiler room, to ensure that all water returning to the boiler did so with the help of the force of gravity. However, nearly all such space heating systems will by now have had a pump fitted, to make the response time less and to save on the cost of the fuel, and also, perhaps, to cater for space heating apparatus and pipes that have been retrofitted and require a circulating pressure that is much greater than could be provided by means of thermosiphonic flow.

In most 'compact' systems, the method of thermosiphonic flow is still used to keep towel rails warm at all times and to enable hot water to circulate from the boiler to the storage cylinder. The latter function is also used as an energy by-pass for solid fuel boilers, directing the heat safely from the burning fuel to the hot water cylinder when it is not required for space heating. It also performs an essential role in eliminating deadlegs in domestic hot water supplies, and for these reasons details of the design method for thermosiphonic flow is given in Chapter 7, section 7.9.

In order to produce the greatest possible circulating pressure around the system, owing to the difference in density between the rising and falling columns of water, the flow pipe should be taken to the highest point possible before it begins its continuous descent through the emitters and back to the boiler.

If the temperature difference between the rising and falling columns of water is 20°C, the value for the circulating pressure/metre height of the flow pipe would be approximately 100 Pa/m and it is this pressure that is available to bring about the circulation of the water. It follows that if the height of the rising column of water is 5 m then the circulating pressure would be 500 Pa.

5.2.5   Pumped flow

When a pump is used to circulate the water in a system, it does so by producing a pressure difference across its inlet and outlet connections. The value of this pressure difference is usually quoted in units of metres of water (assumed to be at the mean operating temperature of the system), instead of the unit of pressure of the Pascal. The duty of the pump would then be described as being capable of delivering the required mass flow rate of water (kg/s) against a head of H (m) of water when running at a speed of N (rev/min). Note that 1 kg of water may for practical purposes be described as having a volume of 1 litre, so that in this case the mass flow rate can also be described as the volume flow rate (litre/s).

A common error is to confuse the head of water produced by the pump with the physical height of some part of the system. The two are not connected. The sole function of the pump is to produce a pressure difference (or difference of head of water H) across the inlet and outlet connections, so that the water will be moved around the system or circulated. It should be noted that pumps used in this way do not 'lift' the water through a height, as the water is already filling the system and, to avoid this confusion, they are referred to in some quarters as 'circulators'. A typical value for H could be 2.5 m; when this head of water is converted into its equivalent circulating pressure at a mean temperature of 70°C using the equation $P = h \times \rho \times g$ we get:

circulating pressure $= 2.5 \times 977.5 \times 9.81$ (Pa)

$$= 23\,973 \text{ (Pa) or } 23.973 \text{ (kPa)}.$$

Comparing this pressure with a circulating pressure of, say, 500 Pa, obtained from a system operating on thermosiphonic flow, it will be appreciated that in this case the pump circulating pressure is approximately 40 times greater, and it follows that this difference will be reflected in the relative size of the pipes required in the two systems.

The use of a pump allows smaller pipes to be fitted, with no restriction on the route taken, making it less costly and potentially more acceptable aesthetically. It is less wasteful of energy than a thermosiphonic system, and it can help with frost control and also with the even distribution of heat around the building. Because of these attributes, pumps (or circulators) are to be found in nearly all space heating systems.

5.2.6 Use of small-bore tubes

The almost universal inclusion of a pump in space heating systems has made it possible to install small-bore copper tubes, which may be loosely categorised as having diameters of 15 and 22 mm, and it has become customary when designing a system for domestic use to try to restrict the sizes of the pipes that are fixed in the living spaces to these two sizes. Using copper is less labour intensive than using steel screwed pipe, and it presents a neater appearance. Steel pipe may, however, be used to advantage where the pipes could be subject to mechanical damage, perhaps in a space or corridor that is used by the public, or where wheelchairs and trolleys are used for the transport of people or goods.

The use of such small-bore tubes makes it essential that during the construction of a building, no foreign materials such as plaster, concrete, cotton waste or rags become sealed inside the pipes; if they are, the performance of the system will almost certainly be adversely affected.

5.2.7 Use of 'microbore' tubes

These tubes are made from annealed copper having diameters of 6, 8, 10 and 12 mm, and they may be used where it is considered essential to make the installation as unobtrusive as possible and where the design of the building makes it feasible. Figure 5.2(a) shows a possible layout using single pipe flow and figure 5.2(b) shows an alternative arrangement, using double pipe radial flow. The valves shown are special-purpose units but, if desired, two conventional smaller valves can be fitted to each emitter instead. The microbore tubes are supplied with water from headers which are connected to small-bore pipes having diameters of 22 mm or 28 mm. It is normal to supply each radiator with separate flow and return pipes, each of which is connected to a conveniently concealed manifold that may also service other emitters.

The limiting parameter in sizing the tubes serving an emitter is the velocity. This should be kept to a figure that is just less than 1.5 m/s in order to avoid 'rushing noises' in the tubes. The tubes serving each radiator are small enough to conceal in a skirting board and are certainly inconspicuous even when they are routed on the surface. It is of course tempting to conceal the microbore tubes in the screed, but new building regulations require that no pipe containing water shall be 'built in'. In some areas this extremely limiting rule is waived and accordingly, it becomes necessary for the designer to make enquiries before he becomes too deeply committed. The cost of a microbore system is meant to be less than one using small-bore tubes, but in the author's experience there is no detectable difference. Microbore tubes can be used with a high- or low-pressure system and must of course include a pump.

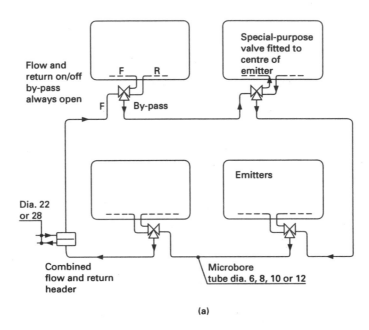

Flow and
return on/off
by-pass
always open

Special-purpose
valve fitted to
centre of
emitter

By-pass

Emitters

Dia. 22
or 28

Combined
flow and return
header

Microbore
tube dia. 6, 8, 10 or 12

(a)

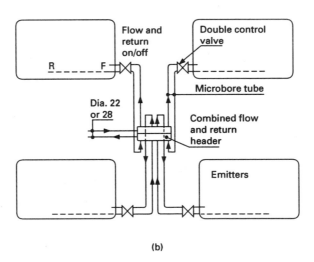

Flow and
return
on/off

Double control
valve

Microbore tube

Dia. 22
or 28

Combined flow
and return
header

Emitters

(b)

Figure 5.2　(a) Microbore with single-pipe flow. (b) Microbore with double-pipe radial flow

5.2.8 Use of macrobore tubes

Extensive thermosiphon flow systems require large-diameter pipes to conserve the very low circulating pressures that are typical of this method of circulation. District heating schemes also require large-diameter pipes to carry the high mass flow rates of water over considerable distances and, in such cases, the pumps used will produce a high head, in order to overcome the very large head loss due to friction as the water is moved around the circuits.

5.2.9 Position of the pump in the circuit

Figure 5.1(c) shows a pump fitted in the hot water flow pipe with the cold feed to the system connected to the suction side of the pump. This method of connection is the preferred one, but the pump, cold feed and expansion pipe may be connected to the boiler in three other ways, each having a different effect on the system.

Figure 5.3 shows the four methods of connection, together with an indication of the way in which the pressure (or head), generated by the pump, is lost around the circuit, owing mainly to the friction set up between the flowing water and the internal surface of the pipes. The pressure lines depicted are referred to as 'hydraulic gradients' and they give an indication of the head of water at any part of the system. When drawing the lines, the starting point on each diagram is the point 1 on the free surface of the expansion tank. The hydraulic gradient can then be imposed on the diagram, provided that the head loss/metre run of pipe is known (see Chapter 7, section 7.9).

When pumps were first introduced to act as circulators in space heating systems, it was considered to be good practice to fit the pump in that part of the system that had the lowest temperature, namely the return pipe. In this way it was reasoned that the motor driving the pump was less likely to overheat, so preventing a possible burn out. As the practice of fitting pumps became more widespread, motors were produced that were rated to run at higher temperatures, and it then became possible to fit the pump in the flow pipe.

Method (a) (figure 5.3(a))

This is the preferred method of connection. Note that the bulk of the system is operating above atmospheric pressure, making venting easier and discouraging the entry of air into the system.

Method (b) (figure 5.3(b))

This form of connection also ensures that the system operates above atmospheric pressure, which is good for venting and prevents air from entering the system. However, in order to prevent continuous discharge

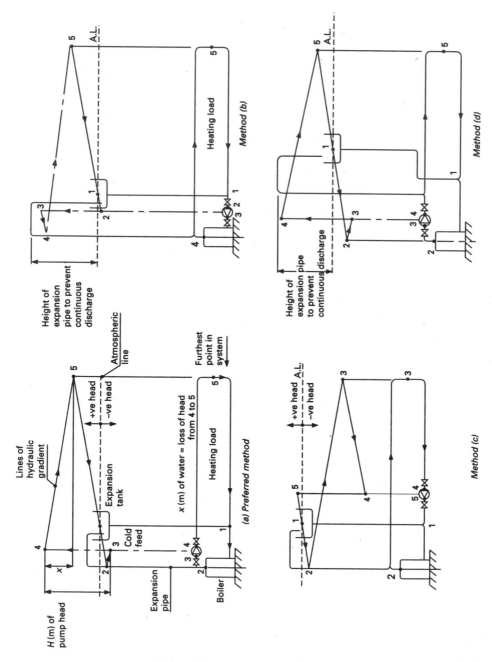

Figure 5.3 Hydraulic gradients associated with four positions of a pump

occurring from the open end of the expansion pipe, it has to be taken to a height that is approximately equal to the head of water generated by the pump, and this is clearly a disadvantage if the headroom above the expansion tank is limited.

Method (c) (figure 5.3(c))

In this case it is not necessary to raise the expansion pipe well above the tank, but the bulk of the system is operating below atmospheric pressure, which encourages air to leak into any small openings around the glands of valves and makes it necessary to stop the pump before any venting can take place.

Method (d) (figure 5.3(d))

In this case the emitters are under positive pressure, which is good for venting, but the height of the expansion pipe has to be extended to prevent continuous discharge due to the impressed head of the pump. Note that the connection of the expansion pipe is not made directly to the top of the boiler but is made downstream of the pump and valves, which may conceivably restrict, or even block off, access to the atmosphere; also in the event of overheating of the boiler this may cause an unscheduled and potentially dangerous rise in pressure in the system. This illustrates an important point of safety: all vent pipes should ideally be connected directly to the top of the boiler and must not have any apparatus that is capable of restricting the flow of water through them at any point along their length or before their connection to the boiler.

Reference to figure 5.3 will show that in cases (a), (c) and (d), the boiler is operating at just below atmospheric pressure. The effect of this on a newly filled system will be to increase the occurrence of cavitation, which involves the release of dissolved gases from the water around small irregularities on the heating surface of the boiler. In the trade, this process is referred to as 'kettling'. With continual heating of the water taking place, the gases dissolved in the water gradually come out of solution and the subsequent occurrences of cavitation then become less frequent.

5.3 Distribution and emission of energy using pressurised water

As the pressure rises, the boiling point of water also rises, so that with a higher pressure it is possible for a specific mass of circulating water to absorb more heat before it boils. It follows that if relatively more heat can be transmitted per mass of water circulating, then the pipe sizes can also be smaller. Also, because the temperature of the water is higher, it is possible to allow for a larger temperature drop across each emitter, enabling these units to be correspondingly smaller than those used in a system operating at a lower pressure. In an extensive system, smaller pipe and emitter sizes reflect a considerable reduction in cost, and for these reasons pressurised systems are used for the higher heating loads. The high temperature of the circulating water makes it essential that all heat emitting surfaces are placed where no part of the apparatus may be touched.

5.3.1 Pressurisation using steam

In the recent past it was common practice to use steam as the means of pressurisation. Water in a conventional boiler was heated until the upper parts of the boiler contained steam at the required pressure. The connection of flow and return pipes below the water line, in the form of 'dip' pipes, made it possible to circulate the hot water safely at the required pressure around the space to be heated. In this system, the water leaving the boiler was mixed with a controlled quantity of cooler return water, in order to prevent it 'flashing off' to steam whenever a slight local pressure drop occurred. Such a reduction in pressure could occur as a result of the effect of pump suction and/or a pressure drop in the pipeline, and because it is not possible to deal with a steam–water mixture using conventional pumps, the circulation of water around the plant (a cyclic process) would be interrupted. The operation of such a system requires skilled personnel.

5.3.2 Pressurisation using air

It is possible to use air as the pressurising fluid, which in an industrial plant would be obtained from an air compressor, from which it is likely that there will be some carry-over of oil into the system. If this oil is deposited on to heat transfer surfaces, it will reduce their effectiveness and may, under some conditions, emulsify with the water, producing foam which cannot be handled using conventional pumps. Usually the air, with its associated oxygen, is in direct contact with the water and it will have a tendency to go into solution, providing an increased propensity for corrosion in the plant and also making frequent topping-up necessary.

5.3.3 Pressurisation using nitrogen

At the present time, the pressurisation of heating systems is more likely to be accomplished using nitrogen. Like oxygen, it too goes into solution in the water, but as it is an inert gas it has no effect on the plant.

5.3.4 Pressurisation of domestic space heating systems

The volume of water contained in such systems is relatively small, so that when it is heated, the volume of expansion can easily be contained in a small pressurised expansion vessel. The vessel is divided into two halves, by means of a flexible butyl rubber diaphragm, which has water in contact with one side and either air, or nitrogen, in contact with the

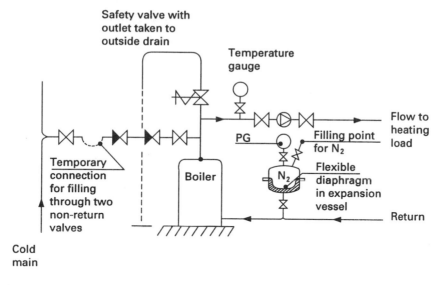

Figure 5.4 Pressurised space heating system for domestic premises

other. Pressurisation is achieved by increasing the pressure of the air, or nitrogen, and the pressure in the system is kept low enough to ensure that the operating temperature of the emitters is acceptable for use in the home, in that skin contact will not result in burns. When air is used as the pressurising fluid, the associated oxygen will not produce corrosion in the plant, but may in time corrode the inside of the expansion vessel.

Such a system is shown in figure 5.4 and is often to be found in those domestic installations where there is no space available for the siting of an expansion tank. For example, it is normal in country cottages for the upper rooms to take up most of the loft space, usually leaving space near the apex of the sloping roof, which may turn out to be just large enough to accommodate a tank, but which does not provide enough headroom above it to carry out essential maintenance.

Water authorities are understandably interested in the way in which the system is filled with water, in order to overcome the risk of back siphonage occurring in the supply mains. There must be no permanent connection between the cold main and any part of the system. Some authorities call for a double check valve assembly to be permanently fitted adjacent to the boiler, at the point where filling is to take place and if the pressure in the main is high enough, then the plant may be filled via a temporary hose connection. In some areas the water pressure is at times too low to fill a pressurised system; in such cases it is possible to introduce water into the system using a break-tank and a pump to raise the pressure.

5.3.5   *Pressurisation of industrial space heating systems*

Large systems may be expected to use the same basic principle of pressurisation, except that the greater volume of expansion of the water is dealt with by allowing it (on warming up) to escape to a separate covered spill tank and when cooling occurs, for it to be automatically pumped back into the system. The larger pressurisation sets dealing with high-pressure hot water systems have direct contact between the pressurisation fluid (usually nitrogen) and the water. As mentioned previously, the nitrogen dissolves slowly into the water and a means for topping up must be provided. Pressurisation sets are available commercially; figure 5.5 illustrates the basic components which are likely to be incorporated. The manufacturers will usually have optional alarm systems available.

When the water from the system enters the spill tank, it comes into contact with the atmosphere and it is at this part of the plant that the oxygen present in the air may dissolve into it, so increasing the risk of corrosion in the rest of the system when the water is reintroduced. In order to minimise this effect it is possible to float a 'ball blanket' on the surface of the spill tank, so limiting the area of the free surface that is in contact with the atmosphere.

5.4   Distribution and emission of energy for district heating using water or steam

District heating schemes usually consist of a number of centrally placed boilers providing space heating and domestic hot water to a group of buildings using a cyclic process. The boilers are usually sited together in one building which houses the pumps, water treatment plant and any flue gas treatment apparatus deemed to be necessary. The advantages are:

(1) Larger boilers can be operated at a greater thermal efficiency.
(2) The choice of an energy source is widened, making it possible to use cheaper heavy oil, pulverised coal, block tariff gas, or methane obtained from a sewage treatment plant or a municipal waste tip. Alternatively, municipal refuse can be either burnt directly, or preformed into pellets and then burnt in the combustion chambers. Waste heat may be used from the condensing cycle used in power stations, from the exhaust of gas turbines, or (depending on the location) from geothermal energy sources (see Chapter 3).
(3) A multi-fuel facility can be made available to exploit any variation in energy costs, and be used if necessary to counteract fluctuations in supply.
(4) The combustion process can be closely monitored so that less smoke is produced.

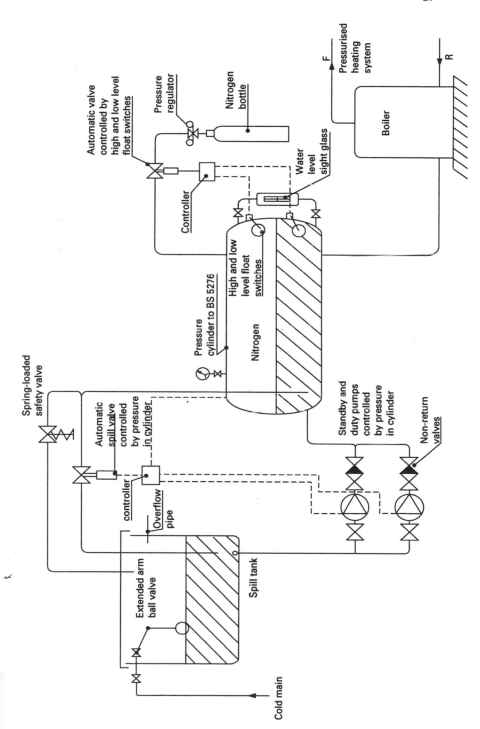

Figure 5.5 Space heating pressurisation set for commercial/industrial premises

(5) The construction of extra tall chimneys can be justified to help disperse the flue gases following treatment in accordance with European Directives.
(6) Such schemes are less labour intensive than when the plant is dispersed over separate sites.

Disadvantages are:

(1) The capital cost is high, so that in a period of high interest rates the development becomes unattractive, unless a Government subsidy is available.
(2) The amount of detailed planning necessary requires the involvement of professionals across many disciplines from the very start of the scheme, resulting in extra pre-planning costs.
(3) In a large development there is always a chance that any errors made in the design or execution of the scheme could be repeated many times, making subsequent attempts to rectify them very costly in terms of both finance and human resources. This can adversely affect designers and increase their professional indemnity insurance.

5.4.1   One-pipe system – a non-cyclic process using water or steam

District heating can be implemented in various ways and it is customary and convenient to classify these in terms of the number and function of the pipes that carry the working fluid. In the countries of the former Soviet Union, the one-pipe system is used where the length of run of the pipe is of the order of 200 km (see [5.3]). Over such a distance it is uneconomic to run a return pipe, making the process non-cyclic. The working substance may be steam which is initially superheated, or typically high-pressure hot water at a pressure and temperature of 1.4 MPa (14 bar) and 200°C.

In those plants using water in the liquid phase, it is first filtered, deaerated and demineralised, using the magnetic ion exchange method; it is then made alkaline by adding sodium phosphate and sodium hydroxide. At the points of consumption some energy is abstracted for space heating purposes and the remainder is then used directly as domestic hot water before being run to waste.

Where the working substance is steam, it is important to the consumer that it is supplied to his premises in a dry and saturated condition, so that he can get the full benefit of the enthalpy of dry saturated steam at the pressure supplied. If the steam enters the premises in a wet condition, the enthalpy is less, even though the pressure and temperature remain the same, producing less heat. To overcome this difficulty, the steam is supplied to the delivery main initially in a superheated condition and with a pressure that is high enough to overcome the pressure drops encountered in flowing through the main. At the consumer's premises it is then allowed to flow through a pressure reducing valve, which approximates to a constant-enthalpy process, so that the condition (or dryness) of the steam is increased as the pressure is lowered, until the steam reaches a condition that is close to that of being dry and saturated.

One advantage of using steam as the working substance for district heating, instead of high-pressure hot water, is that in the case of steam there is no requirement to use pumps to circulate the fluid. In particular, where high rise buildings are involved, it is only necessary to increase the steam pressure at the boilers by a small amount, in order to ensure that the consumers at the highest points are served. In the case of high-pressure hot water, however, it would be necessary to increase the pump head by an amount at least equal to the height of the building, which would involve a large increase in the size of the pumps.

5.4.2 Two-pipe system – a cyclic process using steam or water

This system is common in Europe and consists of a flow and return pipe supplying the consumers in two different ways, described as either 'direct', or 'indirect', entry.

Direct entry

Figure 5.6 shows the flow and return pipes passing through a calorifier housed in a sub-station sited near the centre of a development. The water used for space heating and domestic hot water is town's water, which has been treated with the gas hydrazine (N_2H_4). The gas combines with any dissolved oxygen to produce nitrogen and water with no solid precipitate remaining in the calorifier:

$$N_2H_4 + O_2 \rightarrow 2H_2O + N_2$$

The treated water is used directly in the consumer's heating system, without the occurrence of corrosion in the radiators. The water in contact with the inside of the radiators is then used as domestic hot water and subsequently run to waste.

Indirect entry

This arrangement is possible using steam, or high-pressure hot water; figure 5.7 shows a typical installation using the latter. In this case the domestic hot water is provided for each consumer from a smaller calorifier sited in each dwelling. The pressure of the water is reduced in both the flow and return pipes and in the event of malfunctioning of these valves, additional pressure control valves are fitted to each emitter, so that the danger of the emitters rupturing is minimised. When steam is used in the

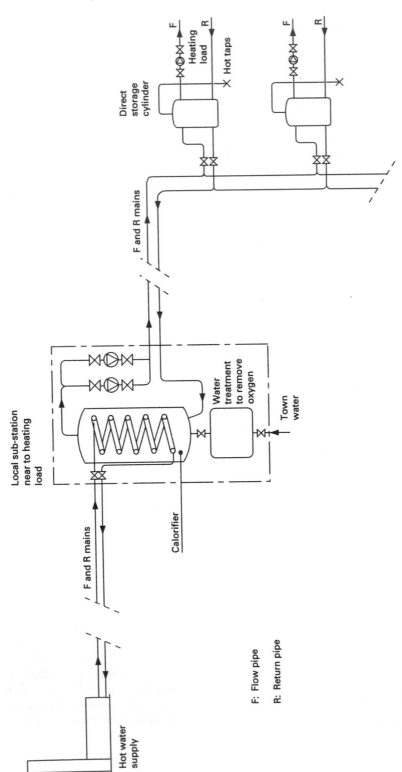

F: Flow pipe

R: Return pipe

Figure 5.6 The two-pipe direct entry system of district heating

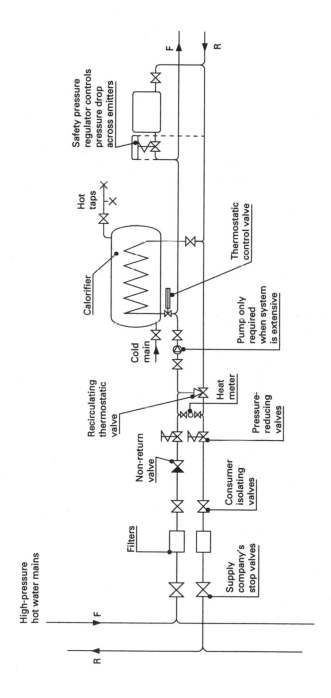

Figure 5.7 The two-pipe indirect entry system of district heating

supply company's mains, two calorifiers may be used. One calorifier will supply the establishment with domestic hot water and the other will provide water for the space heating, and it will be necessary to fit at least one pump on the 'wet' part of the system.

5.4.3 Three-pipe system – a cyclic process using water

This system produces savings in pumping costs during the summer period. The arrangement, shown in figure 5.8, consists of three steel pipes each carrying treated (de-aerated) water. The operating regime changes from winter to summer. In winter, water supplies for space heating and domestic hot water are fed to the plant, via the two right-hand side pipes, and the left-hand side pipe carries the combined return back to the source. In summer, when there is no need for space heating, the right-hand side pipe is isolated and the remaining two pipes serve as flow and return for the domestic hot water.

5.4.4 Four-pipe system – a cyclic process using water

This configuration is included because it was once used in Manchester, where it produced some unscheduled side-effects. Untreated domestic hot water was circulated in two copper flow and return pipes, and water for space heating purposes was circulated in two steel pipes. All the pipes were then encased in a common block of thermal insulation, which, unfortunately, was not adequately waterproofed. The result was that in those parts of the insulation that subsequently absorbed moisture, the steel and the copper pipes behaved as anode and cathode in an electrolytic cell, and the steel pipes became sacrificial.

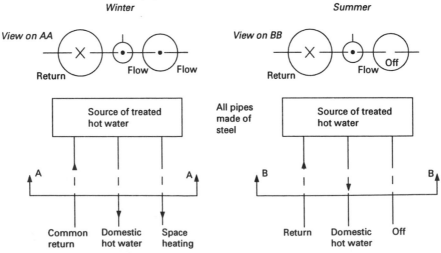

Figure 5.8 The three-pipe system for district heating

5.5   Treatment of boiler feed water

Wherever steam is used as the working substance, it is essential that some form of water treatment is employed. The purpose of this is to limit the potential effects, produced on the heat exchanger surfaces of the boiler and other parts of the plant, by the subsequent precipitation of substances devolving from the compounds contained in the water. The type of treatment, its extent and method of application are best determined by a specialist company. The treatment will depend on many factors, such as the type of salts present in the water supply, the extent to which the water is alkaline or non-alkaline, the pressure at which the boiler is operating and the rate of steam output.

When a mass of water containing dissolved salts is heated and turned into steam, the capacity of the remaining water to hold salts in solution becomes less and the concentration of the remaining salts in the water becomes higher and, as the process continues, some of the salts will be precipitated on the internal surfaces of the boiler. The precipitates of the salts are referred to as scale when they cover the hottest parts of the boiler and sludge when they remain in the mass of the water. The scale that adheres to the metal of the heat exchanger will gradually increase in thickness, and will produce the following two effects to a degree that depends upon its thickness:

(1) a thermal barrier will be set up, which decreases the heat transfer rate, consequently decreasing the thermal efficiency of the boiler by up to 3% (see [5.4]);
(2) the temperature of the metal of the heat exchanger will rise and may reach a value that will affect its mechanical strength, when the metal will deform under the action of the pressure in the boiler.

5.5.1   Internal methods of treatment

This type of treatment refers to the addition of chemicals to the inside of the boiler, to make the salts causing scale and sludge less likely to adhere to the heat exchanger surfaces and also to ensure that when blowdown occurs, the resulting precipitates are easily expelled. An attempt may also be made to control the pH value of the water and the propensity for corrosion due to the possible presence of carbon dioxide CO_2 in solution in the water. For boilers operating at low pressures, the organic materials used to assist the process of blowdown are natural and modified tannins, or starches and alginates. For boilers operating at higher pressures, synthetic materials have been developed, such as the polyacrylates and polymethacrylates, the prescription and application of which require guidance from a specialist company.

5.5.2   *External methods of treatment*

This type of treatment will include all or some of the following processes.

Sedimentation

The water to be treated is allowed to flow very slowly through a tank, allowing material suspended in the water to settle to the bottom. This process may be assisted by the use of coagulating compounds such as aluminium sulphate and ferric sulphate.

Oxidation

The purpose of this is to remove any iron and manganese that may be held in solution in the water. Oxidising agents can be used or aeration methods employed, thereby converting the metals in solution to their insoluble oxides which can then be removed by filtration.

Filtration

The water is allowed to flow through a bed of sand, during which process the larger solids are left behind on the surface and the smaller ones are collected in the spaces between the sand grains. If the volume flow rate of the water is large, pressurising the filter helps to speed up the process.

Softening

One property of water that can be readily detected is its 'hardness'. A hard water requires more soap to produce a lather than does a soft water. It is found that some of the hardness can be removed by boiling the water and this form of hardness is traditionally referred to as temporary hardness. Some authorities now call this alkaline hardness; others refer to it as carbonate hardness, because it can be shown to be caused by calcium carbonate $CaCO_3$, calcium hydrogen carbonate $Ca(HCO_3)_2$ and magnesium hydrogen carbonate $Mg(HCO_3)_2$, which are in solution in the water.

When calcium hydrogen carbonate is heated it produces water, carbon dioxide and calcium carbonate, which is deposited on the heated surfaces. Some hardness cannot be removed by boiling and traditionally this is referred to as permanent hardness. Some authorities now call this non-alkaline hardness and others non-carbonate hardness. This form of hardness is due to calcium sulphate $CaSO_4$ and calcium chloride $CaCl_2$, also magnesium sulphate $MgSO_4$ and magnesium chloride $MgCl_2$. Permanent hardness can be removed by chemical methods.

5.5.3   Categorisation of water softening plants

(a) Lime – soda

Some water contains carbon dioxide in solution and when calcium hydroxide $Ca(OH)_2$ (lime) is added, the products are water and calcium carbonate $CaCO_3$, which is precipated and can then be removed by sedimentation. The water is also likely to contain calcium hydrogen carbonate $Ca(HCO_3)_2$ (a compound responsible for temporary hardness), and when calcium hydroxide $Ca(OH)_2$ is added, the products once again are water and calcium carbonate in the form of a precipitate.

When lime is added to magnesium sulphate $MgSO_4$ in solution in water (a compound responsible for permanent hardness), calcium sulphate $CaSO_4$ is produced and magnesium hydroxide $Mg(OH)_2$ is precipitated and can be removed by sedimentation.

When sodium carbonate Na_2CO_3 (soda), is added to the calcium sulphate $CaSO_4$ (a compound responsible for permanent hardness), sodium sulphate Na_2SO_4 is produced and calcium carbonate $CaCO_3$ is precipitated and can be removed by sedimentation.

Other chemical reactions are possible, depending on the compounds contained in the water and if the process is carried out under hot conditions, with magnesium present, silica will be removed and so prevented from forming on the surfaces of the heat exchanger as a very hard and undesirable form of scale.

(b) Ion exchange

This form of water softening makes use of resins that contain compounds capable of changing places with ions present in the water. There are three variations of this method: base-exchange, dealkalisation and demineralisation.

The base-exchange process removes both temporary and permanent hardness and is ideally suited to providing softened water for low-pressure boilers and for domestic use. Beads of resin contain sodium zeolite Na_2Z, which when brought into contact with calcium sulphate $CaSO_4$ (a compound responsible for permanent hardness) produce calcium zeolite CaZ and sodium sulphate Na_2SO_4, which goes into solution in the water and may be flushed away. Also, when calcium carbonate $CaCO_3$ (a compound responsible for temporary hardness) is brought into contact with sodium zeolite Na_2Z, this produces calcium zeolite CaZ and sodium carbonate Na_2CO_3, which is in solution in the water and may be flushed away. Other reactions are possible, such as that involving magnesium chloride and sodium zeolite.

As the processes continue, the remaining sodium zeolite changes into calcium (or magnesium) zeolite and it becomes necessary for the bed of resin to be regenerated. This is done firstly by back flushing the bed with water to remove any particles of dirt and secondly by bringing the resin into contact with a solution of salt in water. This produces calcium chloride CaCl and sodium zeolite, and finally, after flushing away any excess salt solution, the softening process can be resumed.

The regeneration process is usually accomplished automatically, it only being necessary for the operator to keep the plant supplied with salt. At one time it was thought that the act of discharging salt into a septic tank would destroy the bacteria present. However, in practice this was found not to be the case.

The dealkalisation process uses a resin in acid form, which can deal with temporary hardness only, and regeneration is achieved with the help of sulphuric acid.

The demineralisation process uses a mixture of resins and produces water that is almost free of all salts.

(c) Distillation

This process is reserved for the occasions when only brackish water is available and the cost of using chemical methods becomes too high. The water is evaporated, leaving all the salts behind, and the vapour is then condensed, giving distilled water that is free from all impurities. This method is also used when only sea water is available, in particular to provide drinking water on board ship. Distilled water is found to be not palatable and it has to be aerated before it becomes potable.

(d) Reverse osmosis

This process is currently being used by some of the privatised water authorities in the UK, to assist in the removal of nitrogen-based fertilisers from the available natural water. Some householders who subscribe to the idea (fostered by the advertising agencies) that they can have 'pure' water on tap, have installed their own equipment using this process, in a successful attempt to remove residual traces of drugs from their own drinking water supplies!

The osmotic process is the means by which plants draw water containing nutrients from the soil in which they are planted. The sap contained in the roots of a plant has a higher density than the water in the soil surrounding it. The root wall acts as a semi-permeable membrane and a small suction occurs across it, which, when equilibrium is reached, is referred to as the osmotic pressure.

In the process of reverse osmosis, the osmotic pressure is increased manually until water begins to flow in the reverse direction from the water of higher density, back through a semi-permeable membrane into the weaker solution. This process can be used to extract filtered water from water containing a wide range of dissolved substances, some of which (if taken in large doses) may be toxic. If the application of pressure is high enough (about 2760 kPa), water containing dissolved hardness salts will yield softened water on the downstream side of the membrane, with an efficiency of approximately 90%. The remaining 10% of hardness salts causing hardness (temporary and permanent) may then be removed using a base-exchange process.

5.6   Distribution and emission of energy using low-temperature embedded emitters operating with water

Energy on the move can be recognised as heat when a temperature difference exists and there is no phase change of a substance.

Such heating systems operate using water that circulates to and from the boiler through tubes and panels embedded in the fabric of the building. Because there is a closed circuit of pipes, the system may be described as operating in a cyclic process.

5.6.1   Underfloor heating

Typically the system uses low-pressure pumped water at a mean temperature of around 40°C, with an approximate overall temperature drop of 5°C. When underfloor heating of any kind is used, it provides an overall feeling of warmth, owing to the even spread of heat, of which approximately 50% is radiant, emanating from the whole floor, at a surface temperature between 21°C to 29°C. Because of this relatively high radiation component, comfort conditions are achieved at air temperatures that are about 2°C lower than the values needed with warm air heating, so that with suitable controls it is possible to use less fuel for the same effect. The system also produces a minimal vertical temperature gradient.

Underfloor heating is suitable for areas where, for practical or aesthetic reasons, it is considered unsatisfactory to take up space with conventional emitters or ducts. The system can be applied to any type of building and it will be found to be particularly advantageous in the following cases:

- foyers
- shopping malls
- exhibition halls
- nursing homes and hospitals
- 'safe' rooms for mental institutions and prisons
- swimming pool surrounds
- prestige business centres.

5.6.2   Heating/cooling of ceilings

When pipe coils or panels are built into ceilings, they are frequently designed to receive chilled water during the summer period. One of the limits to the amount of cooling that can be achieved using this method is set by the need to avoid the formation of condensation on the coils and panels. This will be achieved if arrangements are made to cool the surface of the coils to a temperature that is approximately 1°C above the dewpoint of the water vapour in suspension in the air.

5.6.3   Use of plastic tubes

The tubes containing hot water are traditionally made of steel or copper, but with the development of plastics claimed by the manufacturers as being suitable for the simultaneous application of both temperatures and pressures, many installers are now using plastic pipes.

The particular plastics used are known as polypropylene copolymer, or polybutylene, or XLPE polyethylene. One manufacturer using its own patented type of XLPE plastic tube gives a 25-year guarantee on its work and has obtained a British Board of Agrément (BBA) Certificate stating that the particular grade of plastic will last for a minimum of 60 years, or for the life of the building (see [5.5]). The 'XL' of the prefix XLPE refers to the molecular structure of the plastic being cross-linked (as in a chain link fence), so making the walls of the pipe stronger and more durable, and the 'PE' is short for polyethylene. Cross-linking can be achieved by various processes which are subject to patent agreements and one of these involves subjecting the pipe to radiation.

5.6.4   Application of underfloor heating to timber floors

Figure 5.9 shows a cross-section of a concrete ground floor having plastic pipes cast in it for heating purposes. Figure 5.10 shows a plan view of the coil of pipe covering the surface to be heated. When dealing with suspended timber floors, some designers decrease the temperature gradient from the water to the surrounding material by bringing the outside of the pipe into close contact with an extended pipe fin, referred to as a heat emission plate; this is shown in figure 5.11. If in a retrofit involving a suspended timber floor it is not permissible to increase the height of the floor, it may still be possible to apply the system by including the pipe coils and emission plates in the space between the joists. Manufacturers are also able to provide details for the application of the system to floating floors [5.5].

5.6.5   Drying out

In the past, the application of underfloor heating has not always been satisfactory, either to the client or to the occupier. When such systems were installed in floors and ceilings without adequate thermal insulation, the fuel bills were found to be extremely high, particularly when the scheme used more expensive electrical energy and the building was oc-

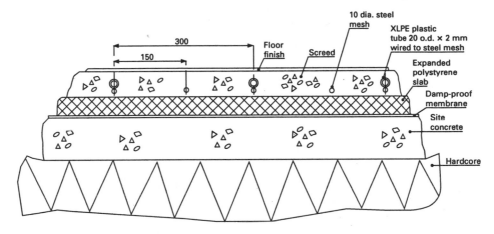

Figure 5.9 Application of XLPE plastic tube to underfloor space heating

cupied and still in the 'drying-out' period. During the initial drying-out period (which normally takes place before occupation), it sometimes happened that heating engineers could be persuaded to use their newly installed system to 'assist' the process and, in some cases, it was not unknown for the water to be circulated at a temperature that was higher than would be expected under normal running conditions.

In such cases, because of the high temperature gradient from the water to the surrounding material, it was not uncommon for longitudinal cracks to appear wherever a heating coil was buried. Clearly, such cracks were not acceptable and, in a building where there was a requirement for the floors to be washed frequently, this resulted in underfloor corrosion and the harbouring of bacteria. Nowadays, it is considered to be good practice to allow the screed to dry out without any assistance from the heating system and, when the plant is commissioned, it should be completed over an extended period, starting with a very low flow temperature and a small temperature difference across the coil, with subsequent increases in temperature taking place only very gradually.

5.6.6 Use of liquid polymer additives

Designers and manufacturers of underfloor space heating systems are at liberty to vary the diameter of the pipes and their pitch, and also the depth at which they are buried. They can also vary the recommended flow temperature and the permissible temperature difference across the coils. Because of the number of variations possible, it is not surprising that recommendations differ with regard to the use of liquid polymer additives in screeds and mortars. It is claimed by some that such additives increase the adhesion between the tube and the surrounding material, so increasing

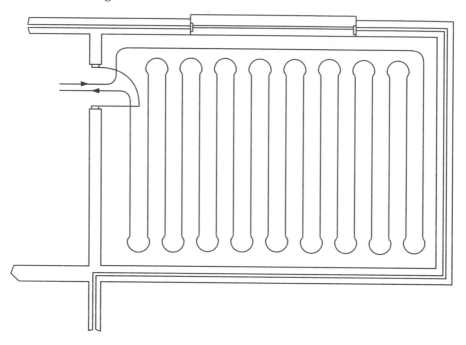

Figure 5.10 Plan of heating coils

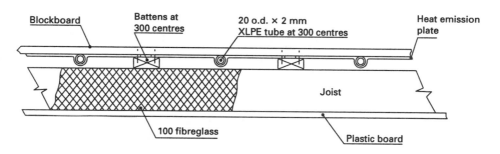

Figure 5.11 Application of XLPE plastic tube to a suspended timber floor

the tensile strength of the concrete and promoting better heat transfer. Other manufacturers maintain that, when using their system, no additives are necessary in a concrete screed, provided that the screed complies with BS Code of Practice 204, Part 2, 1970 (*in situ floor finishes*) [5.5].

Those manufacturers who recommend the use of liquid polymer additives to prevent the formation of cracks in concrete screeds have specifications available to suit screeds that are 'thin' (40–42 mm), or 'thick' (50 mm and upwards). The 'thin' screeds are designed to be suitable for carrying out refurbishment work, in particular when dealing with intermediate floors/ceilings. Generally, it is not good practice to lay PVC tiles (which are waterproof) on to a freshly cast screed before it has had time to dry out.

5.6.7 Permissible floor temperatures

When designing an embedded system it is sometimes found that after using up the option of installing heating elements under the floor, it is still necessary, in order to satisfy all the energy losses, to place additional elements in the walls or ceiling. In an attempt to limit this syndrome, it is tempting to use a smaller pitch when placing the pipe coils (or electrical conductors) in the floor; the result may be floor surface temperatures that are too high and serve only to ensure that people using the building will suffer from aching feet. As a guide, the maximum temperatures for foot contact should not exceed the following: in standing areas, 25°C to 26°C; in sitting areas, 27°C to 28°C; for saunas or bathing areas, 32°C to 33°C; and in perimeter areas, 35°C. A temperature deviation of plus or minus 1°C is permissible.

The Building Services Research Association has produced information that is helpful to potential designers of embedded heating; also when carrying out any design work, it is always advisable to get in touch with at least one of the many companies that have specialist knowledge in this type of heating.

5.6.8 Relative cost of underfloor heating

In the past it was true to say that the cost of installing and running embedded heating (or cooling) elements, in either new or refurbished buildings, was greater than when 'free-standing' emitters were used. Traditionally, in the UK, it was suggested that the first cost was 25% greater; however, the difference in cost is now being narrowed owing to the trend in EC/UK regulations for all floors to have a specified thermal resistance higher than that used in the past. The use of a dense slab insulation under the floor, together with perimeter insulation and a vapour-proof membrane (for ground floors), are essential to make sure that underfloor heating warms only the space within the building and not the space underneath it. Where buildings have a combination of a high thermal capacity and poor thermal insulation, it was found that the response time to any control function was long, making control inflexible and the system costly to run. With the provision of extra insulation, the response time has, predictably, been greatly decreased and the boiler can now be shut down at night and brought on in the morning without having an excessive allowance of extra firing time for warming up.

5.6.9 Positioning of pipe coils in floors

The placing of the pipe coils in the floor can be arranged to suit the client's personal requirements (provided that they are known in time), with regard to the placing of filing cabinets and other furniture. Difficulties may occur when the space is subsequently let to another organisation, which will certainly specify an alternative internal layout that may, or may not, include floor coverings differing from those of the previous tenant. When it is realised that the emission from a floor having a 2.5 mm PVC tile is double that from one that has a carpet with an underlay [5.6], one can appreciate the kind of problems that could arise whenever there is a change in tenancy. The floor may of course be designed originally for the worst known case (carpets with underlay) and the water temperature lowered for a particular zone, where it is known that there will be just the basic floor covering. This approach will naturally be limited by the number of separate temperature-controlled zones that may be incorporated in the pipe circuits.

5.6.10 Temperature control

The controls needed to ensure that the water temperature does not rise above a value of 40°C usually consist of a motorised three-port mixing valve, controlled through a relay by two thermostats, one of which is set to monitor the operating temperature of the water in the flow manifold and the other to react to the air temperature in the room. It is important to ensure that, if there is a malfunction in any part of the apparatus controlling the water temperature, the system will at all times 'fail safe', so that water at a temperature greater than normal will not be allowed to circulate through the buried coils.

5.6.11 Use of a heat pump as a source of energy

The source of energy used to supply water heated to 40°C may conveniently be a heat pump taking energy from the external air, or from an industrial effluent having a temperature greater than the outside air. With such schemes, the initial outlay is always greater than when providing a conventional boiler unit; in fact, if the supply of low grade heat is intermittent and also insufficient to support the thermal load, it may still be found necessary to provide a standby boiler, and the capital cost then becomes even greater. The expected savings in fuel cost are sizeable but, in the experience of the author, the majority of clients allow their feelings on the desirability of energy conservation to be overriden by short-term financial considerations. Enthusiasts should note, however, that such financial restraints are always dependent on the cost of energy (which is always rising), so that a proposed conservation scheme that is considered to be uneconomic at the present time may in the near future prove to be financially viable.

5.7 Heat metering

In those cases where separate buildings on a common site are subject to an energy audit, it is useful to be able to ascertain the amount of energy used by each consumer. This encourages the careful use of the energy supplied and, in order to accomplish these precepts, heat meters may be required. The power transmitted is given by the relationship:

$$P = m \times C_p \times \Delta T$$

where P = power (kW)
  m = the mass flow rate (kg/s) of the working substance
  C_p = mean specific heat capacity at constant pressure (kJ/kg K) of water, at the operating temperature (taken to be a constant)
  T = temperature difference across the flow and return pipes (°C).

A conventional heat meter can consist of two temperature sensors fitted in the flow and return pipes, one small rotating impeller mounted in the return pipe (because the water is cooler) and one electronic 'black box'. A signal proportional to the difference between the two temperatures (the differential temperature, ΔT) is then obtained. Another signal that is proportional to the rate of rotation of the turbine may, when multiplied by a constant, be taken to represent m, the mass flow rate, so that when these two signals are multiplied electronically, the resulting combined signal may be taken as being proportional to the instantaneous power. The sum or integration of this last signal taken with respect to time will then produce an output that is an indication of the amount of energy supplied to the client in 'units'. Note that one 'unit' represents the amount of energy used when one kilowatt of power is received for one hour.

When an incompressible fluid (water) is the working substance, some heat meters make use of a sharp-edged orifice fitted in the pipeline, instead of a rotating impeller, in order to gauge the mass flow rate. The square root of the pressure difference across such an orifice is proportional to the mass flow rate, and an electrical signal whose magnitude reflects the pressure difference is then processed electronically to provide an output, which is directly proportional to the mass flow rate. This signal is then dealt with electronically, in the manner described in the previous paragraph.

The design and subsequent calibration of heat meters require the combined talents of instrument makers and electronic engineers, and many different types of instrument have been produced to satisfy the needs of industry. Some earlier heat meters relied on mechanical means to carry out the integration processes and these require periodic recalibration.

An interesting way of gauging the amount of energy transmitted is to use what is referred to as the 'shunt or inferential method'. In this case a very small part of the whole mass flow rate is diverted from the mainstream return pipe to flow through an electric heater. The heater raises the temperature of the water flowing through it to the original temperature of the mainstream flow pipe, as detected by a sensor fitted to the shunt (or by-pass) and a thermostat fitted to the main flow pipe. The energy required to do this is measured by means of a conventional kilowatt hour meter of the type fitted to domestic premises, and a multiplying factor is then used to 'infer' the amount of energy flowing in the main.

When considering the application of heat meters to district heating, the numbers employed can be minimised by installing them only at key points, such as in the basements of tall buildings, or at sub-stations, each of which serves a group of dwellings. An estimate of individual consumption can then be made by using a volume flow rate meter (which can be inaccurate at low flow rates), or evaporation meters fixed to each radiator. The evaporation rate of the liquid contained in a phial is taken to be proportional to the amount of heat used by the emitter to which it is connected. Unauthorised access to the phial is restricted by a small seal, which of course should be seen to remain intact by the person whose job it is to read the meter (see also [5.3]).

5.8   Further methods of distribution and emission of energy using cyclic and non-cyclic processes

5.8.1   Distribution and emission of energy using steam

Thermal energy (heat), will be transferred when:

(1) there is a temperature difference and no phase change occurs;
(2) there is no temperature difference and a phase change occurs.

Many manufacturing processes require a powerful source of heat to be available for process work and, when there is also a requirement for space heating, it makes good sense to utilise the availability of steam for this purpose. It is possible to pipe steam from a boiler to radiators and/or panels, where after giving off heat to the surroundings, it changes phase (condenses). The condensate is then returned to the boiler, where it is then reheated, changing phase again to steam and repeating the cyclic process.

In order to prevent the possible fracturing of radiators and panels due to the application of excessive steam pressure, pressure reducing valves are fitted at the take-off points from the main supply pipeline. These reducing valves ensure that the heat emitting panels operate at gauge pressures that are no more than 20 to 70 kPa above atmospheric pressure. Engineers frequently refer to 'absolute pressure', which is obtained by adding gauge pressure to atmospheric pressure. The latter has a standard value of 101.325 kPa but it is often taken to be 100 kPa. Thus the heating panels would be receiving steam at a pressure varying between 120 kPa to 170 kPa absolute.

Pressure reducing valves operate by allowing the high-pressure fluid access to one side of a flexible diaphragm, the centre of which is restrained from moving by a spring acting on the obverse side. The fluid does work against the force exerted by the spring, lowering the pressure, and the fluid is then allowed to flow through a valve attached to the underside of the diaphragm, to the outlet. The compression in the spring can be adjusted so that the resulting outlet pressure may be varied. One useful effect of the pressure reducing process is to improve the quality of the steam by increasing its dryness fraction just before it enters the emitters.

Reference to steam tables will reveal that the enthalpy of dry saturated steam at an absolute pressure of 120 kPa is 2683 kJ/kg, while that of water at the same pressure and temperature of 104.8°C is 439 kJ/kg. The difference between the two figures is 2244 kJ/kg and this figure represents the enthalpy of evaporation of the dry saturated steam, which is the amount of heat that will be given off at a constant temperature of 104.8°C when the steam changes phase, that is, condenses, inside the emitter. If the steam has a dryness fraction of 0.9, the amount of heat that can be given off when condensation occurs is reduced to $(2683 - 439) \times 0.9 = 2020$ kJ/kg.

The process of condensation results in a large drop in volume, which allows more steam to flow into the system. Using figures taken from the steam tables, the magnitude of the change in volume can be demonstrated by the analysis that follows.

The volume occupied by 1 kg of dry saturated steam (the specific volume), at an absolute pressure of 120 kPa, is 1.428 m³/kg. When it changes phase (in this case condenses), the condensate has a volume of only 0.0105 m³/kg. So in this case the water occupies a space 136 times less than that taken up by the steam.

Use of steam traps

To ensure that there will continue to be a flow of steam into the system, the condensate must be removed from the emitter; this is accomplished by allowing the steam pressure to discharge the water through a device referred to as a 'steam trap'. There are many different kinds of steam trap, all of which fulfil the following two functions:

(1) they allow condensed steam (condensate) to leave the appliance, but not so-called 'live steam', which has yet to give up its enthalpy of evaporation;
(2) they allow water and air to escape from the unit when the steam supply is reinstated after a shut-down.

A specialised type of steam trap called a 'lifting trap' has a third use, which is to force the emerging condensate to a height that is sufficient to enable it to drain, under the action of gravitational force, through sloping pipes to a receiving tank. This tank is referred to as a 'hot well' and it is sited close to the boiler. It is from this tank that feed water for the boiler is drawn, and it is also in this area that the necessary filtration and water treatment of the condensate and make-up water is carried out, prior to its being reused in the system.

When steam is the working substance, any heat emitter must have a steam trap at outlet and also be fitted with isolation valves and pipe unions on both sides, so that the unit may be easily removed for maintenance purposes. A drain valve should also be fitted to the outlet and a strainer, in a position upstream of the steam trap.

Need to avoid skin contact

For an absolute pressure of 120 kPa, the steam will condense at a constant temperature of 104.8°C and, if in such a system the surface of the emitter comes into contact with skin, for only a fraction of a second, partial thickness burns will occur. The temperature at which condensation occurs inside the emitter can only be varied by changing the operating pressure. This presents a control problem and it is normally accepted that the amount of energy emitted cannot be modulated; it can only be turned on, or off, by means of an isolating valve, which is of course the most basic form of control and is not what is generally expected in a heating system.

The heat coming from a radiator containing steam can be described in non-scientific language as 'fierce'. The author knows of the case of one employee whose place of work was such that the heat from a radiator containing steam was directed on to the small of his back, resulting in his being prematurely retired, with severe back trouble due, it is believed, to the almost constant intense heat. Clearly, if radiating surfaces are to be supplied with steam or other high-temperature fluid, they should be placed so that no one can sit directly in front of them, and they must be guarded if they can be touched.

When it is intended to use steam as the primary working fluid, say to provide space heating and domestic hot water to a small factory, it is very much better to arrange for the steam to pass through a coil of pipes immersed in water contained in a cylinder (a calorifier) and to use the heated water in the conventional way, as a secondary working fluid, supplying heat directly for space heating purposes and indirectly (that is, through another heat exchanger) for hot water. The temperature of the water flowing may then be controlled by means of a mixing valve fitted with a by-pass, and in this way it is possible to modulate the heat output.

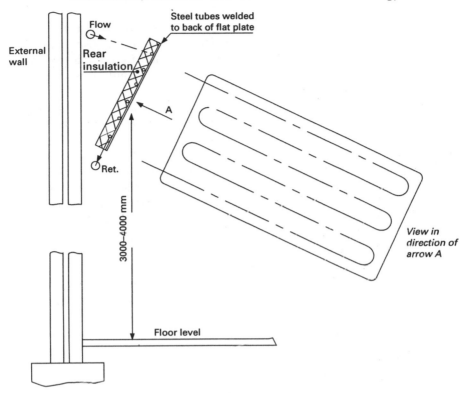

Figure 5.12   Flat-plate radiant panel with snake coil for use with steam or medium/high-pressure hot water

5.8.2 *Emission of energy from overhead radiant panels and tubes, using water*

A useful way of heating the occupants of a series of tall and spacious factory bays, when steam (or high/medium-pressure hot water) is the working substance used in a cyclic process, is to rely on the radiation effect obtained when the working substance is passed through suspended flat plate panels or tubes, which may be insulated at the rear and angled towards the area to be heated. Figure 5.12 shows one such type of radiant panel. The heat convected from the rear of the panel will help to neutralise the effect of any downdraught of cold air coming from openings in the factory roof.

The temperature of the working fluid affects the rate of radiation available per unit area of each emitter. High-pressure hot water would operate between temperatures of 95°C and 150°C, medium-pressure hot water between 85°C and 120°C and low-pressure hot water between 73°C and 83°C. Mounting the panel higher spreads the radiant energy over a wider area (the footprint of the heater) and delivers it to the working level (often taken as 1 m above the floor) at a reduced intensity. When the panel is mounted lower, the converse is true. The net effect of subsequent panels angled towards the same area may be determined by adding the specific intensity of radiation (W/m^2) produced by the new panels at working level, to that of the existing one.

Suitable values for the specific intensity of radiation on the working plane depend on the nature of the task being performed and the net heat loss rate, obtained from the summation of the fabric and air change rate losses. They can vary from 100 W/m^2 for a fully heated, well-insulated building, to 400 W/m^2 for one that is partially heated, lightly constructed and poorly insulated. Designers of such systems should make sure that the net effect of the radiation from all panels aimed at the working area is within the tolerance specified by the manufacturers of the panels, and that there is no likelihood of overheating the heads of the workers.

During the time of warm up of a building, the rate at which radiant energy can be absorbed by a person depends on the magnitude of body area in the path of the beam. When, during this period, the energy emanates from horizontal panels, it is projected vertically downwards and the area presented by the body is that of the head. If, however, the panels are angled and are sited one on each side of the person, the area presented to the irradiating beams is that of both sides of the body. It follows that during the time taken for the building to warm up, the greatest personal heating effect will be experienced when the radiant energy comes from angled panels, placed on either side of the person. However, when the building and its contents approach a steady state of thermal balance, the exchange of radiant energy becomes omni-directional and the difference between the effect of angled panels and horizontally mounted panels becomes less noticeable.

5.8.3 Emission of energy from overhead radiant tubes using hot gases

The directional effect of radiant energy may also be utilised by running pipes, backed with heat reflectors, horizontally above the area to be heated, where the working substance can be hot gases that do not circulate continuously through the system (a non-cyclic process), see figure 5.13.

When considering space heating for medium-sized developments, such as factories having one or two bays, warehouses and garden centres, it can be shown to be cheaper in capital and running costs (when compared with the use of a cyclic process with steam or water as the working substance), to use overhead radiant tubes that contain a mixture of air and burning gases, obtained from one or a series of gas burners.

The air and combustion gases are directed into one end of the radiant tube and drawn through the length of run of tube, by connecting the suction port of a centrifugal fan to the other end. The outlet from the fan may then be ducted through a wall, or to a conventional flue, or in some

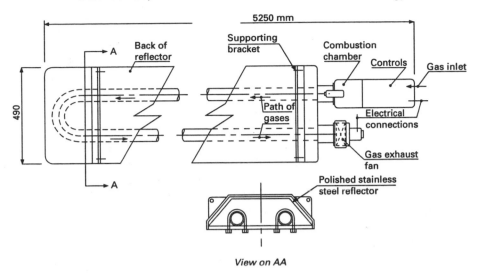

Figure 5.13 Overhead radiant tubes containing hot gases

circumstances discharged into the space being heated. BS 6896 [5.7] gives information on the installation, inspection and commissioning of radiant heaters for non-domestic premises.

The manufacturers of such units produce them with outputs varying from 10 to 50 kW, having recommended mounting heights which range from 2.7 to 3.6 m, if wall mounted, or 2.5 to 18.0 m, when mounted horizontally. From 8 to 10 units, each having its own gas burner, can be connected together in various configurations, with the air and burning gases being drawn through the system by a single fan, prior to discharging the mixture to a flueway. In order to make the combustion process more efficient, the air used for combustion is, in some models, preheated to approximately 80°C, producing (it is claimed) a saving in the consumption of gas of approximately 5% [5.8].

5.8.4 *Emission of energy from low-level 'radiators' using low-pressure water*

Nowadays, the most common way of obtaining the benefits of heat given off by hot water circulating in a piped system (a cyclic process) is to arrange for the water to come into contact with a large surface area, made of a material that is thermally conducting and concentrated in one place. Such units are referred to as 'panels', 'panel radiators', or just 'radiators'. The use of the term radiator, for a device that gives off heat, implies that energy is radiated from the unit, but in fact the greatest proportion of energy emanating from the radiator is convected and the proportion varies according to the type of radiator. The amount of energy radiated depends on the absolute temperature difference between the water and the surrounding air, raised to the power of 4, as well as the shape, emissivity, position and orientation of the heat emitting surface, within the enclosure. Details of the environment surrounding the enclosure must also be specified, if meaningful comparisons are to be made.

Testing of radiators for thermal performance

The performance of emitters is usually determined for the manufacturer by a research group acting on its behalf, which carries out tests in accordance with the recommendations made in BS 3528 [5.9]. The information so obtained is then presented by the manufacturer in the form of tables, enabling the system designer to select the type and size of emitter that best fits the local conditions.

BS 3528 specifies that such tests shall be carried out in an enclosure having external walls that are cooled either by air, or by water, under steady-state conditions, to provide a mean room temperature of 20°C when the mean temperature of water flowing through the radiator is 80°C. Alternative figures are available for when steam is used as the working substance. BS 3528 also gives details of the construction of the test facility and specifies the way in which the tests are to be carried out and the results evaluated.

Figures obtained from such test facilities on the relative proportions of radiant to convective heat emission [5.10] are as follows:

- radiant emission from a single-panel radiator   50%
- radiant emission from a double-panel and two-column radiator   30%
- radiant emission from a four- and six-column radiator   19% and 17%

In an actual enclosure, the normal state is one of thermal imbalance and it is likely that in practice the figures given above will be reduced. For instance, for a single-panel radiator, it is calculated [5.3] that for an outside temperature of −3.3°C, a mean room temperature of 18.3°C and a mean water temperature of 76.7°C, the proportion of radiant heat emission is 38%. It would appear from the previous remarks that the term 'radiator' is to some extent a misnomer and perhaps 'emitter' would be a more accurate description.

It should be noted that the radiant emission of heat from a radiator (or emitter) will be reduced if it is painted with any of the metallic paints, but that other finishes and colours do not affect the magnitude of the radiation component.

Development of low surface temperature emitters to prevent burns

New regulations [5.11] issued by the UK Department of Health and Social Security set out some tenets to ensure the safety of patients, staff and visitors, to all registered accommodation, which includes premises occupied by the elderly, mentally impaired and the young. They suggest that, to ensure safety from burns, flow temperatures should be reduced, guards should be fitted or low surface temperature emitters should be used.

Data given [5.11] show that when a person is in contact for less than a second with a heated surface having a temperature between 80°C and 70°C, partial thickness burns will occur and, if the contact time extends to 10 seconds, that person will receive full thickness burns. It is also stated that the maximum surface temperature of space heating devices in such premises should not exceed 43°C when the system is running at full output.

Compliance with these regulations has been achieved by one manufacturer by adding extra fins to a conventional emitter and by encasing the unit. The casing contains a thermostatic control valve and provides air grilles at the top and bottom. Floor or wall mounted units are available.

Materials used for emitters

Emitters are made from cast iron, steel, aluminium or occasionally copper, and appear in many forms. When they are of cast iron, steel or die-cast aluminium, they are made in short sections that can be assembled to give the emitter the required length. Nowadays, this is normally done at the place of manufacture when the emitter is first ordered, so providing for the requirements of the space in which it is to be fitted. These sections, which form the body of the emitter, are identical and can be screwed together, top and bottom using steel connectors, referred to as 'barrel nipples', each of which has a left- and a right-handed, tapered external thread. Rotation of a nipple by turning a bar wrench, inserted through the hollow centres of the sections and the nipples, draws together adjacent sections, making it possible to produce a watertight joint between them.

Cast iron emitters, unlike those made from steel or aluminium, may fracture if the water contained in them freezes. This condition may occur when the heating plant is shut down for an extended period and the circuits at risk have not been drained. Those readers who have been closely concerned with the task of replacing broken cast iron emitters, sited on the top floor of a development, will always be best qualified to testify as to their weight.

Cast iron and steel emitter sections can be produced in ionic column form, having 2, 3, 4, 5, or 6 columns. They may also be produced to form a section having a smooth surface, which is curved in such a way that it is easy to clean between the sections when they are assembled. This design was developed when cast iron emitters were used in hospitals, at a time when the medical authorities were particularly concerned at the possibility of the streptococcus bacteria being harboured in surface dust. They are referred to as 'hospital' type emitters and are now available in cast iron, or steel.

In the past 40 years the steel panel emitter has become popular and appears mainly in the form of single, or double panels. It is possible to obtain triple and even quadruple panels, but the intensity of heat transfer rate (W/m²) becomes less as the number of panels increases and the effect on the emitter is to make it so wide as to become unsightly.

It is worth noting that the lengths of emitters quoted in manufacturers' tables must be regarded as nominal. In carrying out some developments, it is not unknown for the delivery date for the emitters to be out of phase with that of the proposed installation of the flow and return connections. However, the obvious temptation to get on with the installation of both flow and return connections, using the nominal length of the emitter, should be resisted until the actual length of each emitter is known.

Single-panel steel emitters can be rolled into a curve to suit a bay window or be welded into a series of straight, angled sections to suit a multi-sided angled bay, see figure 5.14.

The use of aluminium as a material, in both extruded and die-cast form, has produced emitters that have a lower water content and are lighter than those made of steel and cast iron. These two factors, coupled with the fact that the thermal conductivity of aluminium is approximately three times that of steel, mean that the thermal capacity of an aluminium emitter is lower than those made of steel and cast iron, so that with suitable controls, a system using aluminium emitters will respond more quickly to changes in the thermal load. In addition, it is claimed that the extruded versions of those emitters made of aluminium weigh approximately one-quarter of that of a steel panel having the same output. Clearly, when such emitters have to be fixed to dry line, or stud walls, their lighter weight will be a bonus. Aluminium emitters are now available in many different forms and colours, obtained by using epoxy resin powder coated finishes. These are applied by the manufacturer, in purpose-built, automatically operated stoves.

Cast iron, steel and aluminium will corrode when they are brought into contact with water containing dissolved oxygen. In a closed heating system, the initial charge of dissolved oxygen in the water will quickly be converted to a finite amount of rust, which will be formed on the internal surfaces. More corrosion due to the presence of oxygen will occur when fresh water is added to the system, or when oxygen is taken into solution from the free surface of a relatively cool expansion tank (oxygen is readily absorbed by water at 'low' temperatures). It may also find its way into the system through seals on valves that are operating below atmospheric pressure owing to the pump being fitted in an ill-chosen position (see figure 5.3).

Whenever dissimilar metals are in contact with an electrolyte, those metals that are high in the table of electro-chemical activity will become

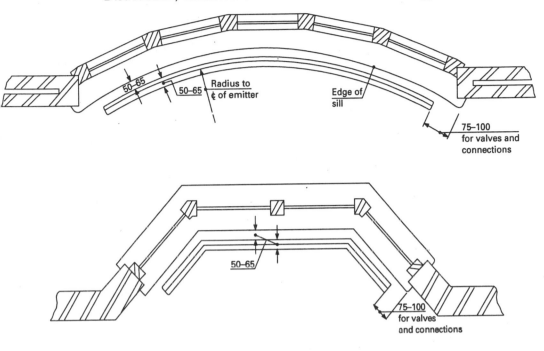

Figure 5.14   Steel panel emitters curved and angled to suit bay

sacrificial, and it is because of this that aluminium has the potential to degrade at a greater rate than steel and cast iron. The electrolytic action set up between the aluminium and the other metals, with the water acting as the electrolyte, can be arrested by the use of a proprietory inhibiting solution. This must always be present in the water contained in the system if degradation of the aluminium is to be prevented.

Positioning of emitters and their connections

Traditionally, emitters (radiators) are positioned under windows where they neutralise cold downdraughts and produce no detectable soot staining. When an emitter is positioned against a wall, the effects of soot staining can be minimised by fitting a shelf above it, such that the space between the shelf and the top of the unit is not less than 50 mm. This is necessary so that the flow of convected air is not unduly impeded. The edge of the shelf in contact with the wall should then be sealed with a foam rubber strip, or equivalent. Siting an emitter against a wall and away from a window is claimed by one established company to reduce fuel consumption. However, the author suggests that any saving can only be at the expense of comfort, as it is known that the slope of the vertical temperature gradient is greater in such a case than if the emitter were positioned under a window.

The transfer of heat from the water inside an emitter is brought about mainly by thermosiphonic flow. This may seem surprising in a system where the circulation is obtained by means of a pump, until it is realised that the cross-sectional area of waterways on the inside of an emitter is very much greater than the area of the inlet connection. Those readers who have a knowledge of mechanics of fluids will know that by evaluating Reynolds number for the water flowing through the emitter, they will be able to check whether the mode of flow is laminar or turbulent. In fact, such a calculation will show that the flow regime is always of a laminar nature, which enables the very small pressure differences developed when thermosiphonic flow occurs to be effective on the internal circulatory system within the emitter.

In order to assist thermosiphonic flow, it was at one time customary to connect the inlet pipe to the emitter, say to the top left-hand side, and the outlet pipe to the bottom right-hand side. This form of connection is known as 'top and bottom opposite ends' (tboe). This configuration facilitates the heat transfer from the emitter but it does not present a tidy appearance in the room. It has now become common practice on pumped systems to connect the inlet pipe to the bottom left-hand side, and the outlet pipe to the bottom right-hand side, that is, both connections to 'bottom opposite ends' (boe). The consequent loss in heat transfer due to this method of connection is believed to be about 10% of the total.

When dealing with column emitters, some manufacturers, on request, will plug the waterway between the first and second columns at the bottom, so that the first column acts as an internal vertical waterway for the entering hot water, and this has the same effect as making an external pipe connection at the top left-hand corner. Fortunately, the matching of emitter sizes to the calculated energy losses in a room is not a critical exercise and it is usually difficult to detect the difference brought about by the type of connections chosen. Hence for all except very long emitters, it is now normal practice to connect 'boe'.

The top connection on the right (or left) of the emitter is usually reserved for the fitting of a concealed valve, the function of which is to allow air to flow out of the emitter when it is being filled and to allow it to flow in when it is being emptied. The same valve is used to vent air and any other gases that may have collected at the highest part of the emitter during use.

In the past it was common practice to fit a 'tongued tee' at the inlet to emitters that were connected to a system operating without a pump. The tee had a metal protuberance facing the direction of flow, the purpose of which was to divert part of the main stream of water into the inlet pipe of the emitter. The effectiveness of such a design was difficult to measure and, because new circuits are now designed (almost invariably) to be operated with a pump, the need for tongued tees has declined and they are no longer manufactured.

Removal and replacement of emitters

It should always be possible to isolate an emitter from the rest of the system by closing the valves fitted on either side of it. Note that some emitters connected to microbore systems have only one connection to the pipework, which operates as a combined inlet and outlet fitting. In a large heating plant, it can take many hours to drain the water from the system and, if alterations are to be made to the small-diameter pipework, on the 'wet side' of the valves, it is sometimes more cost-effective, first to stop the pump and then to freeze the water in the flow and return pipes. This can be done by arranging for liquid carbon dioxide to gasify rapidly, while flowing through an insulated jacket surrounding the pipes that supply the emitter. The consequent change of phase of the carbon dioxide removes heat from the pipes and the water contained in them at a rate that is fast enough to form a plug of ice in each tube. The ice then prevents the escape of water from the pipes long enough for the work to be carried out. The fluid in large-bore pipes can be frozen using liquid nitrogen. Alternatively, a portable refrigerating unit may be used to produce the same effect.

5.8.5   Emission of energy from convectors – natural and forced, using water

A different way of providing surface area for heat emission is to arrange for a tube carrying hot water to be finned. For similar temperature conditions, a 1 metre length of finned copper tube, fitted at the lower part of a cabinet 450 mm high, will give approximately the same output as a single-panel steel or cast iron emitter of the same height and length. The convector has a lower thermal capacity, which means that it can react more quickly to changing temperature conditions.

The cabinet of a convector has openings at the top and bottom, and this arrangement makes use of the stack effect to deliver heat to the space. Such a device is referred to as a natural convector, see figure 5.15. The output of heat to the room may be controlled to some extent by opening or closing a damper (flap), which covers the air outlet to the cabinet. Note that natural or forced convectors do not give their rated output unless they are fitted to a two-pipe system, where the water is circulated by means of a pump (a cyclic process).

When finned tubes are placed inside a cabinet that extends around the walls of the building, this system is referred to as 'perimeter heating' and this is a convenient way of providing an even spread of heat around the outside walls of a building. In order to cater for all possible layouts, the height of the perimeter units varies from normal sill height to skirting height.

In many office blocks the internal space is subsequently partitioned into individual offices, and perimeter heating (provided that the block is

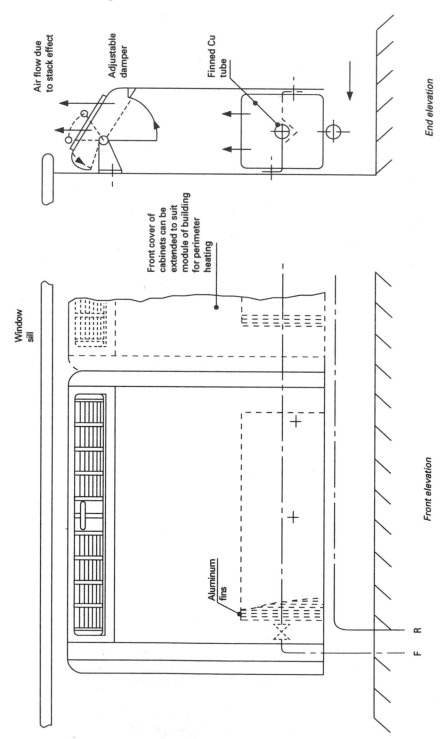

Figure 5.15 Natural convector

not too deep) will continue to provide conditions that are generally satis-factory. For those buildings having a 'deep' aspect to the plan, it is more likely that ventilation and or cooling will be necessary in the centre of each floor.

In an attempt to deliver more heat to the space using the same fin area, a fan (and if the environment warrants it a filter) may be fitted into the unit, so that air is drawn from the room and discharged over the finned tube or tubes, see figure 5.16. This arrangement is referred to as a forced or fan-assisted convector. The temperature of the room may be controlled by a room thermostat with a logic device and a timer if required. The function of the logic device is to prevent the fan from responding to a call for heat from the thermostat when the flow temperature is too low, so preventing the discharge of cool air over the occupants.

One disadvantage of using a fan may become apparent in a quiet room, when the fan bearings have become worn and vibration of the cabinet and trim occurs. Nowadays, small cabinets are fitted with a cross-flow (squirrel cage) fan, but some earlier models had propeller fans fitted and the diameter of the fan was prescribed by the small size of the cabinet. This meant that in order to obtain a suitable output, it was necessary to operate such fans at high speed, which sometimes produced unacceptably high noise levels for 'quiet' rooms.

5.8.6 *Emission of energy from wall-mounted and recessed warm air cabinets using water*

When considering the use of fan-assisted convectors, it is a natural pro-gression to try to take advantage of the economy of scale and to install, say, one large recessed unit in the area to be heated. Such a unit is referred to as a warm air cabinet and, if the size and speed of the fans are chosen carefully, the system (a cyclic process) could be described as being marginally satisfactory. However, when it is proposed to use the same system to heat a lecture room, special care should be taken to choose fans that operate extra quietly, bearing in mind the difficulties encountered by those students sitting at the back of the room when lecturers having a quiet delivery are in active competition with the noisy opera-tion of the fans in the heater cabinet.

The same type of recessed units can be used more successfully to heat a gymnasium. In order to ensure that fresh air is introduced into the space, some of the air inlets to the heating coils must be connected (via a filter) to the outside of the building. In a cold period and during a prolonged shut-down, it is possible for the water in those coils that are in close proximity to the outside air to freeze, possibly causing damage to the pipework. Clearly this is undesirable and three possible ways of avoiding this are suggested.

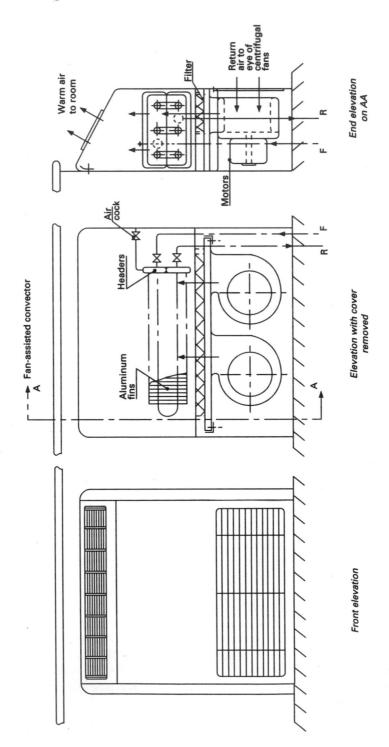

Figure 5.16　Fan-assisted convector

Method 1

This relies on the fact that, in temperate climates, water in motion on the inside of a pipework system that is largely contained within a building is most unlikely to freeze. Hence by allowing the circulating pump to operate when the boiler is shut down, the problem should be avoided. In order that the pump should not run continuously, it can be controlled by means of two thermostats connected in parallel (in case one fails to operate), so that the pump is energised when the outside temperature falls to, say, 5°C. It would not be realistic to make the set point 0°C, as this temperature would not allow for any wind chill factor. When the outside temperature rises above the set point, the pump will be stopped.

Method 2

The boiler and pump can be activated throughout the shut-down period at selected intervals and for predetermined times (assuming that the interpretation of the energy audit will permit this). When the system is fired by solid fuel, this mode of operation is sometimes used to ensure that the fire is never extinguished and is referred to as a 'kindling' mode.

Method 3

A suitable length of electrically operated tubular heater, placed underneath the pipe coil at risk and controlled by air thermostats and the main boiler time switch (or a separate time switch synchronised with the boiler time switch), would prevent freezing. The tubular heater would only be energised when two conditions were satisfied:

(1) when the time switch had shut the boiler down (normally at nights and at weekends);
(2) when either one, or both, of the thermostats arrived at the set point of, say, 5°C.

The heater would be switched off when the outside temperature exceeded the set point, or the time switch controlling the boiler returned to the 'boiler on' mode.

5.8.7 Emission of energy from free-standing warm air cabinets using air

It is now common practice to heat warehouses and factory complexes using a non-cyclic process involving separate free-standing warm air cabinets having self-contained oil or gas burners, firing directly into a heat exchanger, over which the air to be heated is passed, see figure 5.17. Each unit has a flue which, in common with the oil or gas pipes, can easily be disconnected, so that with the provision of lifting eyes, the heater may be

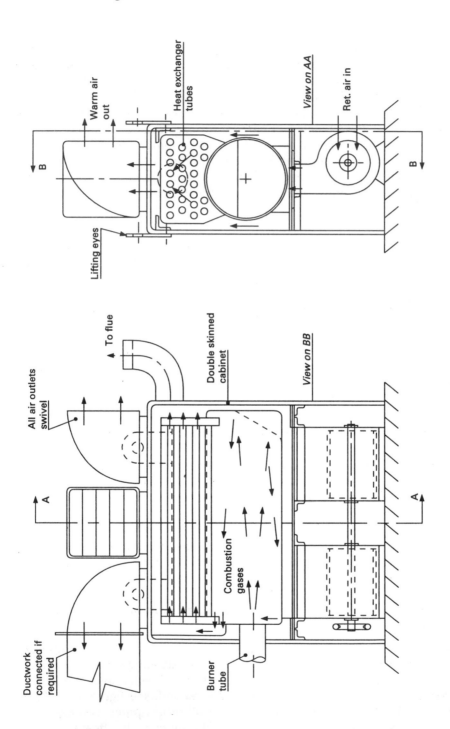

Figure 5.17 Free-standing semi-portable air heater

resited, allowing for any subsequent rearrangement of the working space inside the building.

The warm air outlets are at the top of the cabinet and the air is usually projected horizontally outwards in any chosen direction from 0° to 360°. The magnitude of the air exit velocity governs the possible 'throw' of each air stream, which entrains the surrounding air, warming and circulating it around the space to be heated. The allowable velocity at outlet also depends on the height of the outlets above the heads of the occupants.

In some cases it is possible (if required) to direct all, or a part of, the air output into ductwork, which could have outlet grilles placed in a distant part of the factory, or which may provide warm air to adjacent office space. An outside air intake which is controlled and filtered is usually provided to those heaters that are sited close to the external walls. In order to economise on fuel when operating in the heating mode, approximately one-third of the air is drawn from outside the building and the remaining two-thirds is recirculated. It is found to be counter productive to cut down on the proportion of fresh air provided, as it leads to stuffiness, resulting in the drying of the mucous membranes in the throat and possible sickness among the workforce, with associated absenteeism.

During the summer, the plant may be used for ventilation purposes, and it is usually possible to obtain a limited amount of 'free' cooling for short periods during the day and night. When choosing the number, size and position of the units, it is necessary to make sure that the 'throw' of each air outlet covers the area to be heated and that the height of each outlet is sufficient for the velocity and temperature being used. Figure 5.18 shows a possible arrangement for a factory bay with associated office accommodation.

It is expected that future legislation for the workplace, instigated by the European Union and dealing with ventilation rates used in the built environment, will call for a higher percentage of fresh air intake to recirculated air. This will set new standards for the removal of odours emanating from human occupation and also from furnishings, printing and other processes that take place; it follows that with these proposed higher fresh air intakes, in the future it may be more cost-effective to use heat recovery devices.

5.8.8 Emission of energy from overhead unit heaters using steam or water

A unit heater consists of a heat exchanger, comprising tubes and fins through which either steam or hot water flows, which warms up the air in contact with the outside of the fins using a cyclic process, see figure 5.19. A fan then projects the air into the space to be heated. The principle is analogous to that occurring in the engine cooling system of a motor

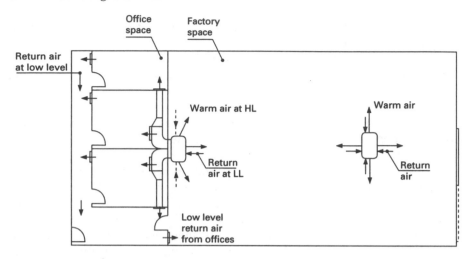

Figure 5.18 Arrangement of free-standing semi-portable air heaters for factory and office accommodation

car using a conventional radiator and a fan, the only difference being the purpose to which the system is put. Unit heaters are often used to provide warm air in large factories. The mounting height has to be chosen to suit the 'throw' of the fan and the temperature of the warm air, as it leaves the heater battery. If the air temperature is too high, the air will rise upwards owing to its excessive buoyancy, and the heating effect will not be experienced by the occupants at floor level. Conversely, if the air temperature is too low, the net effect will be to produce cool and very draughty conditions throughout the space.

In order to avoid discomfort to the occupants, the control of the fans in winter should be such that they can never be activated until the water in the system has reached its design temperature. Those manufacturers of unit heaters who are anxious to have their products installed to their best advantage produce data that, when properly applied, should result in a satisfactory outcome. When applying a number of heaters to a long and narrow factory bay, one method that has proved satisfactory is to place the heaters around the periphery of the building so that the suction side of each heater faces the delivery side of the heater mounted behind it; in this way the air will be moved around the space in either a clockwise or an anticlockwise direction.

Another factor that should be considered is that of the amount of fan noise generated. Such considerations are relative to the type of premises being heated. At one extreme, there is the case of the factory bay, where noisy manufacturing operations, perhaps involving sheet metalwork, are being carried out. In this case it is likely to be considered satisfactory to choose small-diameter fans running at high speed, which are relatively cheap but also noisy. The noise from such fans will barely be noticed in

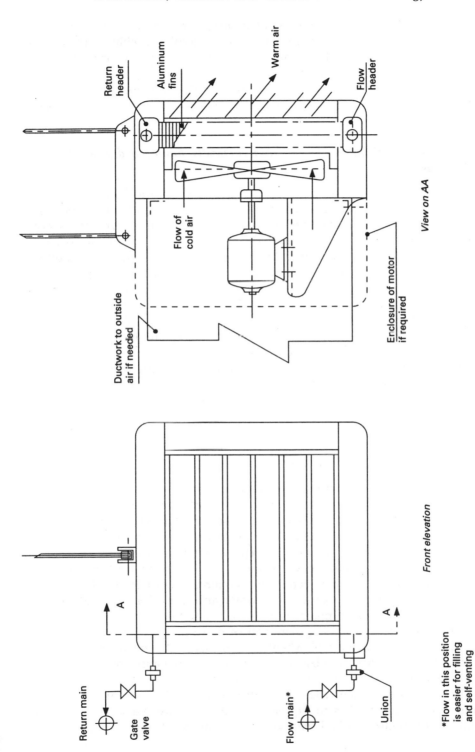

Figure 5.19 Overhead unit heater using high- or low-pressure hot water

the general din set up by the manufacturing process. The other extreme could occur if the same factory bay was undergoing a change of use, to become, say, an open-plan general office. In this case, in order to meet the new lower noise criterion, it would be possible to specify fans having the same duty, but which are larger in diameter and operate at a lower speed. These fans and their associated heaters will be more expensive but will be quieter in operation.

5.8.9 Recirculation of energy from vertically mounted ductwork using air

Whenever streams of low-velocity warm air are used to heat a building, the vertical temperature gradient from floor to ceiling is likely to be high – especially if the temperature of the air coming from the emitter is also high, making the air excessively buoyant. This is not conducive to comfort and is also wasteful of fuel. An obvious way of minimising this effect is to arrange to redirect the warm air that collects under the apex of the roof into a reheater, contained in ductwork, mounted vertically in the space; such a system is referred to as an economy warm air diverter. Figure 5.20 shows the use of such a system with a semi-portable air heater. The heater battery contained in the vertical duct can be connected via a thermostatically controlled valve to a supply of steam or hot water (making the process cyclic) during those periods when heat addition is required, or to a chilled water supply if the company is prepared to pay for cooling the inlet air in the summer.

5.9 Heat recovery

5.9.1 Use of the thermal wheel

There are many cases, in both commerce and industry, where it is economically viable to extract heat from a stream of contaminated hot exhaust gas and then use it to warm a stream of fresh air. For instance, by providing a supply of warmed air for combustion in a boiler, the thermal efficiency of the boiler is increased. The warmed air may also be used in industrial processes and for crop drying, or for space heating. Alternatively, the ventilation air rejected from an air-conditioned building during the summer months is likely to be cooler and drier than the outside air, so that savings may be made by pre-cooling the fresh inlet air, before it reaches the air conditioning plant. All of these processes may be accomplished by using a thermal wheel (or regenerator). The process is noncyclic and the working substances are gaseous. Because the two streams of air can mix, some contamination can occur.

The face area of the wheel is divided into separate sectors of a circle and when the wheel is to be used to recover only the enthalpy of the fluid, each sector is filled with wire wool, made of aluminium, stainless

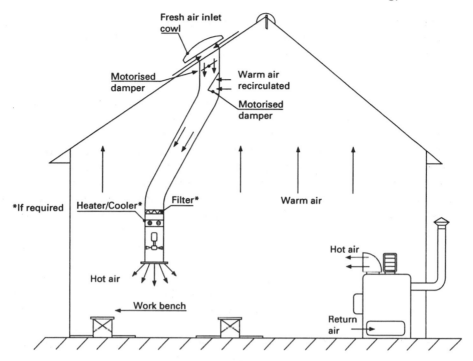

Figure 5.20 Use of economy warm air diverters with semi-portable air heater

steel, or monel metal, depending on the chemical composition and temperature of the exhaust gas. Alternatively, when the wheel is to be used to recover both the enthalpy of the fluid and the enthalpy of evaporation from the moisture contained in the hot exhaust, the matrix may consist of plastic, or textile materials, having a hygroscopic coating, such as lithium chloride, or lithium bromide, which are claimed to be bacteriostatic.

The hot exhaust gas is ducted to approximately one-half the face area of the rotating wheel, where it is allowed to flow through the sectors containing the matrix to a transfer duct on the other side. The direction of gas flow through the wheel is both axial and radial, which is useful when any part of the sector becomes blocked. Both flow components transfer heat and in some cases moisture to the wire, as it is carried across the aperture of the inlet duct. The moisture collected falls under gravity to a collecting tray and is drained away, fortuitously cleaning the wire in the matrix *en route*. According to information obtained from the manufacturers of thermal wheels containing metal matrices, between 10 and 40% of the moisture contained in the hot gas will be transferred to the warmed fresh air.

The cool fresh air is ducted to the opposite half of the wheel, where it passes through the matrix of wire, absorbing the heat from it. The two halves of the wheel are sealed from each other by means of flexible wiping seals, fixed both radially and circumferentially. Additional, peripheral sealing is obtained by means of labyrinth seals, incorporated in the casing surrounding the rotor shaft.

When the face of the wheel moves away from the ductwork containing the exhaust gas stream, the voids in the wire matrix contain elements of contamination and, to overcome this, fresh air is blown through one sector of the wheel and into the exhaust duct, removing the contaminants before the matrix enters the fresh air sector. This process takes place in an area referred to as the 'purge sector'. It is made possible by arranging for the fresh air stream to operate at a slightly higher pressure than the exhaust gas stream and this reduces the degree of cross-contamination to less than 1% by volume, which small amount can usually be removed by means of a standard filter inserted in the warmed air outlet duct.

Depending on the velocity of the air through the matrix, the highest sensible heat transfer efficiency of the heat exchanger (sometimes referred to as the 'system effectiveness') is likely to be about 80%, and this will be obtained when the speed of the wheel is about 20 rev/min. Decreasing the speed of rotation reduces the warmed fresh air outlet temperature and decreases the efficiency of recovery. It is possible (within the range of heat recovery) to maintain a constant, warmed, fresh air outlet temperature. This can be done either by varying the speed of the wheel, by an amount proportional to the warmed fresh air temperature, or by running it at a constant speed and using motorised dampers, diverting part of the hot gas stream around the wheel until the set point temperature of the warmed fresh air is obtained.

When assessing the suitability of any heat recovery system, it is necessary to determine the number of hours for which the plant will be operating, the amount of heat that will be recovered and the amount of electrical energy needed to make the process possible. The ratio of the rate of heat recovered to the rate of heat supplied may be referred to as the 'coefficient of performance' of the plant (compare the COP for a heat pump); this ratio will give an indication of the relative performance of the thermal wheel against any other proposed methods of heat recovery.

The best time to consider the use of a thermal wheel in a project occurs when the project itself is still in the design stage. In this case, any projected power saving may be reflected in a reduction in the size and power of the main plant. However, the power input to both the fresh air supply fan and the hot exhaust gas fan must be increased, to allow for the pressure drop across the matrix of the wheel. The application of thermal wheels to larger projects may require more than one wheel, operating in series, in which case the space required for the ductwork carrying the fresh air and the hot gas (which must all be in close proximity) will be considerable.

5.9.2 Use of a plate heat exchanger

The process is non-cyclic and the working substances are gaseous. No cross-contamination can occur.

The heat exchanger is constructed from a number of thin square hollow boxes, each having two opposite openings formed in the ends of the box having the smallest area. They are assembled one on top of the other, each box being displaced by 90° from the one with which it is in contact, so that the streams of hot gas and warmed fresh air enter and leave the heat exchanger at right angles to each other in cross-flow.

The boxes are commonly made of sheet metal and, when this is the case, there can be no exchange of the enthalpy of evaporation of the moisture carried by the exhaust gas. In order to resist the effects of corrosion, they can be constructed of stainless steel, aluminium coated with vinyl, or epoxy resin. Facilities for draining the moisture from the heat exchanger are provided. At least one company constructs its cross-flow heat exchangers using non-combustible corrugated paper, having a coating of plastic to discourage the formation of mould. The elements have a low thermal capacity and they are tested to the levels specified in BS 476, paragraph 12: 1991.

The heat transfer efficiency of cross-flow heat exchangers lies within the range 30–65% [5.12]. For a given velocity, the pressure drop across the unit depends on the thickness of the gap between the layers of material forming the sides of each box, and it is of the same order as that of the thermal wheel. Control of the unit is brought about by providing motorised dampers which divert part of the hot exhaust gas stream around the heat exchanger, by an amount that is commensurate with the required change in the temperature of the warmed air stream.

Traditionally, ventilation in a conventional house has been achieved fortuitously through cracks around window frames and doors and by means of natural ventilation via chimneys. The modern, well-insulated house has double glazing and draught proofing around the doors, with possibly only one chimney, or none at all. The net effect of these measures is to reduce the natural ventilation rate to a value that is too low to remove cooking and body odours, bacteria and house mites (the latter causing attacks of asthma in those who are susceptible).

At least one company produces a packaged air-handling unit, consisting of a cross-flow plate heat exchanger, complete with fans and heat pump, coupled across the cooled exhaust gas and the warmed fresh air inlet. The unit fulfils the four functions – heating, ventilation, cooling and heat recovery – depending on the requirements of the system to which it is connected.

One of the reasons for having a well-insulated house is to reduce the amount of energy needed to heat it; thus, for smaller houses, it is found that almost all the heat required can be obtained from the installed unit

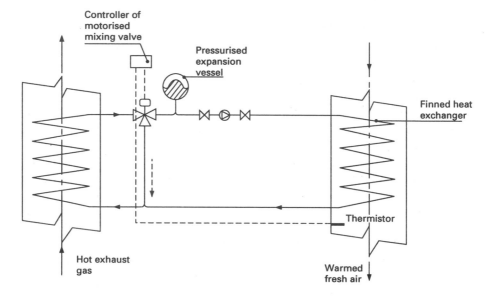

Figure 5.21 Principle of operation of run-around coils

with a small top-up coming from the electrical supply. Similar units, on a larger scale, are produced for use in industrial situations and include humidifiers.

5.9.3 Use of run-around coils

The principle of operation of the run-around coil is shown in figure 5.21. The process is cyclic and the working substance is water which remains liquid. Heat may be recovered from fluids having mixed phases with no cross-contamination.

The hot exhaust gas gives up heat to the finned coil heat exchanger and the heat received by the water contained in the pipework is then transferred, by means of a circulating pump, to the second heat exchanger, over which the warmed fresh air flows. The heat transferred derives from the enthalpy of the fluid and does not include the enthalpy of evaporation. The system is capable of operating with large or small temperature differences and the transfer pipework should be well insulated. The pressure drop across the heat exchange coils placed in the ducts can vary from 100 to 800 Pa, depending on the number of rows of coils required to deal with the expected temperature differences. The heat transfer efficiency covers the same range as the plate heat exchanger, that is, 30–65% [5.12].

When the system is used in conjunction with an air conditioning plant, it is necessary to use anti-freeze solution in the water to prevent icing up. The system is a flexible one, in that it is not necessary to bring the ductwork containing the streams of fresh air and hot gas close together, as is the case with the thermal wheel and the plate heat exchanger. It is possible to introduce additional heat exchange coils into streams of fluid passing through other sections of ductwork by simply extending the pipework of the run-around coil.

Expansion of the water in the pipework is usually taken up inside a sealed vessel, divided into two compartments, by means of a neoprene diaphragm. The expanding water extends the diaphragm against the pressure of the air on the other side of it. On contraction of the water, the air pressure moves the diaphragm in the reverse direction.

The system may be controlled rather crudely by arranging for a thermostat in the warmed air stream to operate the motor of the circulator; or it may be controlled more closely by means of a motorised mixing valve, operated through an electronic control device, which is activated by a temperature-sensitive, solid-state resistor (thermistor) placed in the warm air stream.

Irving and Smith [5.12] state that for low gas and air supply rates, run-around coils take up the least space in the ducts. At higher supply rates, run-around coils take up more space than a thermal wheel and, generally, plate heat exchangers take up the least space in the duct.

A particularly efficient, though more expensive, form of run-around coil makes use of the heat pump principle. The process is cyclic, uses a refrigerant as the working substance and is of course duophaseal. The coil arrangement is the same as that shown in figure 5.21. On the flow line, the motorised expansion valve, by-pass, expansion vessel and circulator are deleted, and a compressor takes their place. On the return line, a pressure reducing valve (expansion valve) is fitted. When conditions are favourable, coefficients of performance approaching 6.0 may be expected.

5.9.4 Use of the heat pipe

The heat pipe comprises a single sealed tube containing a refrigerant and a thin internal circular layer of woven glass fibre, extending along its length. The process is cyclic, the working substance is a suitable refrigerant and it is duophaseal. Heat may be recovered from fluids having mixed phases with no cross-contamination.

When one end of the tube is placed in contact with the hot exhaust gas and the other end is surrounded by the cooler fresh air, the application of heat to the end of the tube causes the liquid refrigerant to boil, changing its phase isothermally to a vapour, which is then displaced to the cool end of the tube. Condensation occurs (also isothermally) and the enthalpy of evaporation is transferred to the incoming fresh air. When the newly condensed liquid refrigerant comes into contact with the annular wick of glass fibre, it is moved by capillary attraction back to the hot end of the tube and the cycle is complete.

The operation of the heat pipe may be controlled, to a small extent, by arranging for the pipe to slope at such an angle that the liquid refrigerant is either helped, or hindered, by the action of gravitational force during its path from the cool end of the tube to the hot end. In the process of heat recovery, a number of heat pipes are arranged in a 'bank' across the respective ducts, or pipes, commensurate with the rate at which heat is to be transferred.

The concept of the heat pipe is very ingenious, requiring no externally applied motive power to enable it to transfer the enthalpy of the fluid isothermally (because of the phase change) from one end of the tube to the other. Unfortunately, the total cost of each unit is very much higher than that of other types of heat recovery equipment, and as a consequence, heat pipes are usually only installed in specialised situations, such as occur in the space and nuclear industries.

When considering the relative efficiencies of the various heat recovery devices so far discussed, the common perception is that the order, from the highest to the lowest efficiency, is as follows:

(1) heat pipe;
(2) thermal wheel;
(3) plate heat exchanger;
(4) run-around coil.

References

5.1.   *Encyclopaedia Britannica,* 1989.
5.2.   Billington N.S. and Roberts B.M. *Building Services Engineering: A Review of its Development,* 1982. Pergamon, Oxford.
5.3.   Diamant R.M.E. and McGarry J. *Space and District Heating (Part 2),* 1968. Iliffe, Guildford.
5.4.   Dearborn Chemicals Ltd. *Basic Principles of Water Treatment for Steam Boiler Systems,* 1986. Widnes, Cheshire.
5.5.   Wirsbo UK Ltd. *Space Saving Space Heating,* 1992. Crawley, West Sussex.
5.6.   Opus. *Building Services Design File,* p. 47, 1991. (Heating and cooling systems, Envirafloor Ltd.)
5.7.   BS 6896: 1987 *Specification for installation of gas fired overhead radiant heaters for industrial and commercial heating (2nd and 3rd family gases).* BSI, London.
5.8.   Opus. *Building Services Design File,* p. 69, 1992. (Benson Heating Ltd.)
5.9.   BS 3528: 1977 *Specification for convection type space heaters operating with steam or hot water* [AMD 5090, 1986, R]. BSI, London.

5.10. Martin P.L. and Oughton D.R *Faber and Kell Heating and Air Conditioning of Buildings*, 7th edition, 1989. Butterworth, London.
5.11. National Health Service Estates Guidance Note. *'Safe' hot water and surface temperatures*, 1992. Department of Health.
5.12. Irving S. and Smith I. It's in the air, *CIBSE Journal*, December 1992.

6 Electrical Energy

6.1 General introduction

It has become possible for man to 'generate' electricity, and conversely, to produce a force from an electrical current, by making use of the information contained in the following statement: 'when electromagnetic fields, such as those emanating from the opposite poles of a bar magnet, are moved by means of a force (acting at any angle), across an electrical conductor, an electromotive force (emf) is set up across the ends of the conductor'. Joining these ends to form a continuous loop will allow an electrical current to flow.

6.2 Direct current (dc)

It is possible to produce a flow of current from a single loop generator by using a single ring, which has been cut in two across its diameter, and this is shown in figure 6.1(a). The split ring is termed a commutator and, if sufficient loops of rotating wire are used, with the ends being connected to the appropriate number of commutator segments, the output from the machine will approximate to a steady flow of current, which is then referred to as 'direct current' (see figure 6.1(b)). It may be collected by using two rods of carbon which are pressed against the commutator, and these are referred to as 'brushes'.

6.2.1 Power loss in a conductor carrying direct current

As power (P) equals amps × volts, then the power loss in a conductor having resistance only and carrying such a direct current is given by the relationship:

$$P = I \times E$$

Applying Ohm's law to this:

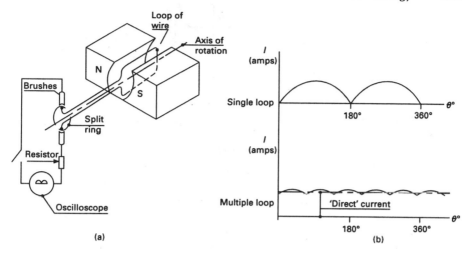

Figure 6.1 (a) Single-loop generator with split ring. (b) Current output from split ring single and multiple loop generator

$(I = E/R \text{ or } E = I \times R)$

then

$$P = I \times I \times R$$

or

$$P = I^2 \times R$$

where I = value of the dc current
    R = resistance of the conductor
    E = voltage.

6.3 Alternating current (ac)

If a loop of wire is connected to two slip rings, as shown in figure 6.2(a), and rotated through 360°, an emf will occur first in one direction, building up to a maximum and decreasing to zero, and then in the opposite direction, building up to a maximum and again decreasing to zero. The change is sinusoidal and this is illustrated in figure 6.2(b).

6.3.1 Power loss in a conductor carrying alternating current

The power loss in a conductor having resistance only and carrying an alternating current of sinusoidal form, the maximum value of which is

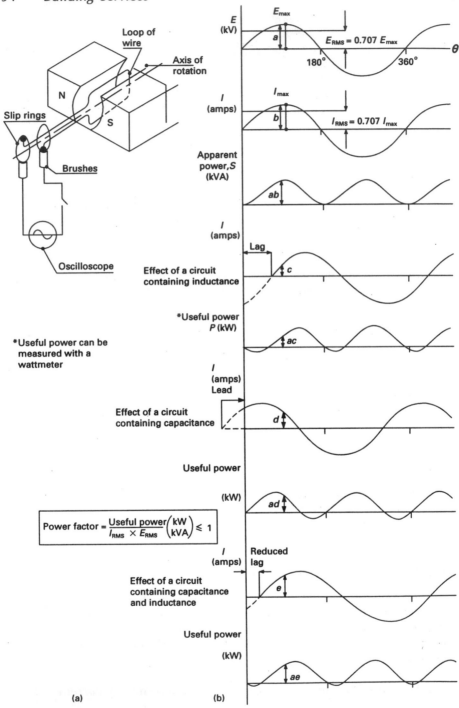

Figure 6.2 (a) Single-loop generator with two slip rings. (b) Generation of alternating current and the effects of inductance and capacitance on the power factor

the same as the direct current in the previous conductor, can be shown to be given by:

$$P = (0.707 \times I)^2 \times R$$

or

$$P = 0.5 \times I^2 R$$

The term $(0.707 \times I)^2$ is referred to as the 'root mean square' (rms) value of the alternating current and, apart from the obvious mathematical connotation, it is also the reading obtained from an ammeter which, if it is digital, may well have been designed to give an rms readout. If, however, the ammeter is of the moving coil type, its scale will have been calibrated to read rms values. When using ac to transmit power, there are other variables that also affect the power loss, but in practice alternating current is used to carry the power to the consumer via the national grid because of the ease with which the voltage may be changed from one value to another.

When the alternator is connected to an external circuit, an alternating current will flow and, depending on the electrical characteristics of the circuit, one of the following three things will occur:

(1) the current will be coincident with the voltage, that is, 'in phase' with it;
(2) the current will lag behind the voltage, that is, it will be 'out of phase' and 'lagging';
(3) the current will lead the voltage, that is, it will be 'out of phase' and 'leading'.

6.3.2 Inductance in an alternating current circuit

Inductance is due to the inertia exhibited by the rapid collapse and subsequent build-up of the lines of magneto-motive force around coils contained in any electrical circuit. Such circuits are said to have 'inductance' and it occurs in particular in the coils of induction motors to a great degree when the motors are over-sized and running on a light load. Inductance is also present in the circuitry of fluorescent light fittings and in the long transmission lines needed to bring electrical power from the generator to its point of use. The effect of connecting a circuit containing inductance to an alternator is to make the current lag behind the voltage (see figure 6.2(b)).

6.3.3 Capacitance in an alternating current circuit

Capacitance is the property of an electrical circuit to store an electrical charge. Such charges may be stored in a capacitor which can consist of two electrically conducting plates, each separated from the other by an insulating substance (called a dielectric); a capacitor can sustain an electrical field and has a low electrical conductivity. The effect of connecting a circuit containing capacitance to an alternator is to make the current 'lead' the voltage (see figure 6.2(b)).

6.3.4 Power factor of an installation

For an alternator, the product of the instantaneous values of voltage and current will give the instantaneous power available. However, when the voltage and the current are out of phase (lagging or leading), the useful power (kW) will be less, because the peak value of voltage and current never occur at the same time (see figure 6.2(b)).

It is found that when an alternating current circuit contains both capacitance and inductance, the capacitor releases energy when the magnetic field due to the inductance is absorbing it, and vice versa. By adding capacitance to an inductive circuit, the inertia effect of the inductance is lessened, decreasing the current lag and making the actual power available closer to that which would be achieved if both voltage and current were in phase.

When dealing with ac, the readings obtained from both ammeters and voltmeters of the digital and moving coil type are rms values, and their product is referred to as the apparent power in VA (more conveniently, kVA).

Hence for an alternator:

Useful power (kW) = apparent power (kVA) × power factor

where the power factor has a value that is smaller than, or equal to, unity.

It is obviously in the interests of any electricity generating company to persuade consumers to improve the power factor of their electrical circuits, so that the existing, or proposed, generation and transmission facilities can serve effectively more customers for a given kVA capacity. For this reason most companies include a tariff having a kVA maximum demand, in which their clients are encouraged to stay within a stated maximum demand for power, by being offered preferential conditions of supply. Predictably, the converse of this arrangement allows the supply company to levy a financial penalty on the customer, which then makes it desirable to seek ways by which the power factor of the installation may be improved.

When the installed inductive load is about half the total load supplied, it is usually found to be economic to install banks of capacitors at the electrical intake position, which can be switched progressively into, or out of, the main feeders, so achieving the best possible power factor with varying loads (see [6.1]). This treatment is usually found to be necessary when dealing with industrial installations and power transmission lines.

6.4 Practical configuration for an alternator

When dealing with the generation of high electrical power, the configuration of a fixed electromagnetic field and rotating coils of wire is not satisfactory. It is found that, owing to the magnitude of the current, electrical discharge occurs between the slip rings and the collecting brushes. The problem may be solved by arranging for the loops of wire to be stationary and for the magnetic field to rotate. The magnetic field does require an excitation current but, as this is small when compared with the current generated, it can easily be dealt with using slip rings and brushes.

6.4.1 Three-phase alternators

Power generating stations make use of rotating magnetic fields which rotate inside three insulated coils of wire placed at 120° around the circumference of a stationary annular ring. Figure 6.3(a) gives a diagrammatic representation of the arrangement. In practice, sections of the coils from one phase overlap those from the other two, and this configuration produces an output of voltage and current from each of the three stationary coils, which approximates closely to the sinusoidal curves shown in figure 6.3(b). The output is described as three phase. It is possible to design a configuration of coils that produces two-phase output, but this is wasteful of the space around the circumference of the alternator that is available for producing electromagnetic fields, and this system is no longer used.

The ends of each of the three coils are connected in a particular way, which is referred to as a 'delta' connection, shown in figure 6.3(c). The vertices of this configuration are then joined to each of three conductors referred to as lines 1, 2 and 3, which are then connected to a transformer before being joined to the distribution grid carrying the electrical energy to the points of use. This three-phase arrangement delivers 73% more power than a single-phase configuration and may be fed to the national grid with the addition of only one extra wire to the transmission towers. For single-phase ac:

Power $= A \times V$

For three-phase load balanced ac:

Power $= A \times V \times \sqrt{3}$

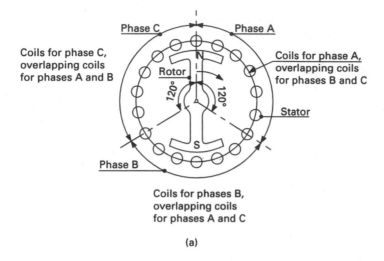

Coils for phase C,
overlapping coils
for phases A and B

Coils for phase A,
overlapping coils
for phases B and C

Coils for phases B,
overlapping coils
for phases A and C

(a)

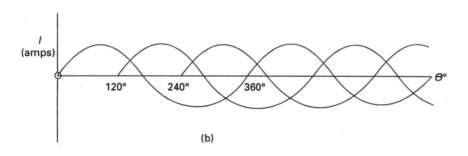

(b)

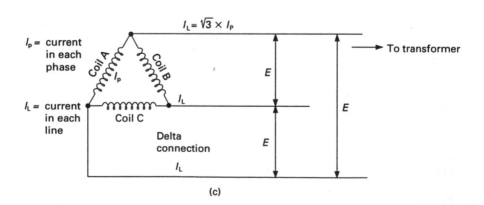

(c)

Figure 6.3   (a) Diagrammatic representation of a three-phase alternator.
(b) Approximate output current. (c) Three-wire, three-phase supply

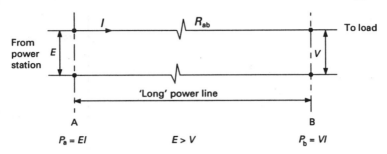

Figure 6.4 Representation of power transmission lines

6.5 Reason for using high voltages for transmission lines

In figure 6.4 the distance between A and B is long and represents power transmission lines having resistance only (R_{ab}).

The voltage drop from A to B is given by ($E - V$). The power available at A is given by $P_a = E \times I$. The power available at B is given by $P_b = V \times I$. The loss of power in the line (P) is then ($P_a - P_b$) or ($E \times I - V \times I$). Hence:

$$P = I(E - V) \tag{6.1}$$

Applying Ohm's law, we have $I = (E - V)/R_{ab}$ or $(E - V) = I \times R_{ab}$. Substituting this in equation (6.1) we have:

$$P = I(I \times R_{ab}) \quad \text{or} \quad P = I^2 \times R_{ab}$$

Hence for an electrical conductor having a fixed value of resistance R_{ab}, the power loss, P, is proportional to the value of the current squared. It follows that, in order to minimise the loss in power in a transmission line, it becomes important to keep the value of the current low; but because $P = E \times I$ for a given input power, this can only be accomplished by making E large.

In order to keep the transmission losses as low as possible, the lines have preferred voltages of 400 kV in the super grid and 275 kV and 132 kV in the national grid.

The upper limit for a transmission line voltage is determined by the combined effects of many factors, some of which are:

(a) cost of the conductors for a given power;
(b) cost of insulation and switchgear;
(c) costs of design, production, erection and maintenance.

It is unlikely that the problem of determining the 'best' voltage can

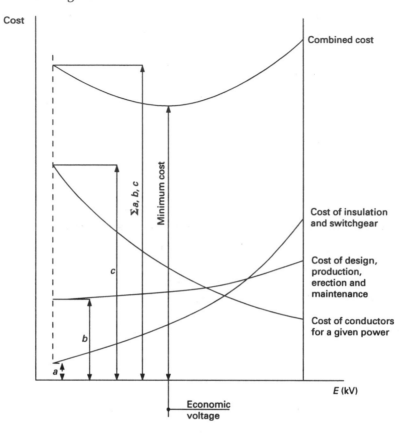

Figure 6.5   Illustration of graphical optimisation

have a definitive solution that will be valid for all times, but in the case of the grid, solutions have been produced that are likely to remain acceptable over a period long enough to justify the overall cost.

Readers may be interested to note that, whenever there are a large number of variables involving costs, a useful way to proceed is to plot them on a graph having common axes (see figure 6.5). Then (with the help of a pair of dividers), each of the ordinates of the curves may be added together to give a point on the total cost curve. After repeating the process for other vertical lines, it will then be possible to draw in a smooth curve through the new points. The economic value of (in this case) the voltage will then be seen to occur at the turning point of the final curve.

Environmental issues are not so easily resolved and it must be noted that recent medical research carried out in St Bartholomew's Hospital, London, has shown that electric and magnetic fields (EMF) of the type set up by high voltages have the effect of increasing the rate of growth of body cells. For those people who live and work underneath high-voltage transmission lines, the results of further research currently being carried out by the hospital and funded by the National Grid Company to the extent of £1 × 10⁶/annum will certainly be of interest. The Company state that, to date (1994), no link has been established between the EMF of the magnitude produced by electrical power transmission lines and the well-being (or otherwise) of humans.

6.6 Use of transformers

The voltage produced by an alternator is limited by the amount of electrical insulation that it is possible to include in the power windings. Power stations incorporating the latest developments have alternators that operate at 20–25 kV (others at 11 kV), and it follows that in order to transmit power over long distances with the smallest possible loss, this voltage must then be increased to the preferred grid voltage. This is accomplished by using a transformer, a device invented in 1831 by Michael Faraday (1791–1867), many years before electrical power was transmitted nationwide.

An increase in voltage can be produced by means of a 'step up' transformer. This is particularly useful after a voltage drop has occurred on a long transmission line. Conversely, a 'step down' transformer is used to decrease the voltage at the end of the line where power take-off is required.

Transformers have a conversion efficiency of approximately 94%. The power loss is due to the magnitude of the eddy currents in the core; these may be minimised by constructing the transformer from a number of thin layers of silicon steel, called 'laminations', each of which is insulated from adjacent layers.

When a transformer is required to deal with large powers, the conversion of 6% of input power to heat presents the designer with a cooling problem. This is usually solved by arranging for the laminated core to be immersed in mineral oil, which acts as both an electrical insulator and as a medium for the transfer of the unwanted heat. This is normally accomplished using thermosiphonic flow (described in Chapters 1 and 5, with the analysis given in Chapter 7). The hot oil, which collects at the top of the transformer, is cooled as it passes outwards and downwards through steel pipes mounted externally, which join into the lower part of the transformer where the oil is coolest. It should be noted that to guard against the possibility of damage occurring to the environment in the event of an oil spillage, these transformers are surrounded by oil-tight walls, of a height sufficient to contain the volume of the oil plus a small margin of about 10%. When transformers are treated in this way, they are said to be 'bunded'.

Figure 6.6 also shows diagrammatically the way in which the three lines (L1, L2 and L3) coming from the alternator are connected through

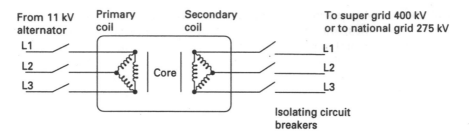

Figure 6.6 Step-up transformer with delta connections

the step-up transformer to the same number of lines that supply the high-voltage super grid and the national grid.

6.7 Transmission of power on a national and regional basis

In the UK in 1920, the National Grid Company was set up to commence operations. It began with 50 different voltages and varying frequencies (25 Hz was used for traction purposes), and initially the power transmitted was smaller than 5 MW.

In 1926 an Act of Parliament set up the Central Electricity Board and the transmission voltage was set at 132 kV with a frequency of 50 Hz. According to [6.2] the effect of this rationalisation of generation and transmission equipment was to cut the price of electrical energy by half. Subsequently, the line voltages for extensions to the grid around large conurbations were increased to 275 kV and the voltage of the supply lines (the 'super grid') was increased to 400 kV. The industry started as a series of private ventures, was nationalised in 1948 and privatised again in 1989.

Currently, 96% of the energy is transmitted by lines that stretch across the country supported on towers, referred to as pylons, and it is known to be approximately 15 times cheaper to do this than to put the lines underground.

6.7.1 Some causes of system failure and their avoidance

Spectacular breakdowns of the electrical insulators may occur owing to the presence of industrial dusts and/or the formation of ice on their surface as a result of freezing fog. Some geographical areas are particularly prone to lightning strikes and, to help protect the current carrying conductors, the topmost line in the cluster of conductors is made the earth line. In other areas, the overhead lines are vulnerable to the mechanical effects brought about by the presence of large numbers of migratory birds (including swans). The boughs of trees adjacent to overhead lines grow ever nearer and can eventually cause a fault current to earth, making it necessary for the supply company to control such growth.

In practice, it has been found that overhead lines operating at 66 kV and 33 kV are less prone to breakdown than lines operating at 11 kV. This is because the lines operating at the higher voltages have higher levels of insulation, greater ground clearance and are mechanically more robust. Experience has shown that most of the electrical faults occurring in overhead power lines are transitory and that isolation of the line, followed by swift reconnection are often all that is required to clear the fault. This fact has encouraged the development of automatically operated protective devices, which operate so that when a fault occurs (say a strike of lightning), the supply is interrupted within 100 ms, to be reinstated automatically in about 8–10 seconds. Any further operation of the automatic trip results in one more try at reconnection, followed by a final shutdown, which must then be investigated, if only to reset the device manually. These devices are generally referred to as 'reclosers', some of which may be adjusted to change the number and times of reclosing. When they are situated in a dusty environment, the mechanism and contacts (hopefully after long periods of inactivity) may become coated with dust, making their operation somewhat problematical. Obviously a suitable maintenance schedule is essential.

6.7.2 Use of power lines for other purposes

An interesting development currently being investigated is the possibility of transmitting telephone and television signals via the power lines, so saving on the existing separate cabling systems now being used. This idea has not been taken up in the past, owing to the difficulties encountered in trying to separate the required signal from its carrier; however, developments in this field are now more promising and further progress is expected.

6.7.3 Breakdown of transmission voltages

400 kV	Super grid (overhead)
275 kV and 132 kV	National grid (overhead)
132 kV	Serving large conurbations and/or heavy industrial complexes (overhead)
66 kV and 33 kV	Small town and/or mixed industrial sites (cables may be underground)
25 kV	Electric traction
11 kV	Villages, hospital and/or light industrial areas (cables may be underground)

Figure 6.7 gives a very simplified view of some of the ways in which the power from an 11 kV intake may be utilised. Note that the earlier three-line system has become a four-line system, feeding into a series of underground cables. The extra wire is added to produce a system that

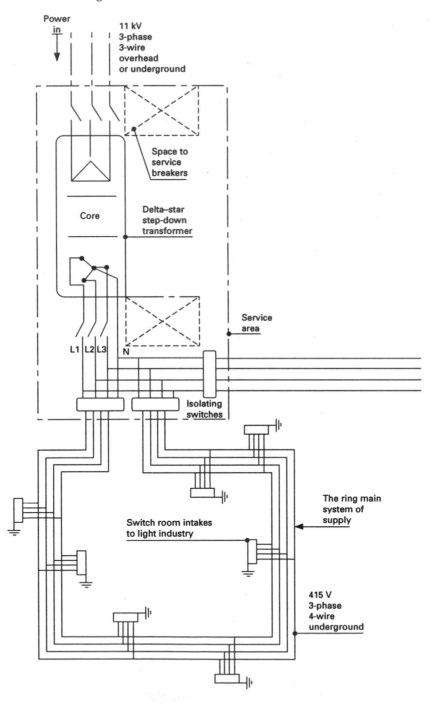

Figure 6.7 Connection of three-wire 11 kV overhead power line to four-wire 415 V underground cables

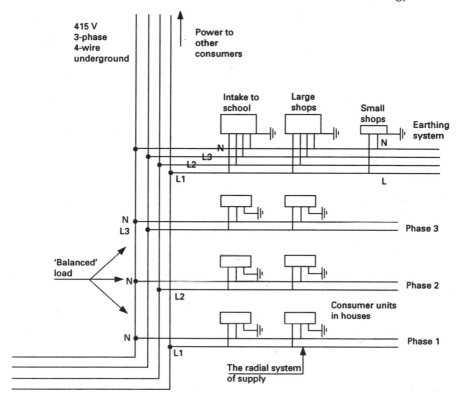

415 V
3-phase
4-wire
underground

Power to
other
consumers

Intake to
school

Large
shops

Small
shops

Earthing
system

N

L3

L2

L1

L

N
L3

'Balanced'
load

N

N
L2

Consumer units
in houses

Phase 3

Phase 2

N

L1

Phase 1

The radial system
of supply

offers a choice of two voltages with three-phase and single-phase options, and is accomplished by connecting the output of the transformer to what is termed a 'star' configuration of wiring.

The consumers can be connected to a radial system of supply or to a ring main system, or to a combination of both the ring main and the radial systems. One advantage of the ring main becomes evident when a fault in the supply line occurs. By appropriate switching, the method of power supply can effectively be turned into one that is radial, so that while repairs are being carried out to the ring main, the consumer can continue to receive power from the 'other side' of the main. This facility is not of course available if the power is taken initially from a radial supply.

6.7.4 Three-phase T T system for installations fed from an overhead supply

This system is usually the method by which small villages and isolated buildings are supplied with power. Figure 6.8(a) shows the way in which domestic and industrial consumers connect their intakes to a system referred to as a T T system. The first letter refers to conditions at the supply and, as it is a T, indicates that one or more points of the supply (in this case at the transformer) are directly earthed. The second letter refers to conditions at the installation and, as it is a T, indicates that all exposed conductive metalwork is connected directly to earth, in this case through the customer's own earth electrode. This may consist of a rod of steel having copper electrolytically bonded to it, which is driven into the ground; or a copper plate, or copper strip buried in the ground. Rods made of stainless steel may also be used.

6.7.5 Three-phase T N–S system for installations fed from an underground supply

This system is used for connecting installations in urban areas and figure 6.8(b) shows the way in which domestic and industrial consumers connect their intakes to a system referred to as a T N–S system. The first letter T has the meanings ascribed to it in the previous subsection. The second letter (as before) refers to conditions at the installation and, as it is an N, indicates that all exposed conductive metalwork is connected directly to the protective earth supply conductor, which in this case takes the form of the metal braid of the supply cable. The third letter refers to conditions regarding the earthed supply conductor and, as it is an S, indicates that there are separate neutral and earthed conductors.

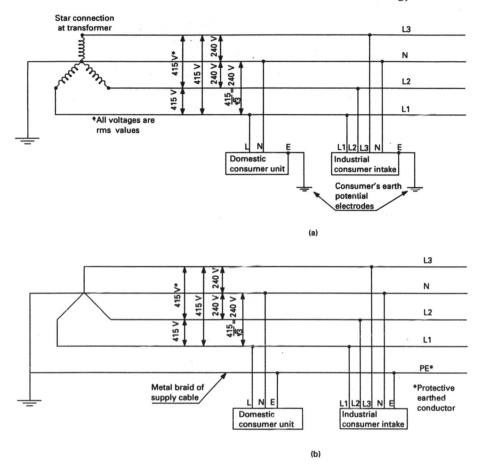

Figure 6.8 (a) Three-phase four-wire T T overhead supply system.
(b) Three-phase four-wire T N–S underground supply system

PART 2: DISTRIBUTION AND CONTROL

6.8 General introduction

Since 1882, the Institution of Electrical Engineers (IEE) has produced (and from time to time updated) a set of regulations that deal in an explicit way with the details of installing electrical wiring in the built environment in the UK. The regulations are not statutory, but in the UK they form the basis for all electrical specifications and, if in any legal action it can be shown that the recommendations have not been taken up by the defendant, this fact would certainly be taken into consideration by the judiciary.

The 16th edition of the *IEE Regulations* (1991) includes the 'form' of Publication 364 'Electrical installations of buildings' of the International

Electrotechnical Commission (IEC), which form has also been accepted for the corresponding work being undertaken by the European Committee for Electrotechnical Standardisation (CENELEC). With the advent of the European Union, attempts are continually being made to 'harmonise' the Regulations, together with the relevant British Standards and Codes of Practice, with those published in other member countries. The current Regulations (the 16th edition) reflect the latest official results of this ongoing process.

As with previous editions of the *IEE Regulations*, their interpretation and application are made easier by the publication of explanatory volumes (see [6.3] and [6.4]).

6.9 Sizing of an electrical supply to a development

It is first necessary to estimate the power required by the proposed development. It is not a task that can be done accurately, but it has to be carried out diligently. In the case of an office block to be constructed as a speculative venture, it is most unlikely that the requirements of the future tenants will be known at the time the design work is being carried out. However, when the development is for a single company it is usually possible to obtain a more accurate assessment of the power that will be needed.

Obviously, not all the apparatus will be in use all the time, and much will depend on the type of building and the kind of electrical equipment installed. The *IEE Regulations* give guidance on the magnitude of the diversity factors allowable, over a wide range of equipment, for buildings having different designated uses. Therefore, by using these tables, it is possible for the designer to arrive at a realistic estimate of the total current, and hence the power required, for any development. Diversity factors are explained later in this chapter.

The local electricity supply company might be expected to supply a single-phase 240 V supply where the current is not likely to exceed 100 A (24 kW). Larger powers (up to about 400 kW) would require a three-phase 415 V supply. Above 400 kW, it is likely that the supply will be taken from a 6.6 or 11 kV supply, which will then require a transformer on site to reduce the voltage to 415 V.

Electricity supply companies obviously reserve the right to decide whether or not a transformer is necessary for any given power, and much will depend on the power already being supplied to other local users. The developer will be expected to pay towards the cost of the supply and if a transformer is required, a suitable space must be provided, with any work necessary being carried out to suit the specifications as prescribed by the supplier.

6.9.1 Entry of power distribution cables to buildings

The cables that enter the switch room of a building usually approach from underneath the ground, and it is important to allow sufficient space at the point of entry to enable the relatively large radius curves required by the cable to be accommodated. The cables are flexible in as much that they can be formed into curves having radii of approximately 20 times their outside diameter. Greater flexibility in cables is achieved by including a larger number of strands of smaller-diameter wire to make up the same area as in the single conductor in the stiffer cable.

6.9.2 Types of power cable

There are many different types of power cable which are best investigated by referring to manufacturers' catalogues. The choice of a particular type of cable for a development considered to be non-hazardous can be made from a very wide selection of cables constructed in different ways and employing different materials, any one of which will usually be satisfactory; such choices, when left to the design engineer, often hinge on his personal preferences. Latterly, aluminium conductors are sometimes chosen for the larger cables, in preference to copper, because of their lower cost.

Traditionally, power cables of minimum rating 100 A are insulated with paper that has been impregnated with a mixture of mineral oil, resin and high melting point waxes. In order to protect the paper insulation from moisture and to some extent from mechanical damage, a continuously extruded lead sheath is formed around the layers of paper. A mastic bedding (or serving) layer is then applied and steel wire is helically wound around this. The bedding layer is made up variously of paper, jute, bitumen and hessian tape, and its purpose is to prevent the steel armouring from cutting into the lead sheath. An external layer of PVC is added and such a cable would be described as a paper insulated lead sheathed steel wire armoured PVC cable (PILSSWAPVC). The addition of high melting point waxes to the oil used to impregnate the paper insulation was found to be advantageous in that it prevented the movement of oil down vertical cable runs into any switchgear fitted below.

The use of polyvinyl chloride as an insulating medium is now common and it is possible to have a power cable constructed of solid or stranded aluminium, insulated with PVC which is then surrounded by a mastic serving layer. This layer is then wrapped with steel wire which is itself then sheathed with PVC to give protection to all the previous layers. This type of construction is coded as PVCSWAPVC.

Some underground cables are manufactured with two or more layers of steel wire armouring, each wound helically in the same direction. The extra layer of steel increases the magnitude of the permissible bending stresses to which the cable can be subjected, and such a construction is used when it is intended to lay the cable directly in the ground, where heave and/or subsidence can be expected. The following sections examine types of cable used for internal distribution.

MICC

Mineral insulated copper covered cable (MICC), sometimes known as mineral insulated copper sheathed cable (MICS), consists of solid copper conductors surrounded by magnesium oxide contained in a homogeneous copper tube. Specifications will be found in [6.5]. The oxide is white and powdery and has the property of being an electrical insulator. It will readily absorb water and, for this reason, special care must be taken to seal off the ends of the cable by means of purpose-made glands at each junction. During manufacture, the powder is compressed inside the copper sheath, so keeping the copper conductors in position and also enhancing the heat transfer characteristics of the powder.

The outside of the copper sheath may then have a thin layer of polyvinyl chloride (PVC) added, which helps to protect the metal from any chemical attack and gives the cable an identity by means of its colour. Such an attack is likely to occur where cables emerge from a concrete floor, where they will almost certainly be vulnerable to the effects of acid from the materials used in the process of floor cleaning. To prevent this, some designers route the MICC through conduit at these levels.

MICC cable is ideal for use in factories, boiler houses and kitchens, where mechanical damage could occur and where the ambient temperature is expected to be higher than normal. It is used exclusively to provide power to fire alarm systems and to firemen's lifts. Some electrical systems require a non-standard number of conductors to be carried in one outer sheath, and such special-purpose cables can usually be provided by the manufacturers.

Mineral insulated cable can be buried directly in the fabric of the building and does not normally need to be protected by the use of conduit. The heat generated by the power loss in the conductor is rapidly dissipated by the metal sheath, without causing any degradation in the materials used in the cable. Accordingly, for those cables that do not have the outer sheath covered with PVC, the IEE has given a higher current rating than other conductors with the same cross-sectional area, thus obviating the need for rewiring for reasons of degradation of the insulating properties.

The use of MICC cable enabled the author to start and virtually complete the rewiring of a technical college during the six-week shutdown period allowed during one summer vacation. A considerable sum of money and time was saved by not having to install the conduit that would have been necessary if conventional wiring had been used; this saving more than offset the higher first cost of the MICC cabling.

MIMS

Mineral insulated metal sheathed cable (MIMS) is a variation on MICC cable in that aluminium may be used instead of copper for both the conductor and the seamless outer tube, or aluminium may be used as a conductor only, with an outer sheath of copper.

PVC

Specifications regarding the manufacture of polyvinyl chloride insulated cables are given in [6.6]. In some electric cables, PVC is utilised both as a material having good insulating properties and also as a material that is generally tough enough to withstand all but the most vigorous mechanical shocks. A typical construction, referred to as 'flat twin and earth', consists of three solid copper conductors, two of which are insulated with PVC, with the third wire being left bare. All three wires are then covered with a thicker layer of flat-sided PVC having rounded edges and referred to as cable sheathing. Such a cable is coded as PVC/PVC. It is widely used in the wiring of houses as well as those installations where surface work is considered to be acceptable. It can be readily fixed to such surfaces using hardened nails held in a suitably shaped plastic cap that fits closely over the external contours of the cable.

Flexible cables

Some appliances require connecting leads that are flexible; typically, these could consist of three multi-stranded copper conductors, each insulated with PVC and coloured red, black and green-stripe on yellow, or brown, blue and green-stripe on yellow, all held in a circular cross-section by means of an outer protective sheath of PVC.

When the need arises for cables to be extra flexible, such as when they are attached to an electric smoothing iron, they can consist of three multi-stranded copper conductors each encased in a seamless colour coded rubber sleeve. The three rubber sleeves are then wrapped in soft cotton yarn, which is then covered by means of an outer sleeve of rubber. This outer sleeve is then finally surrounded by a coloured woven cotton braid.

Some appliances operate in high ambient temperatures; in these cases the conductors may be encased in sleeves of silicon rubber or be wrapped in two layers of varnished glass fibre. For even higher temperatures, the conductors may be covered with a ceramic in the form of fish spine beads. New types of electrical and thermal insulators are being developed to suit environments that are both thermally and chemically hazardous, and it follows that it is always prudent to consult the manufacturers of cable before recommending a solution to a problem for which a better solution may have just become available.

Most commercial developments can now be expected to contain radio and television equipment, and any electrical conductors carrying signals from aerials must be suitably screened so the signals are not corrupted by stray magnetic lines of flux surrounding other adjacent mains current-carrying conductors. The choice of suitable audio and television cables is linked to the electrical characteristics of the equipment they are meant to serve, and the services engineer is best advised to contact the manufacturers of such equipment in order to obtain a suitable specification for any wiring he may be called upon to install.

6.10   Safety considerations and the need to bond metalwork to earth

All electrical apparatus is liable to suffer mechanical damage and, over a period of time, the electrical insulation may also break down. When either or both these defects occur, it is possible that the associated metalwork could come into contact with the line wire, and the metalwork would then take on the value of the line voltage. If a pathway to earth is then provided, a current will flow, the magnitude of which will depend on the resistance offered.

When the only path available for this current happens to be via a human being, he or she will received an electric shock, which in some circumstances could be fatal. If however, the metalwork of the appliance is initially connected to earth by means of a conductor, the resistance of which is less than that offered by any potential human connection, then any fault current will take the former planned route, effectively by-passing the person concerned.

Information about the degree and type of danger involved may be found in Publication 479 of the International Electrotechnical Commission (IEC), entitled 'Effects of current passing through the human body'. In this publication, four 'zones' have been designated (having co-ordinates of time and current for frequencies varying from 15 to 100 Hz), with each zone having a particular physiological effect on human personal life support systems.

It is this syndrome that makes it essential for all metalwork in the installation likely to come into contact with electrical apparatus to be bonded to an earthed circuit protective conductor (cpc). This cpc can take the form of all, or any one, of the following:

(1) a separate conductor with green/yellow insulation;
(2) an earth continuity conductor included in a sheathed cable (twin and earth);
(3) the metal sheath and/or armouring of a cable;
(4) metal conduit and metal trunking;
(5) exposed metal attached to an electrical appliance.

It is clearly important that any metalwork in the building that could conceivably come into contact with any part of the electrical services should offer a low resistance path to earth. Readers should note that with the advent of plastic mains supply pipes for gas and water, it can

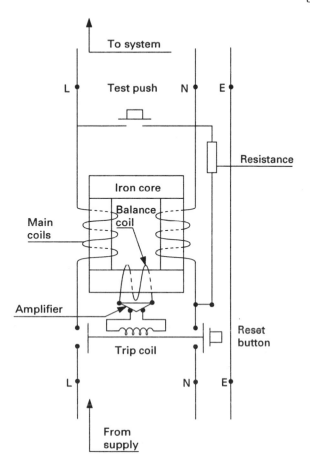

Figure 6.9   Residual current device (rcd)

no longer be assumed that pipes carrying gas and water within the building are *a priori* earthed. It follows that all metal, gas and water pipes (within the building) must now be bonded to an earthed circuit protective conductor.

6.10.1   Use of residual current devices (rcd)

When dealing with the T T system of overhead supply, the consumer has to provide his own earth, the effectiveness of which is known to vary with the type of soil and its moisture content, the latter being clearly outside the control of the electricity supply company. Hence, when such a system is to be installed it has become practice to fit what is referred to as a residual current device (rcd), formerly known as an 'earth leakage relay' (see figure 6.9). Portable versions of the above device, referred to as miniature circuit breakers (mcb), are frequently used in the supply leads to garden tools, which will isolate the appliance whenever there is a leakage current to earth; it is also now common to find an rcd being fitted on individual radial and ring circuits in other systems of overhead supply.

The method of operation is as follows:

(1) When there is no leakage of current to earth, the value of the current in L is the same as that in N, and the opposing windings of the mains coils produce no out-of-balance eddy currents in the iron core. Hence no current flows in the balancing coil and the trip remains closed.
(2) When a leakage to earth occurs, the current in L is greater than that in N. Out-of-balance eddy currents are present in the iron core and a current flows in the balancing coil. This current is then amplified and operates the solenoid trip, which disconnects the installation from the supply.
(3) The operation of the device may be tested by short-circuiting the unit through a resistance from line to neutral, after which it may then be reset for subsequent use.

Rcds have trip ratings of 10 mA, 100 mA, or 300 mA, depending on the type of unit installed. It is known that currents of as little as 1 A sometimes cause an arcial discharge to earth and under these conditions, it is possible for the arc to set light to the surroundings. A residual current device will move to open-circuit whenever the fault current reaches 50 to 100% of its rated trip value, so minimising the chances of a fire being started and also those of receiving an electric shock. In order to restrict the degree of electric shock, it is necessary for the rcd to operate within 50 ms at 240 mA and 150 ms at 80 mA, and an rcd having the rated output of 30 mA will satisfy both these demands.

It is extremely likely that included in the client's electrical equipment will be semiconductors (solid-state devices), which when they fail, could exhibit half-wave-rectified dc current (see Chapter 1, section 1.8 on thyristors). The MK Electric Company Ltd claim that their range of rcds will also operate satisfactorily with this type of fault current.

6.10.2 Safety considerations; limiting the magnitude of current flowing in distribution wires

One effect of adding (or switching in) extra electrical load to a wiring system is to increase the current flowing in the supply wires. Section 6.3 showed that the power loss in a cable is equal to the product of the resistance and the square of the current ($P = I^2 \times R$). Hence any change in the value of the current will produce a change in the power loss which is proportional to the square of the current. The electrical power is dissipated to the atmosphere in the form of heat, the presence of which is signalled by a rise in temperature of the wires and, in order to keep the temperature of the conductors from becoming too high, some form of current limiting device is needed in every circuit.

Traditionally, this has been achieved by inserting into the circuit a short length of wire (a fuse) having a composition and an area such that it would melt when the current passing through it reached a certain value. Higher values of current could be dealt with by using wire having a larger cross-sectional area, and this form of control is still used in many systems. The fusing current is susceptible to the effects of ageing, corrosion and temperature, and the initial fixing tension in the wire, see [6.7]. It is also clearly open to abuse, in that fuse wire of the wrong size may be used to replace a spent fuse, so that the circuit is then protected against a current overload that may be much higher than was initially intended.

In an attempt to overcome some of the above drawbacks, cartridge fuses were produced, with each cartridge having a nominal stated current rating. The value of the current at which the fuse operates is about 40% greater than this rated value (see BS 136). With the subsequent development of other, more accurate forms of excess current protection, it will be seen that the type of current control given by this device and the rewirable fuse may at best be described as 'coarse' – a category recognised by the IEE.

Induction motors can require starting currents that are up to six times the normal running current. This is dealt with by using high breaking capacity (hbc) cartridge type fuses, consisting of specially formulated alloys encased in sand, which allow a higher current to flow for a period long enough for the motor to reach its normal running speed, before the element in the cartridge melts, see [6.8]. This particular type of cartridge offers 'close' control, another category recognised by the IEE, and a conductor having the same physical characteristics is allowed to carry a larger current, when it is fused with a device offering 'close' control, than it would be if it were connected to one giving 'coarse' control.

6.10.3 Miniature circuit breaker (mcb)

The miniature circuit breaker is intended to be used as an overload and fault current protection device to cabling where the load is mainly resistive, and is a very worthy alternative to that offered by rewirable fuses and cartridges other than the hbc type; they are dealt with in [6.9]. Some of the technical details given later in this section have been taken from trade literature supplied by MK Electric Ltd.

The mcb consists of two control devices, each of which operates a solenoid trip in two different modes. The first device operates in a thermal mode and consists of a bimetallic strip through which the circuit current is passed. Values of current greater than the rating of the unit (overload current) heat the strip, so producing differential movement, which is sufficient to trip the solenoid and isolate the circuit. Larger values of overload current serve to trip the solenoid earlier. The second device operates in a magnetic mode and relies on the build-up of magnetic lines of force arising from an excess current (a fault current), which, for a type 2 mcb, is between 4 and 7 times greater than the rating of the unit. The lines of force operate on an armature, which then moves in less than 0.1 second in a longitudinal direction, tripping the solenoid. In the event of a short-circuit current occurring (up to 6000 A), the rapid movement of the armature takes place with such force that it drives a plunger between the contacts of the mcb, safely extinguishing the discharge of the electric arc and isolating the circuit.

The mcb can be tested manually and is easily reset; another useful feature is that it cannot be held in the closed position when there is a fault in the circuit. It will operate with dc, but between lower voltages than with ac; in the magnetic mode, it will require 40% more dc current to trip at the minimum current value. The type of control offered is 'close' and once again the IEE has been able to increase the allowable current ratings of all cables to which the mcb are fitted above those specified for fuses of the rewirable type, which only give 'coarse' control.

6.11   Choosing a suitable cable

It is logical to suggest that a suitable cable would be one in which the product of current and voltage available at the terminals of the fitting match the power needed, and also one in which the voltage at the terminals lies between the limits prescribed by the manufacturer of the fitting. Such considerations take into account the electrical requirements, but ignore the fact that the transmission of electrical power along a conductor always produces a rise in temperature of the conductor together with any associated insulation.

In practice, the current carrying capacity of a cable is limited by two factors:

(1) the perceived 'safe' temperature at which it may be operated, without the insulation being physically degraded;
(2) the magnitude of the voltage drop.

6.11.1   Determination of cable cross-sectional area

Nowadays, most current carrying conductors are covered with polyvinyl chloride (PVC), some grades of which become soft at a temperature of 80°C, making it possible for the conductor to move (migrate) to the outside of the protective covering, where it could short-circuit to earth and/

or come into contact with flammable material. This has in the past proved to be the means by which fires have been started; it is therefore prudent to use the external temperature of the insulation of a cable (however constructed) as a useful guide to its safe current carrying capacity.

Those current rating tables in the *IEE Regulations* that refer to PVC covered conductors have been compiled so that when the quoted value of the current is reached, the maximum temperature of the insulation will not be greater than 70°C. Other tables giving safe current carrying capacities, based on perceived safe temperatures for cables, having a different construction are also provided. The temperature of a conductor will be affected by:

(1) the magnitude of the current;
(2) the construction of the cable and the materials used;
(3) the ambient temperature;
(4) the way in which it is laid, that is, whether it is bunched with other cables in a conduit, or laid on a perforated cable tray, where air can circulate around it;
(5) the way in which any thermal insulation is applied to the cable along its length.

The *IEE Regulations* provide tables of correction factors that reflect the above parameters.

It is necessary to decide on the design current, taking into account the type and method of operation of the equipment being used; it is normally not necessary to assume that all the electrical equipment will be in operation at the same time, and details of appropriate 'diversity factors' are to be found in the Regulations.

The next step is to decide on the way in which the circuit is to be protected from current overload and short-circuiting. As mentioned before, the choice of a mcb and/or a hbc fuse instead of a rewirable fuse results in a relaxation in the size of conductor required.

Using the data from the tables, the required tabulated current carrying capacity may then be calculated by using an equation of the following form:

$$I_t = I_x \times [(1/C_a) \times (1/C_g) \times (1/C_i) \times (1/C_d)]$$

where I_t = required tabulated current carrying capacity (A)
I_x = I_n or I_b (A) as given in table 1.1 of the *IEE Regulations* entitled 'Selection of current and correction factors'
C_a = correction factor for ambient temperature
C_g = correction factor for grouping
C_i = correction factor for conductors embedded in thermal insulation

C_d = correction factor for the type of overcurrent protective device
C_d = 1 for hbc fuses and mcb
C_d = 0.725 for rewirable fuses.

Having determined a value for I_t the appropriate table in Appendix 4 of the Regulations should be examined and a cross-sectional area of conductor selected that has a current value greater than or equal to I_t.

6.11.2   Determination of the voltage drop

Each piece of apparatus will only operate successfully within a declared range of voltage; for this reason, it is necessary to determine the magnitude of the voltage drop along the supply cable. When dealing with long runs (particularly when using single-phase circuits), the fall in voltage that occurs from the intake to the furthest appliance can be significant and, if this is found to be the case, a conductor having a larger cross-sectional area may need to be chosen.

The data published in the IEE tables provide this information for any proprietary cable and method of installation. The appropriate tables list millivolts drop per ampere per metre, when the cable has a temperature equal to the maximum permitted value. What the IEE intend is that its tabulated value (mV/A/m) should be multiplied by the value of the current (A) and then again by the length of the conductor (m) to give the resulting millivolt drop over the whole length of the cable. The resulting figure when divided by 1000 then gives an approximate value for the voltage drop.

The International System of Units (SI) is an elegant and coherent system, which has been ratified for use in the UK by Parliament. It is recognised that the use of the double solidus always results in an arrangement of units that is inconsistent, and it is for this reason that the use of the double solidus (in SI) is forbidden. The problem may be overcome by quoting the units of the quantity in the IEE tables as millivolts per ampere metre (mV/A m), which when multiplied by the value of the current (A) and then again by the length of the conductor (m) can be seen to give a quantity in millivolts (mV), which of course is what is intended.

The voltage drop produced by using the data given in the tables is accurate enough for all except 'borderline cases'. In practice, the operating temperature of a conductor will almost certainly be different from the maximum permissible value of temperature on which the tabulated voltage drop values are based and, for those readers who are required to carry out design work that must comply with the 16th edition of the *IEE Wiring Regulations* and produce an economically viable scheme, [6.4] gives detailed guidance and many worked examples, showing the way in which the necessary calculations may be accomplished.

6.12 Arrangement of the intake panel and other apparatus within a multi-storey commercial development

Figure 6.10 shows one way in which an intake panel could be arranged to deal with the electrical services normally expected in such a development. The incoming cable has its conductors sealed into fuses which are referred to as 'cut-outs'. They are the property of the electricity supply company and are intended to protect the incoming cable from any large unscheduled current. The meter is installed immediately after the 'cut-outs' and this too belongs to the supply company. In figure 6.10, the meter is shown as taking all of the main current, and in a large development this is quite likely to be too large for the size of meter available. In such cases the meter is connected to the secondary coils of a current transformer, with the primary coils of the unit being connected to the mains conductors. The output from the meter may then be subjected internally to a multiplying factor. The meter reading may be accessed electronically and displayed by the central management system or read manually. The wires from the meter (called meter leads, meter bights, or meter tails) are then connected to the main isolator for the building. In the example shown, this takes the form of a triple pole and neutral switch fuse (TP & N S/F) which is the property of the developers.

The switchgear shown in figure 6.10 is usually put together on site, as the installation progresses. It is possible to provide a purpose-made 'cubicle' type switchboard, which is factory-assembled to suit the perceived needs of the developer. Such units are aesthetically more pleasing but more expensive, and some of the subsequent designs make provision for the retrofitting of additional equipment on a module basis, so that the additional work does not look too much like an 'add-on'. However, the potential need for extra equipment must usually be specified at the time of ordering, requiring the designer to be somewhat far-seeing.

One particular manufacturer specialises in producing purpose-made cubicles with the name of 'Powerpod', which may be commissioned about eight weeks before the equipment is required. Such units are very well suited to providing all the electrical needs for an existing (or proposed) building that has been upgraded. The units are normally sited on the roof and in this case a switch room is not required. The equipment is installed using a crane, and the subsequent connections to the uprated (or existing) system are then carried out with the minimum of disturbance to the occupants.

6.12.1 Use of busbars

The conductors coming from the isolator are connected to metal bars made of copper or aluminium, with a typical cross-section of 25 mm × 6 mm and having a length that the designer considers suitable to accommodate the proposed connections for all the switchgear that is subsequently to

Figure 6.10 Proposed layout of switchgear for an 11-storey building

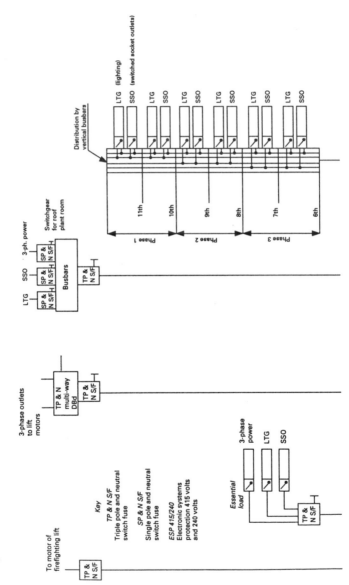

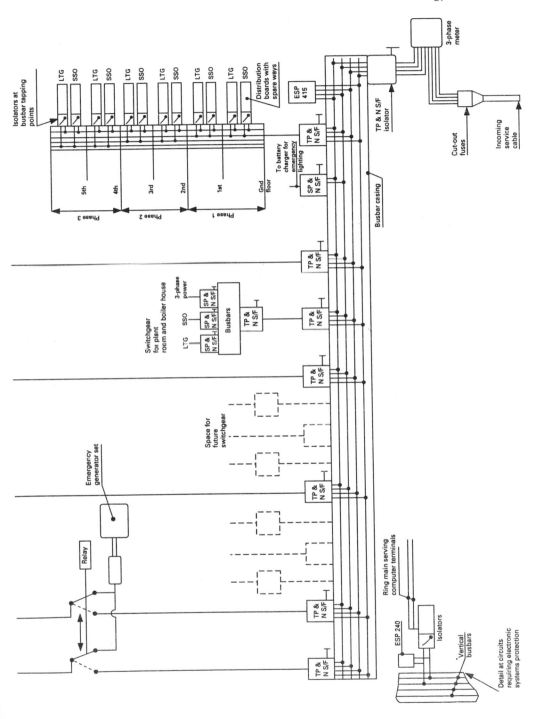

be provided at the intake position. The cross-section of the bars depends of course on the magnitude of current that they will be carrying. These metal bars are referred to as 'busbars', which is an abbreviation from the Latin *omnibus*, meaning literally 'for all' (from *omnis* – all).

The spaces between the bars and the casing in which they are fitted provide the necessary electrical insulation, and each bar must be firmly fixed so that if it is subjected to a short-circuit current, the large forces developed between the bars will not compromise the system. The author of [6.10] calculates that for a short-circuit of 20 000 A, the force of repulsion between two busbars 80 mm apart and 20 m long is 1000 N/m run.

The ampere may be defined as that constant current which, if maintained in two straight parallel conductors of infinite length, of negligible circular section and placed 1 m apart in a vacuum, would produce between these conductors a force of 2×10^{-7} N/m of length.

The use of busbars makes it possible to deal with any extra electrical services that may be needed contemporaneously, because of a change in tenancy, or for future work as yet unspecified. The connections can be made by drilling holes in the bars and then bolting on the take-off conductors, or by using a clamping system.

In a multi-storey development, greater flexibility in the subsequent operation of the building can be achieved by fitting a three-phase busbar system, running in purpose-made enclosed ducts, both vertically and horizontally (if required), through all floors of the building. This ensures that whenever a single- or three-phase service, or extra outlets are required at any location, the connecting points (referred to as tap-off points) can be made on the spot, rather than having to take conductors back to the main supply, which is frequently sited in the basement. The switchgear involved at each tap-off point is likely to comprise a SP & N or a TP & N isolator, or the modern equivalent, a cabinet containing mcbs and an rcd.

Note that the busbars and casing serving each 'lift' of the six floors in figure 6.10 are kept separate, in order to be able to use a smaller section of busbar than would be required if all twelve floors were connected to one vertical run of trunking.

A cheaper alternative to using vertical busbars of a rectangular section is to use lightly insulated circular conductors, fixed inside trunking by means of cleats, which ensure that the conductors are kept a set distance apart. The insulation consists of a thin layer of PVC, to protect anyone working in close proximity from receiving an electric shock. The take-off points required at each floor are likely to consist of the same switchgear described in an earlier paragraph, and these generally lead to isolators, one for lighting and one for power. Each isolator then supplies a distribution board having enough ways to deal with the present requirements, as well as providing a few extra ways for use in the future.

Note that, in order to comply with fire regulations, it will be necessary at those points in the building where the trunking containing either busbars, or circular insulated conductors, passes through each wall or floor, to provide protective insulating sleeves around each bar and then to seal off the remaining cavity with concrete, to the full depth of the barrier. In this way, in the event of a fire, the passage of hot gases and flames from one space to another will be delayed.

6.12.2 Use of conventional PVC insulated cables

Polyvinyl chloride (PVC) insulated wiring may be laid loosely (that is, not fixed) in vertical trunking having horizontal stub ducts at each floor, leading to the services required. This form of wiring can be used to advantage when it is known that the amount and power of the apparatus connected are likely to remain reasonably static for the life of the building. However, when extra equipment is required, such as, say, the provision of forced air coolers on the roof of a multi-storey building, it will be necessary to run new cables from the busbars at the intake position to the roof, and this may be facilitated by the initial provision of a larger than usual allowance of unused space inside the cable ducts serving the existing installation.

Figure 6.10 shows the connections from each busbar in the main switchboard, leading to a number of TP & N S/F, one for each outlet requiring a three-phase supply; as well as a single pole and neutral switch fuse (SP & N S/F) for each outlet needing a single-phase supply.

6.12.3 Use of distribution boards

Each floor will require an electrical supply to connect to a number of lighting circuits. In older properties, this is achieved by providing an SP & N multi-way distribution board, containing rewirable fuses (with the option of an isolating switch). In new or proposed developments, it is more likely that an SP & N multi-way distribution board will be provided that is fitted with miniature circuit breakers and a switch disconnector or a residual current device.

For the supply of power, each floor will also require an SP & N multi-way fused distribution board (switching optional) dealing with separate ring mains containing socket outlets, or an SP & N multi-way distribution board containing mcbs and a switch disconnector or an rcd. The building is also likely to require three-phase supplies to the roof, perhaps for lifts and air conditioning equipment, and also to the boiler house and plant room. The plant room in an eleven-storey development is likely to contain pumping equipment for domestic hot and cold water supplies. These three-phase services can be provided by using multi-pole distribution boards (which may also be multi-way), and can include either fuses or mcbs and either a switch disconnector or an rcd.

In order to limit the effects of voltage drop in the final supply circuits, it is good practice to arrange for the distribution boards serving the apparatus and fittings to be sited at their geometric centre. The distribution boards serving the socket outlets are usually sited adjacent to the lighting distribution boards, and it is common for the socket outlets to be connected to a ring main system. In both domestic and commercial/industrial practice, each ring main should serve no more than 100 m² of floor area; in addition, the total connected load on the ring must not exceed the capacity of the controlling mcb.

6.12.4   Use of conduit

Traditionally, steel conduit was used exclusively to protect the sub-circuits serving the luminaires, socket outlets and other apparatus; specifications for steel conduit are given in [6.11]. In new construction work, the conduit is buried in the fabric of the building, but in retrofits it is often surface mounted. Steel conduit, when new, usually provides a good path to earth, but in older installations where some of the screwed joints have rusted and others perhaps become slack, the continuity to each can be impaired.

The use of plastic conduit is now growing and specifications for this are to be found in [6.12]. An installation using plastic conduit is cheaper than steel and, if the work has to be done on the surface, the effect is aesthetically more pleasing. Because plastic is not a good electrical conductor, it is necessary to provide a separate earth continuity wire running through it, to ensure an easy path to earth for any stray electric currents that may be present owing to faulty appliances.

6.12.5   Provision for three-phase motors

When a three-phase supply is required for a motor, this can be supplied by means of a TP & N S/F and a no-volt and overload release starter, or by using the appropriate triple pole mcb, to protect the conductors of the circuit, together with a four-pole rcd chosen to suit the characteristics of the motor. The use of the 'no-volt release' capability contained in the motor starter is a necessary safety measure, so that in the event of an inadvertent cessation of the supply, the motor will automatically be disconnected from the mains and will not start up as soon as the supply is reinstated.

6.12.6   Provision of a 'balanced' load

The switchgear shown in figure 6.10 shows an eleven-storey building having two sets of vertical busbars, with each set dealing with six floors and, in an attempt to create a 'balanced' electrical load, two floors are connected to each phase. The exercise can only be approximate, as there can be no absolute control over the number and size of the appliances that are connected to the switchboard at any one time.

When trying to assess the state of balance across each of the three phases, a tong-test ammeter is a useful instrument to have available to be able to spot-check on the relative magnitudes of the currents in each conductor.

6.12.7 Avoiding 'mixed' phases on a single floor

Safety considerations make it necessary (whenever possible) to try to ensure that the electrical outlets on each floor are not served by 'mixed' phases. This is because the potential voltage difference between one line on one phase and the line on another phase is 415 V, whereas when all connections originate from the same phase on each floor, the greatest possible potential between lines can only be 240 V. Such considerations become important where there are expectations that portable apparatus attached to running leads will be used, particularly when the leads are long enough to bring the user into close proximity with other electrical equipment connected to a different phase. Whenever mixed phases do occur within 2 m of each other, the owner is required to display a warning notice.

6.12.8 Using 'mixed' phases on a factory floor

Notwithstanding the comments made in the previous section, it has been found to be beneficial to use mixed phases when supplying fixed and adjacent luminaires, in those parts of a factory where rotating machinery is being used. This ensures that there will be no unwanted stroboscopic effects which may lead an operative to conclude that a rotating piece of machinery was stationary when in fact it was rotating.

6.12.9 Distribution of electrical energy using floor and skirting ducts

In an office space it is to be expected that the desk at each workstation will be connected to the power supply, the in-house data network system and the internal and external telephone systems. There are several ways in which this can be facilitated apart from festooning the wires around the internal surface of the building.

Perimeter system

For an office block having individual partitioned offices and a rectangular plan area that is not too deep, it is possible to provide for all three services to be carried in separate electrically insulated sections of proprietary-made skirting ducts. The electrical conductors belonging to each service may then be terminated at any number of chosen points along the length of the duct, to suit the proposed positions of the desks, ensuring in each case the minimum length of trailing wires (see figure 6.11). Telephone wires are insulated to operate between 50 and 60 V, and must

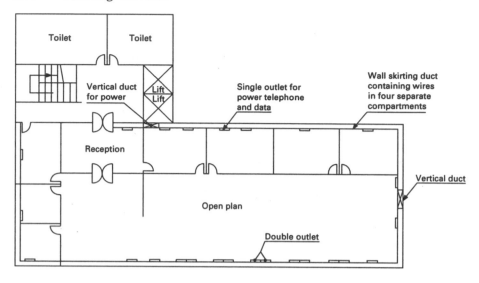

Figure 6.11 Perimeter system of distribution of outlets for power, telephones and data using skirting ducts

be kept separate from power cables which in the UK operate at 240 V. In addition, there is always the possibility of electromagnetic radiation from 240 V cables interfering with the telecommunication signal. The skirting duct system is somewhat inflexible if the duct has to negotiate door openings.

Branching system

For the same type of office block, it may be possible to use shallow ducts having separate compartments for each of the services (see figure 6.12). The ducts are contained within the floor screed and lead to combined power, data and telephone outlets at chosen points on opposite walls. This system is slightly less flexible than that given by perimeter distribution using a skirting duct, but it will not be affected by the positions of doors and walls.

Grid system

The contemporary office is frequently open plan and often has an area that is sometimes square and deep. In such buildings, the disposition and permanency of workstations can reliably be assumed to be variable and temporary and, in order to provide the flexibility required in the placing of desks, the outlet terminals must extend across the floor (see figure 6.13).

Each terminal is served with cables that are routed in separate compartments contained in shallow ducts embedded in the floor screed in

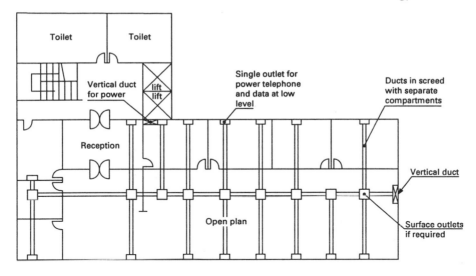

Figure 6.12 Branching system of distribution of outlets for power, telephones and data using ducts in the screed

the form of a 1.5–2.0 m square grid (the dimensions being chosen to suit the module of the building). The corners of each square contain a junction box, fitted with a hinged cover. Each junction box contains flush-mounted terminal outlets, which supply the services in the preferred position to the underside of each desk. Another advantage of using a grid system is that the ducting may also be used to contain the wiring for the luminaires on the floor below.

Suspended floor/ceiling systems

The branching and grid systems for routing of cables are both contained in shallow ducts, the depth of which depends on the thickness of the floor screed. This screed depth may at times be limited and, if the number of outlets requiring full computer and printer services is large, it is sometimes not feasible to accommodate all the cables, some of which have moulded plugs attached to them, in the limited vertical height available. In such cases the provision of a suspended floor having removable access panels may be considered advantageous.

In the case of the suspended floor, all the above-mentioned services are routed under the floor to the perceived points of use which, depending on the requirements of the client, may or may not conform to a square grid. With this system, it is obviously possible to cater for the contemporaneous installation of all, or some, of the following services, with the facility that others may easily be added at a later date. Some of the services mentioned may sometimes be more conveniently routed in a false ceiling (these are shown with an asterisk):

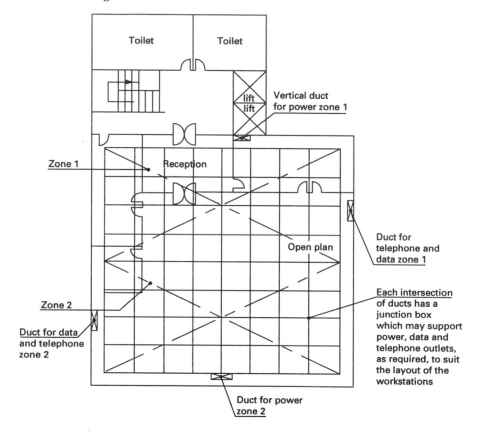

Labels within figure:
Toilet
Toilet
Vertical duct for power zone 1
lift
lift
Zone 1
Reception
Open plan
Duct for telephone and data zone 1
Each intersection of ducts has a junction box which may support power, data and telephone outlets, as required, to suit the layout of the workstations
Zone 2
Duct for data and telephone zone 2
Duct for power zone 2

Figure 6.13 Grid system of distribution of outlets for power, telephones and data using ducts in the screed or underfloor

- in-house data system, with associated printers;
- management control system;
- teleprinters;
- facsimile;
- internal and external telephone systems, nowadays more often combined, so that a single telephone outlet serves both functions;
- radio and television cable distribution;
- security circuits, including closed circuit television;
- fire alarm system, which will include fire detectors*;
- paging system*;
- public address system*;
- slave clocks, each linked to a master clock in the control room of the building*;
- system of bells and calling systems for luxury flats, hotels, hospitals, homes for senior citizens, schools and corrective institutions*.

6.13   Requirements for telecommunication systems

All commercial buildings had at one time a private manual branch exchange (PMBX) requiring an operator to deal with telephone numbers and all incoming and outgoing calls. The private automatic branch exchange (PABX) then followed, which still needed an operator to deal with incoming calls and typically for 100 subscribers and 10 exchange lines required a space of 7 m × 3 m. A subsequent development of the system led to the private automatic exchange (PAX) which for 150 subscribers needed a much smaller space of 3 m × 2 m. The latest development involving the use of computer-controlled switches is described as a 'call connect system' and for 134 subscribers and 24 exchange lines requires a space of only 0.6 × 0.7 × 0.98 m

6.14   Data security

Nowadays, commercial organisations are becoming increasingly aware of the ease with which 'sensitive' electronic data can become known to their competitors. Word processors and printers emit electromagnetic radiation, which can be detected outside the building in which they are used, and an unscrupulous competitor, with suitable equipment, can become privy to all the data currently being processed. In view of this, it is now considered prudent for the managers of companies to quantify the effect of the occurrence of any 'leakage' of data on the commercial viability of their organisations, so that the directors of the companies may then sanction any extra expenditure required.

Depending on the results of such a survey, the implementation of security safeguards may, or may not, prove to be cost-effective. If a company decides to initiate security measures, it will be necessary to operate selected word processors and printers from power circuits controlled from a switch fuse contained in a locked cupboard. Also the word processors, printers and all leads have to be screened against the self-emission of electromagnetic radiation. This may be accomplished by enclosing the circuit boards in copper or aluminium cans, which are cross-bonded electrically and earthed, by the provision of either fibre optic leads (light does not radiate electromagnetically), or electromagnetic filters. These are inserted adjacent to the source of power and prevent the supply leads from transmitting the radiation to all parts of the wiring system.

The task of screening word processors and their associated equipment is usually carried out by the manufacturers, resulting in acquisition costs of the equipment being typically two or three times the normal cost. Quite clearly, the personnel involved will have to conform to an agreed regime of handling, duplicating and storing (and destroying) the tapes and discs (both hard and soft) that are used for the electronic storage of data.

The technology dealing with security is changing at a great pace, and suitable specifications for making the equipment that handles classified information 'safe' should only be drawn up in consultation with those who are expert in the latest developments.

6.15 Control of electrical energy used for artificial lighting

With the continuing development of integrated circuits to carry out specialised tasks, it is now possible to introduce an elegant centralised regime, consisting of automatic lighting controls over all or any luminaire, used in any part of a building. The objects of such controls are to assist with energy conservation, to further the convenience of staff, to contribute to their safety and to facilitate building security.

6.15.1 *Artificial lighting, energy conservation and staff convenience*

The effects of variable cloud cover and the diurnal variation in the apparent position of the sun ensure that the light intensity incident upon the different facades of any building will always be different. Energy may be conserved by arranging for the luminaires that are fitted at the perimeter of a development to operate at a lower intensity when light sensors detect an overall lightening of the sky, while those luminaires situated nearer to the core of the building continue to operate at full intensity. Any changes that may be required in the sensitivity of the apparatus can be carried out by the operators of the building management system.

The convenience of staff may be taken into account by the provision of light switches, through which they can exercise progressive control over the lighting intensity in their particular zone. This can be done by making the light switch react to sequence switching. The first short movement of the pushbutton operates the lights at a reduced intensity, perhaps to enable the use of a word processor; the next slightly longer push will increase the lighting intensity sufficiently to permit the use of, say, a drawing board.

Such is the elegance built into contemporary integrated circuits, that it is possible for personnel to use the internal telephone (if required) for the purpose of switching local lighting.

6.15.2 *Artificial lighting, energy conservation and staff safety*

These dual aims may be achieved conjointly by introducing a timed and controlled pattern of deactivation of the luminaires for those times when the building becomes only partially occupied and then finally unoccupied. Members of staff who may be working late in a zone scheduled for shutdown are given a visual indication that shutdown is about to occur, and they are then able to control the lighting in their part of the building, using local controls over individual luminaires (if needed), rather than by means of the use of a multi-gang switch controlling many fittings. In this situation, the escape routes must obviously remain illuminated. Infra-red sensors fitted at the place of work detect when the space has been vacated and after a timed interval, the progressive shutdown of all the luminaires may be recommenced. For further details on lighting, see CIBSE publication LG7 (1993), 'Lighting for Offices'.

6.16 Assessing the need for rewiring

Rewiring is often carried out because of a change in use of the building, when the opportunity may be taken to install a system that embodies the results of the latest developments in the use of electrical services.

Rewiring may also be carried out because the insulation of the existing wiring is seen to be literally cracking up. In the past, buildings were wired using vulcanised rubber insulation (VRI or VIR). The cables consisted of tinned copper conductors, covered with two layers of rubber, the second layer being vulcanised to improve its elasticity and strength. This outer layer of vulcanised rubber was wound with a fine braid, which was then covered with a more coarse coloured braid. The copper conductors were tinned because of the corrosive effect of rubber on bare copper. The insulation is flammable and the insulating properties of this kind of cable are much reduced when the braid becomes damp. Whenever oil is absorbed by the braid, the rubber insulation is degraded.

Over a period of 20 to 30 years, the cable may have been operating at temperatures above those for which it was intended. In the trade, this syndrome is referred to as 'cooking', and in this case it produces two effects: the tensile strength of the yarn making up the fine braid becomes less, and the rubber insulation becomes brittle. Provided the wiring is left undisturbed, it usually functions satisfactorily but, as soon as it is subjected to even the smallest mechanical shock, the brittle insulating rubber falls away from the conductor (usually at the bends), leaving it bare.

Whenever new wiring is added to older properties having an original VIR system, the new work will almost certainly disturb the existing wiring, in which case, in order to avoid a potentially dangerous situation, it is prudent to recommend a rewire. Such a rewiring scheme is supposed to be made easier if conduit was used in the original installation. However, it is likely that the threads on the existing conduit will be rusty, making the task of obtaining earth continuity more involved. In such circumstances the use of MICC cabling could be considered, particularly as it is easier to bury in the fabric of the building and it does not need subsequent rewiring.

Nowadays PVC sheathed wiring is used extensively in all sectors and it has physical properties that are generally superior to those of VIR. It is chemically unreactive, tough and not flammable. However, it deteriorates if it is subjected to ultra-violet light, and if the temperature is greater than 80°C it becomes soft, causing the embedded conductors to migrate, producing the possibility of a short-circuit, as described earlier. The frequency with which systems have to be rewired, owing to deterioration of the electrical insulation, has been reduced because of the widespread use of PVC and the general ongoing effects of adherence to the constantly updated *IEE Regulations*. It follows that the case for habitually installing conduit, to provide a completely rewirable scheme, is not nearly as strong now as it was 30 years ago. It is often found to be satisfactory to use PVC sheathed cable, routing it through voids and cavities and using conduit only for switch-drops and the crossing of inaccessible places, and for those places where mechanical damage, or exposure to ultraviolet light, may be expected.

Traditionally, the instrument used to obtain an indication of 'insulation worthiness' is the 500 V megger. This can measure the resistance between lines, between line and neutral, and between line and earth. When using the instrument, there should be two separate modes of testing. One mode is to test with all the apparatus isolated (with switches open), and the other is to test with all switches closed (but with the ends bunched together). A reading that is smaller than 1 Megohm (1×10^6 ohm) would generally indicate that the electrical insulation of the system is in a poor state and rewiring should be considered.

In the case of an installation that is in a permanently damp environment, some caution must be exercised when interpreting the results from a series of megger readings. Such an environment can often be found in pumping stations, parts of which are constructed below ground level. The author has knowledge of a pumping station that had just been rewired and in which all the megger readings taken on the wiring were below 1×10^6 ohm, and it had to be realised that, because of the dampness, such readings were the best that could be achieved.

6.17 Temporary distribution of electricity on the construction site

On every site, builders have to create their own 'workshop', and this implies that a temporary source of power must be provided to facilitate the proposed construction. The requirement may be met by the provision of a portable generator, but most main contractors arrange for a temporary supply to be made available from the local electricity company. The greatest electrical load is likely to be that of the tower crane and, for this reason, the position of the crane on site always has some bearing on where the supply unit dealing with the temporary electrical supplies is placed.

When faced with the task of selecting the most advantageous position for a crane, an attempt has to be made to reconcile the effects of a large number of sometimes conflicting variables and, in order to obtain the optimum solution, it is inevitable that compromises have to be made. A list of some of these variables follows:

- shape and size of the site;
- expected shape and height of the buildings;
- ease with which the crane may be constructed and finally dismantled;
- 'reach' of the crane;
- positions for off-loading incoming supplies;
- preservation of lines of sight for the crane operator; and
- position of the greatest workload during the course of construction.

Having decided on the position of the crane, the incoming underground armoured supply cable may be routed towards it, taking care to avoid those areas where future excavations are planned. Commonly, energy is required for the following facilities:

- tower crane (three-phase 415 V, 50 Hz);
- heating and lighting of site huts and floodlighting the site (single-phase 240 V, 50 Hz);
- provision of mains power to portable equipment (three-phase 415 V, 50 Hz);
- lighting of construction work (single-phase 110 V, 50 Hz);
- power for portable tools (single- and three-phase 110 V, 50 Hz);
- portable lighting for hazardous and damp areas (single-phase 50/25 V, 50 Hz).

The main contractor must provide a waterproof, heavy-gauge steel enclosure, referred to as a 'supply incoming unit' (SIU), so that the electricity supply company can terminate the cable in the usual way, with cut-out fuses, a multi-meter and an isolator. The adjacent crane and any hoists, fixed pumps, mixers and compressors will require three-phase 415 V supplies, and these may be obtained from a mains distribution unit (MDU), which provides for the appropriate number of waterproof socket outlets. The MDU is made of steel and has lifting eyes and skids, and it may be attached to the incoming supply unit, when the combined unit becomes known as a supply incoming and distribution unit (SIDU), or it may stand on its own (see figure 6.14).

In order to cater for portable three-phase 415 V motors, which may be powering submersible pumps, screw cutting machines or compressors, it is usual to provide a separate EMU connected to the MDU. The energy flowing in all the temporary circuits on site is controlled using residual current devices (rcds), which monitor any earth leakage currents, and miniature circuit breakers (mcb), which protect against both over-current and heavy short-circuit currents.

The flexible cables providing three-phase 415 V electricity supplies to portable plant such as submersible pumps, screw cutting machines, compressors and mixers are particularly vulnerable, and it is essential for the safety of personnel on the site that the earthing circuits (protective

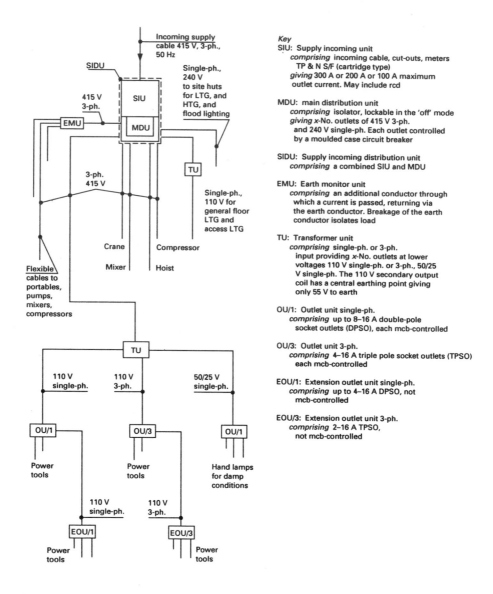

Figure 6.14 Distribution of electrical energy on site

conductors) always have continuity. Accordingly, provision is sometimes made to incorporate an EMU into the system. In these cases, an additional conductor is connected to the protective earth conductor and a small current is then allowed to flow. If this flow of current is interrupted, the equipment being used is then automatically disconnected from the supply.

The provision at the places of construction of a 110 V supply, serving lights and portable tools, and a 50/25 V supply, serving inspection lamps for underground ducts, requires a transformer unit (TU), which is connected to the mains distribution unit. The transformer unit has a number of outlets, connected to separate outlet units (OU), which may be sited on each of the floors under construction and provide for a number of 110 V single-phase or three-phase socket outlets for power tools. Extension outlet units (EOU) may also be used to reach those places furthest away from the source of supply. Figure 6.14 illustrates and details the function of each of the aforementioned units.

When the incoming supply unit is sited near the boundary of the development, the PVC insulated/PVC sheathed sub-distribution cables to the site accommodation can conveniently be slung from an independent catenary wire, mounted between poles and carried around the perimeter of the site to the points of use. If, however, it is decided to position the incoming supply unit in the centre of the site, the sub-distribution cables serving the temporary accommodation may have to be routed in heavy gauge steel trunking to avoid mechanical interference from site traffic.

When the cables have to cross a roadway, they should be laid in drain pipework 600 mm below the surface with markers at each side of the road. If it is decided to route the power cables over the roadway, the recommendations given in [6.13] should be followed.

There are many ways in which the temporary supply of electrical power can be provided and, whatever methods are used, it is important that they are carried out by suitably qualified personnel who are charged with the tasks of periodical reassessment of the layout of the temporary plant and the instigation of regular checks on its operational safety.

6.18 Provision of emergency power

BS 5588 deals with fire precautions in the design and construction of buildings in the UK and, for those buildings that are taller than 18 m, it is specified that at least one firefighting lift be provided, having a dedicated electrical power supply. The lift can be completely separate, or it may be selected from a group of lifts. During an emergency it is intended that the lift will be under the control of a firefighter from within the car, and that its use will be for the evacuation of personnel and for the conveying of firefighting equipment to the appropriate floors.

In the case of a hospital building, in order to ensure the continuing operation of life support machines in intensive care units and other essential apparatus in operating theatres, a dedicated electrical power supply is an absolute requirement. In such cases, it is necessary to install a generator that will self-start whenever the mains voltage falls below a selected value, automatically disconnecting the main supply and taking up the load that is connected. Figure 6.10 shows the way in which a firefighting lift and any other essential load within the capacity of the emergency generator could be connected. On resumption of the main electricity supply, the process of reconnection is then carried out automatically.

Generators currently available have specifications that are capable of taking up 70% of the load in 10 s, and 100% of the load in 20 s, while keeping the output voltage to plus or minus 1.5% of the declared value. Experience has shown that, under emergency conditions some facility for sustaining an overload is often required, and this can be made possible by specifying that the emergency generator must be able to operate with a 10% overload for 1 h.

The amount of standby generator capacity that is to be provided depends heavily on the perceived importance to the client of the continuance of the work function during a power cut and bearing in mind the cost of the equipment, such considerations are not always treated as rationally as they should be. An affluent client will often insist on providing sufficient generating capacity to deal with the total electrical load of the establishment, believing that by making such demands he is serving the 'corporate image' best.

6.19 Use of a standby generator during periods of maximum demand

Some industrialists operate on a maximum demand tariff, which gives the electrical supply company the right to levy an extra charge if the demand for power exceeds a stipulated amount. Under such conditions it is sometimes economic for the factory owner to install his own generator, the operation of which can be automatically triggered by a rise in the reading of a kVA meter above a preset value. The generator will then remain in operation until the peak loading condition has passed and a fall in the amount of kVA is detected.

Recent legislation in the UK has now made it possible for those companies that have spare generating capacity to put power into the national grid system, by arrangement with the appropriate power utility.

6.20 Provision of emergency lighting

In buildings that are open to members of the public, it is necessary to provide an electrical supply system that will operate the emergency lights whenever the normal electricity services are interrupted. Traditionally, emergency lighting systems use incandescent filament bulbs. For prestige installations, however, where the sight of a line of bulbs may be considered to be aesthetically inappropriate, it is possible, with suitable control equipment, to install fibre optics which may be concealed in a false ceiling to supply each flush fitted lighting outlet.

The electrical load needed to supply emergency lighting is small when compared with that required for the operation of a lift, or other equipment requiring power, and it is possible to arrange for this to be supplied by batteries of the traditional lead–acid, or alkali type, which take a trickle charge from the electricity supply main when it is functioning normally. If the emergency lighting load for a particular building is small, it can sometimes be shown to be economic to use the more expensive nickel–cadmium batteries, which have a longer life than the lead–acid type and require less maintenance.

Readers should note that according to BS 5839: Part 1: 1988 section 2, lead–acid batteries are not considered to be suitable for use with a fire alarm service.

6.21 Provision of an uninterruptible power supply to computers

Whenever a computer is shut down it should not be done so by disconnecting it abruptly from the mains supply. It is necessary to follow a procedure whereby blocks of data necessary to the functioning of the internal programs are cancelled, or withdrawn, from their points of location in the memory. If this process is not carried out diligently, subsequent operation of the computer in the manner intended eventually becomes impossible, and to restore its operational capabilities it may be necessary to reprogram the hard disc.

In addition, whenever an unscheduled shutdown occurs, there will be an obvious loss of data that is currently being processed; it is therefore essential that the computers (and any peripheral equipment at risk) should be connected to an uninterruptible power supply (ups). This unit consists of a battery charger, batteries and an inverter. The function of the inverter is to convert the dc output from the batteries into an ac input to the computers. In normal operation, the power taken by the computers is just less than the power taken by the battery charger, so that the batteries are always fully charged and may be described as 'floating'. When the mains supply is disconnected, the battery charger no longer operates but the batteries continue to supply the connected apparatus for a period of up to 30 minutes. During this time it should (hopefully) be possible for the operators to finish their immediate work and then to shut down the apparatus in the normal way, thus avoiding any 'glitches'. For maintenance purposes, a by-pass to the system is normally included.

If it is found to be necessary to continue operating the computers and the establishment also happens to have a standby generator in operation, it may then be possible to arrange for the batteries contained in the ups to be recharged from the generator supply.

6.22 Provision of a battery room

Details of suitable accommodation for batteries are given in BS 6132 or BS 6133.

Lead–acid batteries are bulky and when they are being charged give off hydrogen which, if mixed with air in a concentration greater than 4% by volume produces a potentially explosive mixture. It follows that the battery room must have no spaces where hydrogen could accumulate, and a policy of 'protection by continuous dilution' should be used, whereby fixed natural ventilation (having no electrically operated fans) is provided at high level. The provision of luminaires should be restricted to appliances that are wall mounted.

Generally, no space heating is required, as the electrolyte used in the batteries will not freeze until the temperature approaches about –30°C. Space is usually provided for the storage of the acid needed for topping-up purposes, and all fixtures and fittings should be acid resistant. The battery room must be designated as a no-smoking area, and the use of electrically operated hand tools and portable lamps which are not flameproof should only take place under strict supervision. Clearly, when designing such an installation, compliance with the relevant safety regulations is essential and details about the use of flameproof equipment are given in section 6.23.

6.23 Use of electrical equipment in a potentially explosive atmosphere

A potentially explosive atmosphere is one that is not normally explosive but which may become so, owing to the unforeseen or abnormal operation of equipment. Examples of such malfunctioning could include a rise in local and/or ambient temperature and the escape of flammable gases, liquids and/or powders. Potentially explosive atmospheres are to be found in the mining industry and throughout industry generally, and in particular inside a battery room, where the charging/discharging of lead–acid batteries takes place.

In the UK, the first certificates of flameproof construction for electrical equipment were issued by Professor Stratham of Sheffield University [6.14]. Subsequently, the Mining Equipment Certification Service (MECS) was set up in the decade beginning 1920 to give practical application to the results of the Safety in Mines Research Board, which was dealing with pit explosions triggered by electrical equipment. The certificates issued to equipment that complied with the requirements of the MECS, resident in Buxton (Derbyshire), were known worldwide as 'Buxton Certificates'. Currently, the electrical equipment for use in British mines is classified under 'Group 1'.

Flameproof equipment for use in areas other than mining and covering all flammable substances other than methane is classified as being under 'Group 2', and the certification is carried out by 'The British Approvals Service for Electrical Equipment in Flammable Atmospheres' (BASEEFA). Both MECS and BASEEFA have now become part of the Research and

Laboratory Services Division (RLSD) of the Health and Safety Executive (HSE). The Electrical Equipment Certification Service (EECS), see [6.15], is involved in European Union activity in the explosion protection sector, and it is also linked to the RLSD.

6.23.1 Subdivision of Group 2 substances

Because of the large number of flammable substances, the Group 2 classification has been further subdivided into Groups 2A, 2B and 2C, with each substance being placed in a particular group, depending on:

(a) the smallest gap in a flameproof enclosure allowing an internal explosion to trigger off an external explosion; and
(b) the energy required to initiate the process.

Group 2A contains substances that will tolerate the largest gaps and require the greatest input of electrical energy, while Group 2C substances require the smallest gaps and the least input of electrical energy.

The complete groupings, together with ignition temperatures, are to be found in table 7 in BS 5345: Part 1: 1989. Some examples of the ways in which the substances are categorised are as follows:

Group 2A

  Propane, butane, pentane, hexane, methanol, petroleum, cellulose solvents

Group 2B

  Hydrogen cyanide, methyl acetate, tetrafluoroethylene

Group 2C

  Acetylene, carbon disulphide, hydrogen

6.23.2 Classification of hazardous areas for flammable gases, vapours and mists

The British Standard divides these areas into three zones:

zone 0 – that in which an explosive atmosphere is continuously present
zone 1 – that in which an explosive atmosphere is likely to occur in normal operation
zone 2 – that in which an explosive atmosphere is not likely to occur in normal operation but, if it should occur, it will exist for only a short time.

6.24   Type of protection

Detailed specifications of each of the types of protection currently completed are to be found in the relevant parts of the newly harmonised BS 5345: 1989 Parts 1 to 8. There are currently nine types of protection that are recognised. The details for types o and q are in preparation (Part 9 of the British Standard), and those for type m are not as yet (1994) completed.

6.24.1   Type d – Flameproof enclosure (Part 3)

The switchgear is constructed in such a way that it will withstand the internal explosion of a gas that has entered it, and will subsequently allow the products of the reaction to be exhausted to the surrounding atmosphere without an external explosion resulting.

The construction of a container for a piece of switchgear requires the use of machined joining surfaces (referred to as flanges), which have a depth and a clearance that can be determined in the design and manufacturing processes. In practice, it has been found that there is a maximum experimental safe gap (MESG), which will not allow the products of an explosion inside the apparatus to initiate an explosion outside it, and over many years the values of the MESG have been obtained for a wide range of flammable gases and vapours. In practice, in order to allow a margin of safety, flameproof enclosures are constructed having flange gaps that are smaller than the MESG.

It is a common misconception (even among those who deal in flameproof equipment) that external explosions are prevented solely by effective heat transfer in the flange. Phillips [6.16] and others have now shown that the process depends on other factors, one of which requires that the escaping jet of hot gases has a velocity that is high enough to bring about entrainment of the surrounding, cooler, flammable gases. The effect is to lower the temperature of the gases that make up the jet to a value that is below the ignition point of the flammable mixture in the enclosure.

6.24.2   Type i – Intrinsic safety (Part 4)

This is applicable to those electrical circuits that operate at specified low-energy levels and includes telecommunication and data transmission circuitry. The total energy contained in the system, including that given out following a change in inductance or capacitance, or the heating up of components, should be insufficient to initiate an explosion. Apparatus awarded an 'i_a' classification is suitable for all hazardous areas, but that awarded an 'i_b' is suitable for zone 1 and less hazardous areas only.

6.24.3 Type p – Pressurised apparatus (Part 5)

This refers to the process of ensuring that any space or enclosure containing equipment that arcs and sparks, such as a large electrical generator, is initially purged of any ignitable gaseous mixture and then pressurised to a level at least 50 Pa above atmospheric pressure. It is possible to use an inert gas for this process but, if the space is to be inhabited, it is safer, cheaper and much more convenient to use air as the pressurising medium.

A process of continuous dilution is required and this is defined in BS 5345: 1989 as 'the technique of preventing the formation of an explosive gas/air mixture in an enclosure, by the supply of a protective gas, at such a rate that the concentration is always kept below the lower explosive limit'. The committee responsible for this standard has provided sufficient data (together with a worked example) to enable the determination of a suitable volume flow rate for different flammable mixtures. The document goes on to recommend that up-to-date records should be kept of the integrity of the system and that all operations should be subject to a 'Permit to work'.

Part 5 of the same standard gives details of the extent and type of instrumentation required for zones 1 and 2 (the code does not apply to zone 0), together with the actions to be initiated for each zone to ensure that the system will 'fail safe' at all times.

6.24.4 Type e – Increased safety (Part 6)

This concept applies only to those parts of electrical equipment which, when they are in operation, are not expected to produce sparks, arcs, or high temperatures, such as the windings of three-phase and single-phase motors. The use of such circuitry becomes acceptable when extra mechanical, thermal and electrical protection is employed.

6.24.5 Type N – given as Type of protection N (Part 7)

The apparatus is not capable of igniting a surrounding explosive atmosphere and a fault capable of causing ignition is not likely to occur. This type of protection is mainly employed in the petrochemical industry, based in the UK, and it is in some respects complementary to type e protection and in others more stringent. Both types e and N are the result of the development of two separate standards.

6.24.6 Type s – Special protection (Part 8)

The construction of the apparatus is such that it does not comply with other established types of protection but can, by means of tests, be shown to be suitable for use in hazardous areas in prescribed zones.

6.24.7   Type o – Oil immersion (Part 9, in preparation)

Electrical contacts are submerged in oil to a specified depth and when the contacts are operated the oil 'quenches' the sparks so produced. The quenching process brings about a deterioration of the oil and also releases hydrogen, which can represent a fire hazard.

6.24.8   Type q – Powder/sand filling (Part 9, in preparation)

In this method, the electrical apparatus contains a granular material, which fills the enclosure to a prescribed depth so that, if arcing occurs, the flammable atmosphere outside the enclosure will not ignite.

6.24.9   Type m – Encapsulation (no part number allocated)

In this method, electrical components that could cause sparks, or heat, are completely enclosed by a compound that prevents the ignition of any external flammable mixture of gases.

6.25   Selection of type of protection

Each zone has assigned to it the type of protection that is permitted, and these will differ depending on the country of use:

zone 0 (UK) – i_a and s (specifically certified for use in zone 0)
zone 1 (UK) – any explosion protection suitable for zone 0 and d, i_b, p, e and s
zone 2 (UK) – any explosion protection suitable for zone 0, or zone 1 and N, o and q.

6.26   Selection of enclosure according to temperature classification

It is possible for any flammable gas to be ignited by coming into contact with a surface the temperature of which is greater than, or equal to, the ignition temperature of that gas. This will not occur if the highest possible equilibrium temperature of the external parts of the enclosure is always less (by a safe margin) than the ignition temperature. Unless special conditions apply, tests are carried out at an ambient temperature of 40°C and the temperature classification is as follows:

T class	Maximum surface temperature (°C)
T_1	450
T_2	300
T_3	200
T_4	135

$$T_5 \qquad\qquad 100$$
$$T_6 \qquad\qquad 85$$

6.27 Illustration of the selection of suitable apparatus

A telephone is to be installed in a zone 1 area, where propane gas having an ignition temperature of 470°C may be present.

For zone 1, any one of the types of protection i_a, d, i_b, p, e and s could be used. Propane is classified under group 2A and, from the previous table, a temperature classification of T_1 would be suitable. The final coding could be as follows:

EExi$_b$ 2A T_1

'Ex' is an internationally recognised abbreviation that denotes explosion protection, and 'EEx' indicates that the apparatus complies with one or more types of protection as detailed under the auspices of the European Commission for Electrotechnical Standardisation (CENELEC).

If the apparatus specified was not available, it would be possible to use equipment designated for more stringent conditions, which for example could have alternative codings of:

EExi$_a$ 2B T_2 or   EExi$_a$ 2C T_3

Note that the temperature classification values T_2 to T_6 are all lower than T_1 and would therefore be acceptable in any of the codes given.

6.27.1 Marking of equipment

Flameproof equipment carries markings which correspond to the codes given in the previous paragraph, and the way in which the details are presented is prescribed in BS 5345: Part 1: 1989.

6.27.2 Some important provisos

When dealing with electrical equipment for use in potentially explosive atmospheres, the person responsible must ensure that the choice of equipment for the intended country and the way in which it is proposed to be used comply in every respect with the appropriate and current regulations. It should also be emphasised that would-be practitioners in this field must also be seen to be guided by those recognised experts who operate in the potentially hazardous domain of flameproofing.

Readers who are approaching the subject of flameproofing for the first time should note that the information given in the previous sections should only be regarded as a preliminary introduction to what is a very large and involved topic; particularly as the process of 'harmonisation' within the European Union has now begun. This will become apparent as soon as any of the references listed at the end of the chapter are accessed.

PART 3: EMISSION AND CONTROL OF ELECTRICAL ENERGY FOR SPACE HEATING

6.28   Description of apparatus

6.28.1   Emission of short-wave energy from a quartz lamp/heater

In 1985, a breakthrough in high-temperature radiant heating was made possible by the development of the quartz lamp/heater.

When considering the transfer of electrical energy from a power station to the position where the change from electrical to thermal energy takes place, the process does not require the use of a 'working substance', and the terms 'cyclic' and 'non-cyclic' in this case have no meaning. However, when considering the way in which the thermal energy derived from the electrical energy is subsequently processed, the term 'working substance' and the description of the handling process as being 'cyclic' or 'non-cyclic' remain valid. In the case of a quartz lamp/heater, which is used to warm an enclosure containing people, the working substance is air and the process may be categorised as being non-cyclic.

The lamp/heater consists of a sealed, clear quartz tube, containing a tungsten resistance element that operates in an atmosphere made up from the halogenous group of gases, which mixture is generally referred to as 'halogen' gas. The chemical reaction that takes place when the tungsten filament is brought to incandescence is referred to as the 'tungsten–halogen regenerative cycle', and the effect of this is to reduce the net evaporation rate of the tungsten, so that the filament can be operated at an elevated temperature of 2200°C. Blackening of the inside of the tube is said not to occur and, in addition, there is an increased efficacy of light output (18–22 lumen/W) over that of a tungsten filament operating in an inert gas (11–17.3 lumen/W).

The tube produces a relatively intense output of short-wave radiation, with the maximum wavelength occurring between 1 μm and 1.5 μm (see figure 6.15). Approximately 90% of the heat output is radiated, with the balance being convected, and if the quartz tube is mounted at the focal point of a parabola, it is possible to make the output from the tube directional. The energy radiated is unaffected by humidity and, when it impinges on the skin, produces a feeling of comfort akin to the sensation of being in the rays of the sun.

When a radiant beam falls on an object in its path, some of the heat is absorbed and the temperature of the object rises. Convection currents in the surrounding air then distribute the heat to other parts of the space. Some of the radiant energy falling on the object is reradiated at a longer

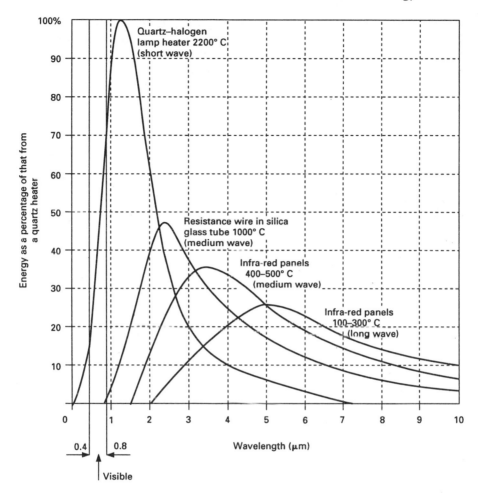

Figure 6.15 Characteristic curves of the intensity of radiation against wavelength for different types of electric heater

wavelength and when this in turn impinges on other objects, the process is repeated. It should be noted that the space is heated without the need for any forced circulation of air, the movement of which may entrain surrounding dust particles, and this facility is important for those industrial processes that require a dust-free atmosphere.

If it is intended to use the lamp for heating purposes only, then the light output is usually unwanted and, by adding a ruby coloured filter to the tube, just a small amount of light will pass through the filter, giving an attractive cherry red glow.

Tubes are manufactured to give outputs that commonly change in multiples of 500 W, and casings are manufactured to contain the required number of tubes for a particular location. If required, it is possible to use an appliance that contains a bank of tubes, some of which are unfiltered, providing light and heat, and some are filtered, giving heat only. The tubes emit their full heat just seconds after being switched on, and they are ideal for use in buildings having an intermittent pattern of use, such as churches, village halls and gymnasia.

Quartz tubes must not be installed in enclosures where there could be contact with flammable materials or flammable gases, or dusts, as the operating temperature of the outside of the quartz tube (750°C to 800°C) is high enough to ignite flammable mixtures and materials and, if the conditions are critical, might initiate an explosion. In the UK, information about such hazards is readily available from the Health and Safety at Work Area offices, see also [6.15].

Installers should note that:

(1) the switches used to control the tubes must have the correct current rating; also, the thermostats used should have the capability of 'zero voltage switching';
(2) the tube should be fixed in a horizontal position, to prevent migration of the halogen gas to the lower end of the tube, resulting in failure of the filament;
(3) the surface of the quartz tube should be kept clear of grease to avoid the development of fine cracks (referred to as 'devitrification'). Normally each tube has a life of about 5000 hours.

6.28.2 Emission of medium-wave energy from a portable electric fire operating at 1000°C

The heat is obtained by means of an electrical resistance wire having a high resistance per unit length. It operates at a temperature that produces a bright cherry red colour. The energy radiated is more diffuse than in the case of the halogen tube, see figure 6.15, with the maximum wavelength occurring between 2 μm and 3 μm. In the past, the resistance wire was routed through grooves cast into blocks, or wound around rods made of refractory material. Nowadays the heating element in an electric fire is contained in a tube of glass made from almost pure silica and referred to as quartz glass. The surface of the tube is smooth and may readily be kept free of dust, so helping to avoid the odours associated with a dusty ceramic in contact with a hot wire.

Conventionally, the element is fixed at the focus of a polished parabolic reflector, so that part of the energy output which is initially projected towards the surface of the reflector will be reflected forwards in the form of a parallel beam. The use of this type of emitter is almost wholly confined to domestic premises, where great care must be taken to avoid setting fire to any fabric on which the radiant beam may fall.

6.28.3 Emission of medium-wave energy from overhead heaters operating between 400°C and 500°C

The heat is obtained by means of an electrical resistance wire, contained in a steel tube and having a lower resistance value/unit length than in subsection 6.28.2 above. It is placed at the focal point of an overhead, horizontally mounted, parabolic reflector. The lower operating temperature produces more diffuse radiation than in the previous case, with the maximum wavelength occurring between 3 µm and 4 µm, see figure 6.15.

The complete unit is not unlike a conventional overhead fluorescent tube lighting fitting, having the heating element in place of the tube. It may be used to provide 'spot heating' for members of staff and customers congregating around a counter, situated in one part of a large store or warehouse. The effect is localised and is experienced only when personnel are in the path of the radiant beam.

This type of heater is sometimes referred to as an 'infra-red' heater, meaning that all the wavelengths of the radiated energy are beyond those of visible light, and therefore invisible. The energy given off is sometimes described as 'black' heat. Reference to figure 6.15 will show that all the heaters described give off some of their energy in the infra-red region.

6.28.4 Emission of long-wave energy from panel heaters at 100°C to 300°C

In this case, the heat is obtained by means of an element around which no air can circulate and which has an even lower value of resistance per unit length than that described in subsection 6.28.3. By preventing the circulation of air, the energy removed by convection is reduced and the operating temperature of the element is increased, resulting in an increase in the value of the radiation component. The radiation is given direction by applying suitable thermal insulation to one side (the rear) of the unit. Such appliances are referred to as 'high' (or 'low') temperature (infra-red), radiant panels; the energy is even more diffuse than the previous heaters (see figure 6.15), having a maximum wavelength occurring between 4.5 µm and 5.5 µm.

One version of the 'high' temperature radiant panel is shown in figure 6.16(a). The front part is covered by a ceramic tile or a vitreous enamelled steel plate, and the casing is equipped with adjustable angle brackets, so that the unit may be either wall or ceiling mounted. The appliance is robust in its construction, easily cleaned and eminently suitable for use in places that are unsupervised, such as school washrooms and corridors, or in industrial situations.

Another form of panel, also operating at a temperature of about 300°C, takes the form of electrical resistances fixed on the centre-line and to the rear of two elongated rectangular fins. An external, thermally insulated

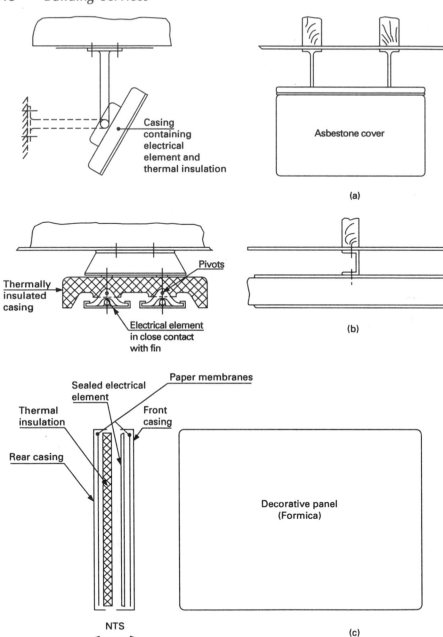

Figure 6.16 (a) 'High'-temperature radiant panel (300°C). (b) Linear radiant panel with adjustable fins (300°C). (c) 'Low'-temperature radiant panel (70–100°C)

casing supports the fins which may be angled in the direction required, see figure 6.16(b).

A 'low'-temperature radiant panel operating at temperatures of about 70°C to 100°C is shown in figure 6.16(c). An electrical resistance is sandwiched between a mesh of synthetic fibre which is covered by a suitable polymer. The matrix of fibres is then sealed by polyester film and brought into contact on the rear side with a thermal insulator, and on the front side with a laminated plastic panel. It can be built into a wall, or partition, to provide 'local' or spot heating for an industrial operative, who may be sitting at a workstation where the ambient heating at times proves to be minimal.

6.28.5 Distribution and emission of electrical energy using embedded low-temperature emitters

When using embedded electrical conductors for space heating, they may be inserted in 'D'-shaped conduits to facilitate subsequent access for repair, if required, or they may be buried directly in the fabric of the building. Mineral insulated copper covered conductors (MICC) are ideal for this purpose and are often specified for use outside the building, when the surface of a roadway has to be kept free from frost. For surfaces that are not subjected to road traffic, it is possible to cast XLPE coated cables into the screed.

The technical sales offices attached to the electricity companies have a great deal of experience in the design and use of such systems, and they will also be able to advise on the specifications for suitable concrete and plaster mixes, which may be needed to cover the conductors satisfactorily and to prevent any subsequent cracking of the final finish. The use and application of underfloor heating elements is discussed more fully in Chapter 5.

6.28.6 Emission of long-wave energy from electrical convectors – natural and forced

Natural convectors using electrical energy are similar in operation to those using a coil containing water or steam, except that an electrical resistance is fitted in place of the pipe coil and the air control damper is usually omitted (see Chapter 5 and figure 5.15). Normally, the electrical resistance is designed to operate in the infra-red region, so that the colour of the element is always 'black' rather than 'cherry red'. In the case of conventional nickel–iron resistance wire, this has been shown to result in a longer life for the element. Nowadays, the electrical element sometimes consists of an electrically insulated resistance wire contained in a tubular metal sheath.

The room temperature may conveniently be controlled using a room thermostat to open or close the electrical circuit. Control of the rate at which heat is put into the room is sometimes accomplished by manually switching an additional resistance coil into or out of the circuit and, predictably, the use of a time switch will control the periods of operation of the complete unit. The design of the casings in which the elements are housed can range from utilitarian to ornate, to match the decor of the surroundings.

Because the element is continuously cooled by air passing over it, the convection effect is maximised at the expense of the radiation effect, and the units should only be used to heat spaces where the warming-up time matters little and where the space to be heated is not too lofty. It is most likely that the convector will be operating during the hours of peak energy consumption, so that the energy will be bought at peak prices, and for this reason, it becomes important to make sure that the enclosure is insulated to a high standard.

The height of the casing can be varied to suit the sill heights of any room, so that the units may easily be fitted under the windows. The casing can also be made to be continuous around the room, so providing perimeter heating; by reducing the height, the units can provide skirting heating if required.

Forced convection is accomplished by arranging for a cross-flow (squirrel cage) fan to draw cool return air from the room across the electrical elements and to discharge it upwards, with a suitable velocity, into the space being heated.

6.28.7   *Emission of long-wave energy from electrical tubular heaters – mainly convective*

These heaters take the form of 50 mm diameter circular tubes, mounted horizontally and containing electrical resistances, which are surrounded by air at atmospheric pressure. The power input is at a standard rate of 180 W/m run, and this produces an even surface temperature of approximately 80°C. Because the surface is too hot to touch, the tubes should have a guard fitted to prevent accidental skin contact.

Some manufacturers produce an elliptical tube, which, if fixed with its major axis vertical, results in an increased volume flow rate of convected air over its surface. The required heat output can be obtained by specifying the length of tube and by mounting a number of tubes one above the other. These can be controlled in the same way as the convector heaters described previously and, because of the relatively high cost of peak electrical energy, a high standard of insulation is again desirable.

6.28.8   *Emission of long-wave energy from electrical overhead unit heaters – mainly convective*

Unit heaters have been discussed in Chapter 5, where figure 5.19 provides an illustration of a unit heater using hot water as the working substance.

In the case of a unit heater using electrical power to provide the heat output, the heat exchanger is replaced with a bank of electrical resistance elements. A specialised form of heater, which can be operated electrically, can be found sited over external doors, where it is used to produce a downward flow of warm air across the full width of the opening, resulting in a warm air 'curtain' to prevent draughts.

6.28.9 Emission of long-wave energy from electrical oil-filled emitters – mainly convective

The apparatus in this case takes the form of a conventional steel panel emitter, filled with oil and heated by an immersion heater that is fitted at the bottom of the panel. Operation of the unit is controlled by a thermostat and the surface temperature reaches approximately 70°C. The energy output, in common with water-filled emitters, is mainly convective. Such panels can be fixed or free-standing; they can also be made portable, by mounting them on castors, to provide space heating (up to 1.5 kW/unit). Those panels that are wall mounted, or free-standing, may usefully be employed in small offices and in the small enclosures usually occupied by receptionists.

6.29   Description of apparatus using 'off peak' power

6.29.1 Storage heater (non-cyclic)

These heaters consist of refractory bricks assembled in an insulated steel casing, which has been designed to blend easily with most colour schemes. The heaters can be free-standing or wall mounted. Air circulation channels formed in the bricks contain electrical heating elements which, by means of a remote time switch, are automatically connected to and disconnected from, the power supply during off-peak periods. The overall temperature of the refractory bricks increases throughout the charging period, absorbing the heat given off from the electrical resistance.

During normal operation there is a continuous emission of heat from the casing, which reaches a maximum at the end of the charging period and falls to a minimum at the start. When air is allowed to flow naturally over the refractory bricks, the unit is referred to as a 'static' heater; in this case, the mode of heat output is approximately 50% radiated and 50% convected. When air is drawn over the bricks by means of cross-flow fans, the unit is called a dynamic heater; here the mode of heat output is about 25% radiated and 75% convected (see figure 6.17(a)). By controlling the fans by means of a time switch, it is possible to arrange for more of the heat output to be delivered when the occupants are in residence, or planning to be, than would be the case with a static storage heater.

It has long been recognised that the occupants of rooms heated by

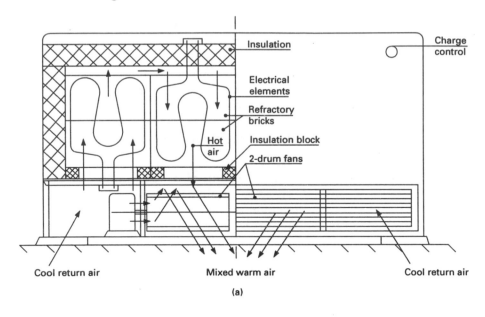

(a)

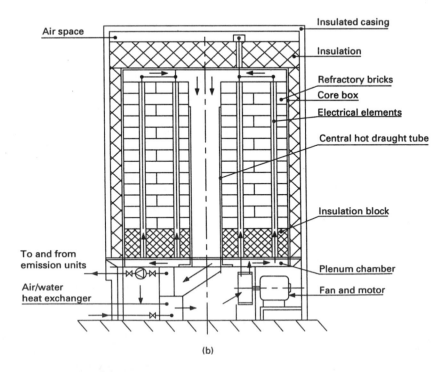

(b)

Figure 6.17 (a) Half-sectional view of dynamic storage heater. (b) Sectional view of dry core boiler

block storage heaters can often benefit from an extra input of energy to the room, especially towards the end of the day, but this time is usually several hours before the off-peak period is scheduled to commence. Manufacturers therefore now produce 'combination' storage heaters, which contain a smaller quantity of refractory bricks and a built-in on-peak forced convector so that, irrespective of the state of charge of the heater, a suitable heat output is available at all times. By adding an air-mixing chamber and a thermostatically controlled air flow damper to the unit, the air temperature can be kept to a value within plus or minus 0.5°C.

For those who have to design a scheme using such storage heaters, [6.17] gives a useful set of tables, offering a series of correction factors, to be applied to the basic equation:

$$R = 24 \times Q/n$$

where R = the rating for the appliance (kW)
  Q = the total design heat loss (kW)
  n = the number of off peak hours available for charging purposes.
The correction factors reflect the order of the design heat loss, the thermal capacity of the building and the level of heat gain from the lighting and the occupants.

Clients who use storage heaters of this type nearly always benefit from having some extra input of energy which, if it is electrical, will normally be taken during the peak period. Reference [6.17] also gives a table of multiplying factors that allow for the proportion of annual energy consumption used off-peak. Provided that the storage heaters are properly sized, they are eminently suitable for use in the enclosures generally used by the old, infirm and nursing mothers, all of whom have a requirement for 24 hour heating. However, the heaters must be guarded against accidental skin contact, as surface temperatures in excess of 100°C have been measured on the front surface of the cabinets.

6.29.2 Central storage warm air heaters (non-cyclic)

Such units concentrate the storage medium into one large appliance, which, because of its weight, must be positioned with care in the building. A heater having a rating of 15 kW and a charge acceptance of 104 kWh has a mass of 754 kg. The heat from the unit is removed by means of a fan discharging into a duct system, the outlets of which terminate in registers fixed in each space being heated. The heater has a return air grille with a filter and a filtered fresh air inlet, which may be adjusted to provide the required volume flow rate of fresh air.

Because all the heat storage medium is contained in a single casing, the system has a smaller casing loss per kW of charge. This is because it can be assembled into a shape that can be more effectively insulated than is possible with separate storage units, which have to be long and relatively thin to fit as unobtrusively as possible into living rooms. In order to provide a reserve of power, for ease of warming of the spaces being heated, the rating of the central unit is generally made greater than the thermal capacity of the refractory bricks. The temperature of the air leaving the unit is normally kept to a constant value by means of a thermostat-controlled damper which mixes the hot air with that returning from the space being heated.

6.29.3   Dry core boiler heaters (cyclic process with water)

The dry core boiler consists of refractory bricks, interlaced with electrical heating elements spaced around a central hot draught tube, operating at a mean temperature of about 750°C. These are contained in a single casing which is so well insulated that one particular manufacturer claims that the casing retains unused heat for up to 10 days. An integral fan circulates hot air through the heated core and across an air-to-water heat exchanger. The heated water is then circulated by means of an integral pump through a conventional low-pressure hot water radiator heating system, see figure 6.17(b). This clearly provides a system that is more responsive to the heating needs of the occupants of the building. Also, because the working substance is recirculated (a cyclic process), the residual heat is kept within the boundary of the apparatus and is not lost to the atmosphere, as would be the case with a ducted warm air system.

6.29.4   Thermal storage cylinder (cyclic process with water)

This method of utilising off-peak electricity is the precursor of all the systems discussed so far, and the huge size of the water storage cylinder has sometimes necessitated special attention from the transport authorities, to enable it to be delivered safely to the site without causing too much interference to other road users. Such storage cylinders are insulated to a high standard and are usually installed in the basement of a building. They rely upon the head of water available from the roof to the basement to increase the temperature at which boiling of the water could occur, and in this way more heat can be stored in a given mass of water, before a phase change to vapour occurs.

The stored water is heated during the off-peak period, by means of electrode boilers for the larger cylinders, or by means of banks of immersion heaters for the smaller ones. (The principle of operation of electrode boilers is discussed in Chapter 4.) Figure 6.18(a) shows a typical arrangement of the plant required to transfer the heat from the cylinder to the building, during the period of occupation, using conventional emitters. These may be carefully controlled to provide a suitable quantity of heat, when and where it is required, using an inside/outside compensatory control unit, the operation of which is described in subsection 6.29.5.

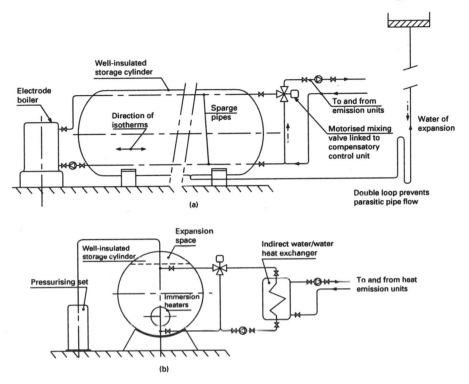

Figure 6.18 (a) Thermal storage system pressurised by means of a standing head of water. (b) Packaged and pressurised thermal storage system

In modern practice, it is likely that the large storage cylinder would be replaced by a number of smaller packaged and integrally pressurised insulated storage cylinders. Such an approach avoids the obvious delivery problems associated with a very large cylinder, and also makes it easily possible to vary the amount of hot water that is stored to suit the current and projected needs of the client.

The hottest water stored in the cylinders is the least dense and is automatically displaced to the upper part of each container by the cooler and more dense water, which falls naturally to the bottom. When this process is complete, the water is said to have 'stratified'. Clearly, the hot water supplying the emitters must be drawn from the top and the cool return water introduced at the bottom of each cylinder; furthermore, in order to preserve the stratification, both the suction and return pipes must be connected to 'sparge' pipes. The sparge pipes extend for the full length of the cylinder and have radial holes drilled in them along their length so that the passage of the water occurs as evenly as possible along the extended length of each pipe, see figure 6.18(a).

Typically, the smaller pressurised hot water storage cylinders operate at an absolute pressure of 450 kPa (350 kPa gauge). The boiling point of water at the absolute pressure is 148°C and, when the sets are operated at a maximum storage temperature of, say, 138°C, this allows a margin of 10 K to prevent boiling taking place. Water is circulated from the cylinder to a water/water heat exchanger using a pump and a mixing valve connected to a by-pass, see figure 6.18(b). In this way, water from the external circuits is kept separate from that contained in the cylinder, so reducing the possibility of corrosion occurring in the cylinders.

6.29.5 Cold thermal stores

Cold thermal stores can be particularly useful in providing a supply of chilled water for use in air conditioning plants, to supplement that provided by the refrigerators during peak load periods. By taking advantage of off-peak tariffs and by using the thermal capacity of the pipework installed in the building complex, it is possible, by operating the coolers during the off-peak period, to precool the water sufficiently to overcome any temporarily large increase in the cooling load, which may occur during the remainder of the day.

Particular care must be taken to ensure that the water is not cooled to such an extent that it turns to solid ice. With a pumped system there are always parts of the plant where the flow velocity is reduced to a relatively low value and it is at these parts that the change in phase will first become noticeable, making it a matter of fine judgment as to when the cooling process should be stopped.

6.30 Use of a 'compensator' control unit to overcome effects of thermal lag

It is possible to control the amount of energy supplied to a heating system by means of a preset thermostat, sited in a room judged to be representative of the conditions required for the whole system. In an attempt to make this method of control less coarse and more representative of the needs of the occupants, additional thermostats controlling motorised zone valves, or valves fitted directly to the emitters, are frequently installed.

Implicit in this method of control is the assumption that the change in the outside temperature is small and that therefore the change of heat flow rate from the enclosure, through the insulated boundary to the outside, will also be small. In fact (particularly in the UK), the change in the outside temperature is frequently large and it may swing in either direction, resulting in correspondingly large changes in the magnitude and direction of heat flow rate across the boundary. Depending on the construction of the boundary, the thermal effects of such changes are delayed in time (typically by 30 minutes to two hours) and the length of the delay is referred to as the thermal lag.

The effect of thermal lag is such as to ensure that the perceived thermal conditions inside the enclosure are likely to be out-of-phase with conditions outside it, so that upon a sharp rise in outside temperature, heat will continue to be introduced into the space for a period equal to the time of the thermal lag, before the effect of the changed external conditions has been detected. The obvious result is that of wasteful overheating. Conversely, the effect of a sharp fall in outside temperature is to produce a state of under-heating, for a period equal to the length of time of the thermal lag.

The compensator control system avoids the problem of thermal lag. It detects a change in outside temperature and varies the flow temperature of the working substance by a proportional amount, the magnitude of which may be altered to be 'in tune' with the type of building construction used. At least one other control function is provided, which sets the value of the flow temperature corresponding to the outside temperature, for the particular type of heating system to which the control is being applied. An installation using convectors as emitters requires a relatively high flow temperature, one using radiators a somewhat lower temperature, and embedded emitters the lowest temperature, in order to avoid the occurrence of cracking of the fabric of the building. Further controls make it possible to have a 'boost' period and a 'set back' period, both of which can be operated automatically or with a manual override. The operation of the compensator relies on a Wheatstone bridge, described in the next section.

6.30.1 Wheatstone bridge

The English physicist and inventor Sir Charles Wheatstone (1802–1875) invented a network of resistors that can be used to determine the value of an unknown resistor. Figure 6.19(a) shows the arrangement and the result of its analysis, when the bridge is 'balanced', that is, when the flow of current between A and B is zero. The bridge may be used to produce proportional control in the following way.

If one of the resistors is heated (or cooled), the value of its resistance will change and an out-of-balance current will flow between A and B, the magnitude of which is proportional to the change in value of the resistor. By inserting a specially sensitive solid-state resistor (a thermistor) into one of the arms of the bridge and by bringing it into contact with the fluid whose temperature is to be sensed, any change in its resistance will produce a proportional change in the current flowing between A and B. The magnitude of this current may then be used to produce a proportional movement in the motorised valve controlling the flow temperature in the by-pass system, see figure 6.19(b), which shows just one way in which the compensatory control may be applied to a heating system, using water as the working substance. Figure 6.19(c) shows some typical

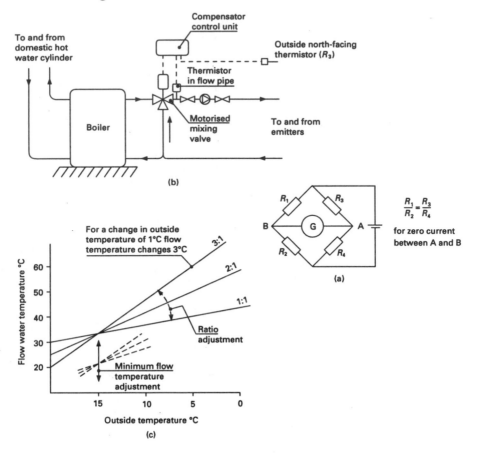

Figure 6.19   (a) Arrangements of resistors to form a Wheatstone bridge.
(b) Installation of a compensator in a heating system using water
as the working substance. (c) Typical characteristics of
compensator

characteristics of the control unit. The control box incorporates solid-
state electronics and, as such, can be expected to be reliable.

6.30.2   Use of intelligent controllers

The up-to-date version of the compensatory controller is referred to as an
'intelligent controller' and it is increasingly being linked to a building
management system (bms), which may either be on site or at head office.
The latest units can control space heating and cooling, ventilation plant
and domestic hot water supply. The facilities available can include the
following:

(1) the control of functions linked to both the internal and external temperatures;
(2) frost control;
(3) automatic 'set-back' and 'boost';
(4) the remote rescheduling of the functions of the unit from the bms control, or from head office, or (if required) from a mobile telephone;
(5) the automatic disconnection of the telephone line after information has been transmitted;
(6) on-site manual control;
(7) optimum start-up and shut-down times.

This last facility brings forward, or puts back, the time of starting up the heating plant, in accordance with the outside temperature, so that the period of warming up always has an optimum value. Such systems have a realistic 'pay-back' time and they can be shown to reduce energy costs, while providing consistent comfort control, in a wide variety of buildings.

Generally, the success or failure of any thermal storage system depends on the difference in environmental temperature between that at the end of the charging period and that at the beginning; [6.17] suggests that for the scheme to be viable, the difference should not be greater than 3.5 K.

PART 4: CONTROL OF ELECTRICAL ENERGY GENERATED IN THE ATMOSPHERE

6.31   Electrostatic charges and their connection with lightning strikes

The phenomenon of electrostatic charges has been experienced from very early times and is the result of frictional forces acting between materials in relative motion.

Benjamin Franklin (1706–1790) first proposed that buildings could be protected from lightning strikes, by attaching vertically mounted earthed iron rods (called Franklin rods), at their extremities.

Michael Faraday (1791–1867) was an English physicist and chemist, who by his work has succeeded in becoming known as the first discoverer of the phenomena of electromagnetic induction. He pioneered insulated Faraday cages, which are used today for maintenance work on live overhead power transmission lines. One method providing lightning protection for a building having a height of 4.0 m is shown in figure 6.20(a), and for a four-bedroom house in figure 6.20(b).

6.32   Need for protection

The establishment of the need for protection for any structure can be clear-cut, such as in the case of a tall and isolated building situated on the equator, or a building where explosives are to be manufactured. On

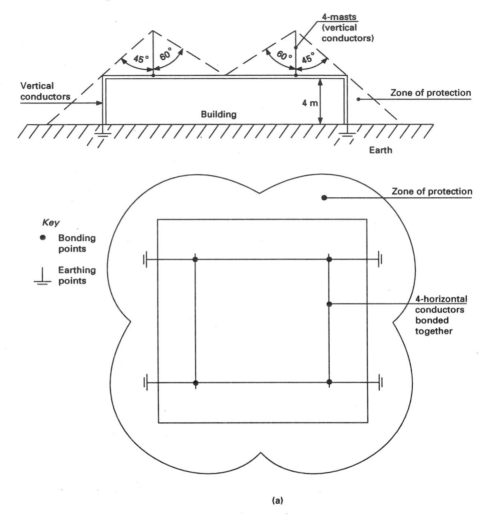

Figure 6.20 (a) Lightning protection for a building 4.00 m high

the other hand, a four-bedroom detached house situated in Northern Europe or North America could be categorised as a fairly low risk. Decisions concerning the need for lightning strike protection depend on many factors and only some of these are quantifiable. Some considerations will hinge on the degree to which those who occupy the building need to 'feel protected' and of course on the terms that are on offer from the insurers of the building.

The 'harmonised' British Standard *Code of Practice for the protection of structures against lightning* [6.18] points the way to making a realistic assessment of the need for protection by first determining the probable number of lightning strikes/year (*P*) for the locality. This value is then modified by the application of a series of multipliers referred to as 'weighting

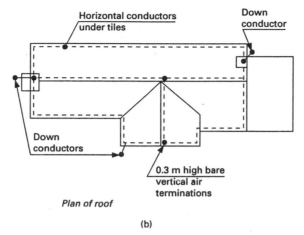

Plan of roof

(b)

Figure 6.20 (b) Lightning protection for four-bedroom detached house

factors'. The probable number of lightning strikes/year (P) = the lightning flash density (N_g) multiplied by the effective collection area (A_c), hence:

$$P = N_g \times A_c$$

The overall risk factor is then given by the product of P and the weighting factors.

6.33 Production of electrostatic charges in a thunder cloud

The random movement of rising and falling currents of air within a cloud containing particles of dust, water vapour, water droplets and ice crystals produces a positive charge around the top of the cloud and a negative charge around its base (see [6.19]). The effect of this negative charge is to induce a positive charge on the uppermost surfaces of any adjacent building, and the process continues until the voltage between the opposite sets of charges on a particular building is large enough to permit the initiation of a downward moving 'leader' of invisible ionised particles, from the cloud towards the positively charged surface. This 'leader' moves downwards in a series of intermittent steps of the order of 50 m.

Some points on the surface of the building will have a greater positive charge than others, and it is from the point of highest intensity of charge (often the highest part of the building) that a positive upward moving 'streamer' of invisible ionised particles will emanate. This streamer can leave the building in any direction and the distance it has to travel before it joins up with the downward leader is referred to as the striking distance, or 'last step' (in the process of ionisation). When the leader and the streamer meet, a current of electricity flows downwards along the conducting path formed by the ionised particles, so producing a visible flash of lightning referred to as the 'return stroke' or lightning strike.

BS 6651: 1990 [6.18] gives details of the way in which lightning protection schemes for tall office blocks, single-storey factory bays and other buildings, both permanent and temporary, should be designed and specified. Details of the zones of protection offered by air termination points are given, together with details of the way in which tall buildings with complicated outlines should be assessed by means of an imaginary rolling sphere, having a radius that approximates to the striking distance, or last step of ionisation.

6.34   Earthing of steel-framed buildings and reinforcing bars

Traditionally, advantage has been taken of the steel framework and the large amount of reinforcing bars cast into piling systems, foundations, columns and floors of a building to provide a 'built-in' Faraday cage and earthing system; BS 6651: 1990 [6.18] recommends that this practice should continue. When this method is chosen, care must be taken to ensure that there is electrical continuity between the encastré steel and the lightning protection system and all other earthing connections. It follows that when it is intended to use this method of earthing, it is best to introduce it during the design process.

6.35   Effectiveness of earthing systems

In practice, the earth on which buildings are constructed is non-homogeneous and often consists of rock with only a light covering of topsoil. BS 6651: 1990 recommends that at least two strip electrodes should be used, but also suggests that a suitable earth termination may be obtained by drilling the rock to a diameter that is not less than 75 mm and filling the space around the electrode with an electrically conductive cement, made with graded granular carbonaceous aggregate (trade name Marconite), or by backfilling with a material capable of retaining moisture (trade name Bentonite).

The depth to which such earth terminations should be taken must depend on local readings of the resistivity of the ground. For structures built on ground that is not rocky, the overall resistance from any part of the protection system to earth must not be greater than 10 ohm. For structures built on rock, this value is not applicable.

6.36   Inspection testing and records

BS 6651: 1990 states that all lightning protection systems should be visually inspected at regular intervals (not greater than 12 months apart) to check that they are in accordance with the recommendations given in the code, and the results from testing (see [6.20]) should be recorded in a lightning protection system log book.

6.37   Protection of power supplies and data communication lines against lightning

The growth in the use of computers in commerce and industry has been of an exponential nature, and it follows that the continuing control of plant and the accumulation of stored data, in such systems, are becoming increasingly important to the owners and operators of the organisations concerned. The equipment used to process the information is particularly sensitive to transient over-voltages and it has become painfully obvious to the insurers of the contents of buildings that their business portfolios include an ever-increasing factor of risk involving the possible loss of data and electronic equipment due to this cause.

Transient over-voltages (referred to as 'peaks' or 'spikes') can occur as a result of internal switching events releasing stored electrical energy from apparatus of an inductive nature, such as motors, fluorescent lights and transformers, and also as a result of external events, such as the secondary effects from a lightning strike. When such a strike occurs, power lines and conductors carrying data are subjected to the effects of magnetic induction, which can cause transient over-voltages up to about 6 kV to appear in the conductors. Particularly susceptible are those (external) communication lines connecting one building to another.

In order to protect data lines between buildings, it is possible to insert transient voltage limiters at each end of the line transmitting the data. When connected in series, these units have a low resistance and a specified 'let through voltage', which is restricted to less than twice the stated voltage of the lines.

The protection of the equipment connected to the power lines may be accomplished by installing, in parallel, a device that mimics the effect of a Zener diode. The output voltage across the lines increases by only a small amount, when the induced voltage due to the secondary effects of a lightning strike is applied to the system, so that the current flowing in the conductors is also kept low. Figure 6.10 shows such a unit connected to the mains input of a building and also shows smaller electronic systems protection units, connected in parallel with key electronic equipment, on each floor as required. The author understands that such devices are designed using 'non-linear technology', the details of which are too involved to be included in a book of this nature.

It should be noted that the increasing use of fibre optic cables for the transmission of data has eliminated the problems associated with electrically induced interference. An Appendix is scheduled (1994) to be added to the British Standard dealing with methods of protecting electronic equipment against lightning strikes.

References

6.1. *CIBSE Guide*, Vol. B, 1986, sections B10–11. CIBSE, Delta House, 222 Balham High Road, London SW12 9BS.
6.2. Lazell P. *Power distribution on the national grid*. Lecture given to an area meeting of the Institutions of Electrical and Mechanical Engineers, Chelmsford, 6 November 1991.
6.3. *Handbook on the 16th edition of the IEE Regulations for Electrical Installations*, 1991. Blackwell Scientific Publications, Oxford.
6.4. Jenkins B.D. *Electrical Installation Calculations*, 1991. Published on behalf of the Electrical Contractors' Association by Blackwell Scientific Publications, Oxford.
6.5. BS 6207: 1991 *Specification for mineral insulated copper-sheathed cables with copper conductors*. BSI, London.
6.6. BS 6346: 1987, 1989 *Specification for PVC insulated cables for electricity supply* (under revision). BSI, London.
6.7. BS 3036: 1992 *Specification for semi-enclosed electric fuses (ratings up to 100 A and 240 V to earth)*. BSI, London.
6.8. BS 88: 1988 *Specification for cartridge fuses for voltages up to and including 1000 V ac and 1500 V dc* (under revision). BSI, London.
6.9. BS 3871: Part 1: 1990 *Specification for miniature and moulded case circuit-breakers* (under revision). BSI, London.
6.10. Niedle M. *Electrical Installation Technology*, 1988. Newnes–Butterworth, London.
6.11. BS 4568: Part 1: 1988 *Specification for steel conduit and fittings with metric thread of ISO form for electrical installations*. BSI, London.
6.12. BS 4607: Part 1: 1988 *Specification for non-metallic conduits and fittings for electrical installations*. BSI, London.
6.13. Building Research Station. Digest 179, *Electricity distribution on site*, 1975. HMSO, London.
6.14. Fordham-Cooper W. *Electrical Safety Engineering*, 3rd edition, 1994. Newnes–Butterworth, London.
6.15. Dickie P.M. *Electrical Equipment Certification Service*, 1992. Harpur Hill, Buxton, Derbyshire SK17 9JN.
6.16. Phillips H. *Safe Gap Revisited*. Health and Safety Executive, Buxton, Derbyshire, UK, 1988. (Published by the American Institute of Aeronautics and Astronautics, 1 Nc, 370 L'Enfant Promenade, SW, Washington DC 20024–2518).
6.17. Martin P.L. and Oughton D.R. (Faber and Kell) *Heating and Air Conditioning of Buildings*, 7th edition, 1989. Butterworth, London.
6.18. BS 6651: 1990 *Code of Practice for the protection of structures against lightning* (under revision). BSI, London.
6.19. Trade literature supplied by W.J. Furse and Co. Ltd, Nottingham, UK, 1991.
6.20. *BS Code of Practice 1013: 1965 Earthing*. BSI, London.

7 Water Services

7.1 Effects of the rationalised water bye-laws on the design and installation of water services

Those organisations responsible for the supply of potable water to buildings in the UK have the statutory power to make and enforce water bye-laws, which are designed 'for the prevention of waste, undue consumption, misuse or contamination of water supplied by them'. Building Regulations deal with 'aspects of safety' and British Standards deal with 'fitness for purpose'.

In England and Wales alone there are 10 water authorities and 28 water companies, and predictably, this statutory power has in the past resulted in a proliferation of water bye-laws that differed from one authority to another and were a veritable jungle for designers, installers and clients alike.

In 1985, in order to rationalise and bring about a uniformity of approach to the bye-laws in the UK, and at the same time produce a document that could be kept up-to-date and would be suitable for training purposes, the water authorities jointly produced and published a *Water Supply Bye-Laws Guide* [7.1]. This publication took into account the guidance given by the Department of the Environment in [7.2] on the categorisation of the severity of risks due to back siphonage, together with methods for its prevention, which are currently in use both in the USA and in Europe and which, it was deemed, could with advantage be introduced into the UK.

In the following section, the principal changes made to the bye-laws using the new approach have been identified, so that when new work is being contemplated, the changes may be incorporated and advantage may be taken of the latest developments; these include an introduction to the concept of different classes of risk to be expected for any given installation.

The role of the local inspectors is such that they can be relied upon to offer helpful advice whenever practical work or design work is being carried out. It is obviously cheaper and more satisfactory to make changes to a drawing than it is to carry out changes on site, when the designer's credibility may then be 'on the line'. Essential reading for designers, surveyors and installers will be found in [7.3] and [7.4].

For those readers who wish to obtain knowledge on the way in which plumbing is arranged in Continental European developments, much information will be found in [7.5].

7.2   Effects of back siphonage in water service pipes

A large part of the content of water bye-laws is concerned with the prevention of the passage of water from consumers' apparatus back into the mains. This is called back flow and, by definition, if water has entered consumers' premises, it is assumed to be contaminated. If no precautions are taken, back siphonage will be possible, when the pressure inside the water supply main falls below atmospheric pressure.

Figure 7.1(a) shows the way in which the contents of a fish pond could be drawn into the town main, owing to the formation of a vacuum in the supply pipe. This situation may be avoided by fixing the end of the hose pipe above the final water level of the pond, so that there is an air gap between the end of the hose and the water in the pond.

7.2.1   *Changes to the bye-laws to prevent back siphonage (bye-laws 11 to 29)*

In practice, one cannot rely on the ability or the willingness of individuals to ensure that an air gap, as just described, would always be provided; hence the need for bye-laws specifying the inclusion of mechanical devices in the system to prevent back siphonage.

Under bye-law 18 in the *Guide* [7.1], the connection of the hose pipe on domestic premises will only be allowed to those draw-off taps that are provided with an approved double check valve assembly, or some other no less effective back flow prevention device, see figure 7.1(b). The servicing valve shown is provided to facilitate maintenance without disturbing other parts of the system (see bye-laws 68–71).

7.2.2   *Back flow prevention devices*

A back flow prevention device may consist of: the provision of air gaps, of which there are two types, A and B, related to the diameter of the inlet pipe, as illustrated in figures 7.2(a) and (b); or any one of the following – a check valve; a check valve and a vacuum breaker; a double check valve assembly; a pipe interrupter; or some other approved combination of fittings. Vacuum breaker valves and pipe interrupters are not used in the UK as commonly as they are in Continental Europe. Experience has shown that the former require periodic maintenance and the latter may under certain circumstances discharge water, so that installers in the UK prefer to fit a double check valve in place of both units.

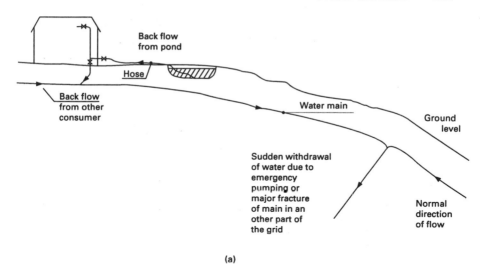

(a)

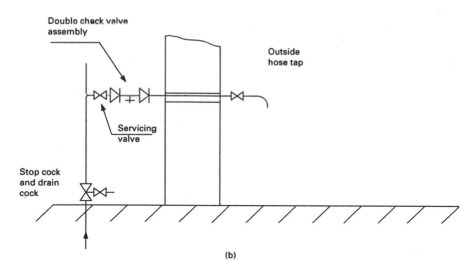

(b)

Figure 7.1   (a) Example of back flow. (b) Preferred method of connection of hose tap to main

7.2.3 Classification of levels of risk

The water undertakers have categorised the levels of risk posed by different types of water systems into three classes, some of which are quoted as examples, with Class 1 risk being the most stringent.

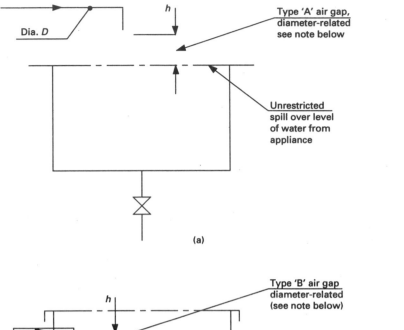

Type 'A' air gap, diameter-related see note below

Unrestricted spill over level of water from appliance

(a)

Type 'B' air gap diameter-related (see note below)

Overflow

Steady-state water level when float is punctured or removed

$D \leqslant 14$ mm	$h = 20$ mm
$D > 14$ mm but $\leqslant 21$ mm	$h = 25$ mm
$D > 21$ mm but $\leqslant 41$ mm	$h = 70$ mm
$D > 41$ mm	$h = 2D$

Valve closed gives worst-case scenario

(b)

Figure 7.2   (a) Type 'A' air gap. (b) Type 'B' air gap

Class 1

Occurs where there is serious contamination, that is likely to be harmful to health; for instance, where hospital apparatus is being used, industrial processes are involved, or where a hose is used commercially. For this class of risk, all the above must be supplied from storage, or in the case of hospital apparatus and industrial processes, a type 'A' air gap can be used.

Class 2

Occurs where there is risk of contamination by a substance that is not continuously or frequently present, but which may be harmful to health; that is, where commercial water softeners, vending machines with CO_2 injection and domestic hose union taps are being used, and also when unvented heating systems are being filled. For this class of risk, all the above must be provided with a double check valve assembly and in the case of the charging of the heating system, there must also be a temporary connection (flexible hose), which is removed after the filling is completed.

Class 3

Occurs where the risk of contamination by a substance is not likely to be harmful to health, but is undesirable; that is, where domestic water softeners, vending machines without CO_2 injection and unbalanced mixer taps are being used. For this class of risk, the first two cases mentioned must be provided with a single check valve and the last must have a check valve fitted on each pipe.

7.2.4 Prevention of secondary back flow (bye-law 26)

Secondary back flow describes any flow that may occur from one installation to another in buildings having multiple occupancy. Figure 7.3(a) depicts the way in which protection of the supply pipe is assured and 7.3(b) shows the way in which protection of the distributing pipe is to be arranged.

7.3 Changes to bye-laws

7.3.1 Dealing with the prohibition of the use of lead (bye-law 7)

When lead is ingested by the human body, it cannot be excreted and becomes a cumulative poison that is known to impair the mental abilities of children. The new bye-laws have now prohibited its use where water is used for domestic purposes. Accordingly, the composition of the solder rings in capillary fittings and the solder used to 'feed' such joints must be lead free. Bye-law 7 also deals with the suitability of other materials for use in water systems.

7.3.2 Dealing with storage cisterns (bye-law 30)

The storage cisterns used either for drinking water or for secondary supply purposes must now have a cover that is securely fixed and fits closely to the cistern.

When the cistern stores more than 1000 litres of water, the lid must be

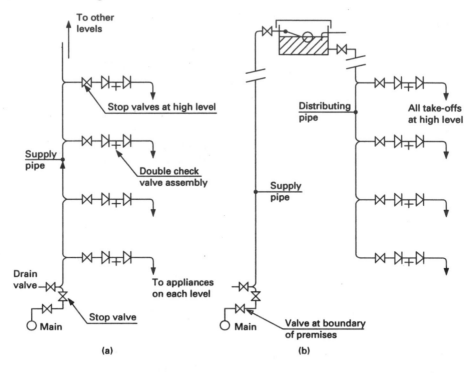

Figure 7.3 (a) Secondary back flow protection of supply pipe. (b) Secondary back flow protection of distributing pipe

of such a size that, when essential maintenance is carried out, not all of the top of the cistern is uncovered.

The lid of any cistern must be fitted with an insect-proof vent, consisting of a 0.65 mm corrosion resisting mesh, a sleeve and tight fitting grommet for the vent pipe, and a screened (insect proof) warning pipe assembly, complete with a dip tube. The cistern must be lagged against frost and should be so positioned that, when it contains potable water (usually in a multi-storey development), the temperature of the contents will normally be less than 20°C.

7.3.3　Dealing with the identification of underground pipes (bye-law 27)

When dealing with underground services, it is not unknown for operatives to mistake the black outersheath of an electric cable for the black unplasticised polyvinyl chloride (PVC-U) wall of a water main. In order to help minimise the number of such mistakes occurring in the future, all underground pipes are to be made of blue medium-density polyethylene pipe (MDPE), although this may give rise to problems where radon is present.

7.3.4 *Dealing with the provision of servicing valves (bye-laws 68 to 71)*

The purpose of these valves (which can be operated with a screwdriver) is to be able to isolate an appliance for maintenance or repair, without interrupting the operation of associated equipment and without draining down and wasting relatively large volumes of water, which may also have been heated. The valves are fitted to cold or hot water, cisterns, cylinders, or tanks, having a capacity greater than 18 litres, as close to the appliance as possible. The bye-laws also state that a servicing valve shall be fitted (as near as is possible) to every ballvalve.

7.3.5 *Dealing with float-operated valves (bye-law 42)*

Piston type valve

Over the years, the piston type float-operated ball valve has been the subject of continual development. There are two types of piston-operated valve: the Croydon type, in which the piston, activated by the lever arm attached to the float, moves vertically to shut off the water supply; and the Portsmouth type, in which the piston moves horizontally to shut off the water supply. Figure 7.4 shows a cross-section of the Portsmouth type valve. The earlier versions of both valves were noisy in operation and both suffered from a build-up of salts around the moving parts. Also, the construction of the valves was such that the moving parts had sufficient inertia to prevent an early reaction to the occurrence of back siphonage in the main.

The noise emanating from a ball valve exhibits different characteristics during different phases of its operating cycle, and in a successful attempt to attenuate the external noise due to the splashing of the incoming stream on to the water surface, a vertical discharge tube (a silencer tube), made of easily collapsable plastic, is fitted to the Portsmouth valve. This directs the water coming from the underside of the valve to a position below the water level. In the event of back siphonage conditions occurring in the main, the walls of the flexible silencer tube collapse and prevent the back flow of water from the cistern. Because of the form of its construction, a similar modification could not easily be made to the Croydon valve.

In addition to the problem of external noise in the Portsmouth valve, it was found that erosion was taking place on both the seat and valve surfaces, resulting in a leaky closure. Subsequent work carried out by the UK Building Research Establishment showed that this was due to the phenomenon of cavitation, which was shown to be occurring at different parts of the assembly, at different times during the cycle of operations.

Cavitation is the term used to describe the emergence of dissolved gases from a liquid, when the pressure of the liquid is reduced to the vapour pressure of the gas. In the case of potable mains water having a temperature

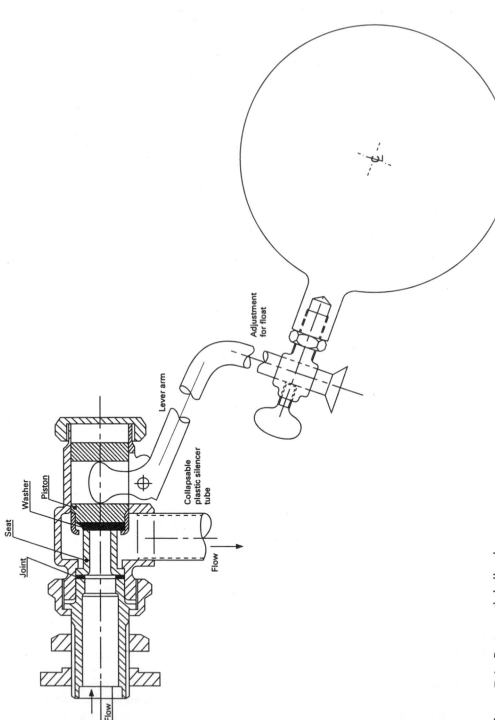

Adjustment
for float

Lever arm

Collapsable
plastic silencer
tube

Piston

Washer

Seat

Joint

Flow

Flow

Figure 7.4   Portsmouth ball valve

of 17°C, the dissolved gases are those of air, and the vapour pressure is of the order of 0.02 bar (2 kPa) absolute.

Wherever there are abrupt changes in the internal contours of a fitting, a drop in overall pressure occurs, accompanied by much larger local pressure drops at points downstream of the change of shape, and it is at these positions that conditions suitable for cavitation can arise. When the small bubbles of air so formed eventually come into contact with the higher mainstream water pressure, they collapse and because water is virtually incompressible, the resulting impulses are transmitted to the surrounding surfaces, which become eroded. The collapsing air bubbles also produce noise, which is transmitted by the water, both downstream and through the material of the valve, which is not to be confused with water hammer.

Most of the earlier versions of Portsmouth valves did have sudden changes in their internal contours and subsequent development was then channelled into making all the changes in cross-section smooth and gentle, in order to avoid the occurrence of cavitation at these points. The latest specification of form and materials of the Portsmouth type valve is illustrated in BS 1212: Part 1: 1990, which also gives details of a suitable way of adjusting the position of the float, relative to the fulcrum point of the lever, without having recourse to bending the arm. Under the new bye-laws, a Portsmouth float valve may be used provided that it is installed with a type B air gap.

In the process of selecting a Portsmouth ball valve for a required volume flow rate, when the local pressure in the main is known, tables 11, 12 and 15 of BS 1212 are relevant. Tables 11 and 12 together show that each nominally sized valve may be fitted with variously sized seats, made of metal or plastic, to suit certain body pattern numbers. In addition, table 15 of the standard quotes the estimated volume flow rate that may be expected for a series of orifice diameters when the inlet pressure varies from 0.039 to 14.1 bar gauge. Alternatively, the manufacturers of such fittings also provide pressure/flow data for their particular valves.

Diaphragm type valve

The Building Research Establishment (BRE) designed and developed an alternative float valve (the BRE ball valve), using a diaphragm in place of the piston of the Portsmouth type valve (see figure 7.5). The force required to accelerate the diaphragm is much less than that needed to accelerate the piston, and the force needed to flex the diaphragm is also less than the friction force between the piston and its cylinder. These considerations make the diaphragm valve more responsive to the conditions encountered with back siphonage.

In addition, the 'heel' of the lever arm may be positioned so that it operates under dry conditions, so preventing the encrustation of this part

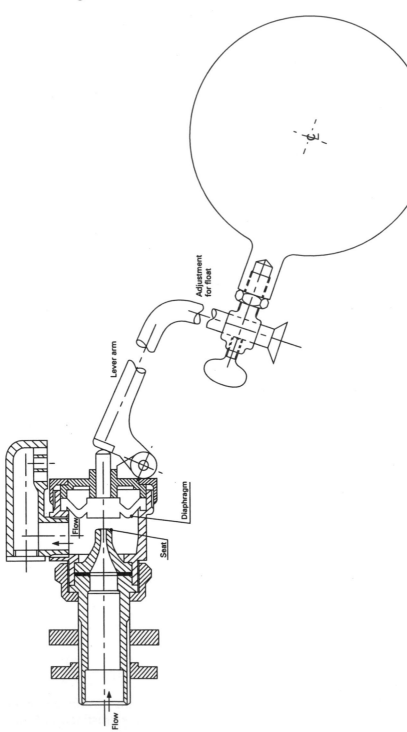

Figure 7.5 Diaphragm valve

of the assembly with salts precipitated from the water. The outlet from the diaphragm valve faces upwards, so that the orifice of the seat runs 'drowned'. This ensures that external air cannot become entrained in the flow of water, thus causing unwanted noise.

Further silencing is achieved by the connection of a sparge pipe to the outlet. The holes drilled along the longitudinal axis of the tube are positioned so that the outward flowing water jets of small diameter are directed against the side of the containing vessel, avoiding the generation of noise due to the splashing of a single jet of water on to the free surface.

Tests for back siphonage have shown that the performance of the piston type float-operated valve (Portsmouth ball valve), as detailed in BS 1212: Part 1: 1990, is inferior to the diaphragm type float-operated valves as detailed in Parts 2 and 3 of the same standard. This suggests that Part 1 (piston type) ball valves will require back siphonage protection, whereas Parts 2 and 3 diaphragm type float valves will not.

Generation of noise caused by self-induced oscillations in a ball valve

The author's attention was drawn to a situation that involved a Portsmouth type ball valve, fitted to a water storage tank, in a six-storey office block, which was continually generating an unacceptable amount of noise that could be heard all over the building. The valve was exhibiting a forced vibration at the natural frequency of the system, which was reinforced by the incidence of ripples on the surface of the water in the storage tank. The system may be 'detuned', and the syndrome prevented, by changing any part of the mechanism; such as the length of the lever arm, the size of the float, the size of the orifice fitted to the ball valve, or (rather less conveniently) the overall length of the tank.

Equilibrium type ball valve

The pressure in a water main varies diurnally, in some cases over a wide range, and this means that in any ball valve, the length of the lever arm and/or the diameter of the float must be designed to deal with the highest possible pressure in the main. This necessarily results in lever arms that are extra long and/or float valves having large diameters.

The equilibrium ball valve is designed to make the movement of the valve independent of the variation of pressure in the main, resulting in a relatively much shorter lever arm. Figure 7.6 shows the cross-section of such a valve as manufactured by Pegler Ltd of Doncaster, in which the mains pressure is transmitted to both ends of the piston, so that the force required to make it move in a longitudinal direction is balanced only by the friction force. By making the area of the piston at the end remote from the seat larger than the area of the seat, the resulting longitudinal force due to the difference in areas helps to close the valve.

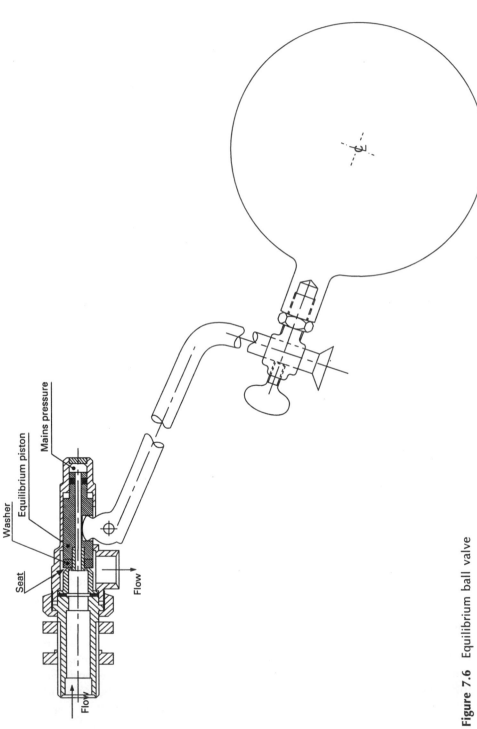

Figure 7.6 Equilibrium ball valve

7.3.6 'Water Fittings and Materials Directory'

There are many different variations in the design of ball valves and other water fittings, though not all of them have been approved for use in water installations by the Water Research Centre.

Every six months, the Water Bye-laws Advisory Service publishes a *Water Fittings and Materials Directory*. This is an updated list of those fittings and materials that are acceptable for use by the Water Authorities under the conditions specified in the Directory.

7.4 Explanation of terms categorising hot and cold water systems

(a) *Vented*: The system is open to atmospheric pressure.

(b) *Unvented*: The system is closed to the atmosphere and operates under pressure, usually above atmospheric.

(c) *Direct*: A direct hot water system is one in which the water contained in the boiler, or other heater, may also be drawn from the taps.

(d) *Indirect*: An indirect hot water system is one in which the water contained in the boiler, or other heater, is kept separate from the water drawn from the taps.

(e) *Primary circuit*: That part of a system in which the water circulates from the boiler, or other water heater, and the heat exchanger contained inside a hot water storage vessel.

(f) *Secondary circuit*: An assembly of pipes and fittings in which water circulates in distributing pipes to and from a water storage vessel.

(g) *Secondary system*: That part of a hot water system consisting of the cold feed pipe and any storage cistern, water heater and flow and return pipes used to deliver hot water to the taps.

(h) *Distributing pipes*: Such pipes can convey either hot or cold water, and they are frequently referred to as either 'secondary hot', or 'secondary cold'.

7.4.1 Changes in bye-laws dealing with the connection of unvented hot water systems (bye-laws 90 to 95)

Unvented hot water systems may now, with certain precautions, be connected directly to the water main, making storage and feed and expansion cisterns in the loft space unnecessary. However, provision must still be made for the expansion of the water in both the primary and secondary hot water pipe circuits.

In such systems, the mains pressure is common to both hot and cold water supplies. This means that pipe sizes can be smaller, for both hot and cold water services. Thermostatic mixing valves used for washing/showering purposes can operate more effectively, because the pressure is higher and any differential pressure caused by the simultaneous operation of other outlets forms a relatively smaller proportion of the total pressure. This results in a reduced requirement for overly sensitive automatic adjustment of the temperature of the mixed water, so reducing the need for complexity in the design of the thermostatic mixing valves.

Prior to 1 January 1989, nearly all properties had to install a storage cistern, of a capacity agreed by the local water authority. The amount of storage was usually based on the amount of water likely to be used over a period of 24 hours (table 1 of [7.3] gives details of the recommended minimum storage of cold water for domestic purposes for both hot and cold outlets). This ensured that, in the event of any cessation of mains water supply lasting less than 24 hours, or the occurrence of a temporary fall in mains pressure, the user would not be unduly inconvenienced.

A system with storage can cope with an instantaneous demand for water from the outlets connected to it, having only a limited effect on the mains distribution network. This was important for those water authorities operating in areas where property development was ongoing and which found that their existing mains distribution networks were no longer capable of sustaining the duty required. In addition, in such overloaded systems, the pressure would have been low and the chances of back flow occurring would have been high.

Note that, in towns, water authorities should always be consulted as to the likely effectiveness of a proposed installation of hose reels used for firefighting on a given floor, and if the scheme is sanctioned, it would then seem logical for the water supply company to recognise that it has a duty not to allow the pressure in the mains to fall below the level at which the use of the hose reels became ineffective. In development areas incorporating low rise buildings, the effects of the diurnal variation in water pressure in the mains is not so critical.

However, whenever it is proposed to install an unvented hot water system, enquiries should always be made about the likely pressure and volume flow rate available, to sustain the combined instantaneous demand of both hot and cold water outlets at the development.

7.5 Hot and cold water and cyclic process space heating

Figures 7.7 to 7.14 show some of the ways in which the services of hot and cold water and space heating may be provided to a dwelling, using vented and unvented systems, all of which will be found to be in current use in the UK.

7.5.1 Method 1

Figure 7.7 shows a traditional fully vented system having indirect secondary hot water storage supplied by a conventional boiler, and also providing space heating. This is typical of the type of installation likely to

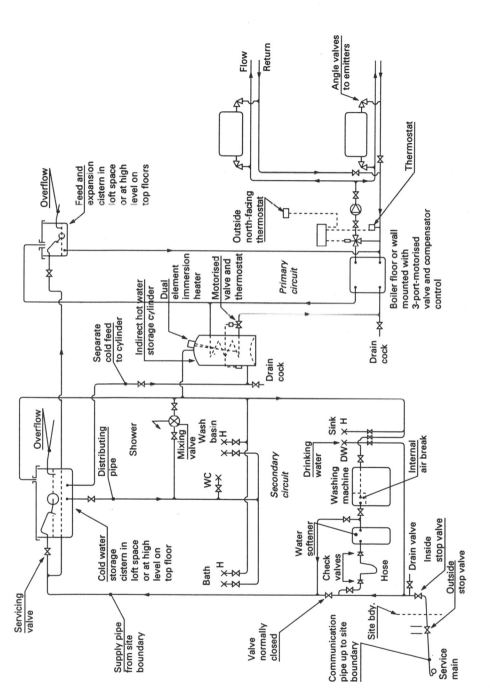

Figure 7.7 Vented system, having indirect secondary hot water storage supplied by a boiler, also providing space heating

have been fitted throughout the UK prior to the time of the change in the bye-laws in January 1989. In some areas only one connection to the main was allowed, that being to the drinking water tap in the kitchen, while in other areas the bathroom services indicated could also be connected to the main.

Advantages

(1) Storage capacity provided by the cistern minimises the effects of low mains pressure and provides a reliable source of supply.
(2) The hot water storage facility ensures that adequate simultaneous draw-off rates are possible, from more than one outlet.
(3) Minimal maintenance is required.
(4) Installers do not have to be certificated by the British Board of Agrément (BBA).

Disadvantages

(1) Pressure of water at a shower is limited by the height of the storage cistern.
(2) Instantaneous use of the shower and any other fitting alters the differential pressure at the thermostatic mixing valve by a relatively large amount, causing a noticeable change in the temperature of the mixed water.
(3) In order to achieve a satisfactory volume flow rate from each fitting, using the head available from the cistern, relatively large-diameter pipes must be used.
(4) Space is required at high level, within the living space, or in the loft, to accommodate the storage and feed and expansion tanks.

7.5.2   Method 2

The incorporation of a 'Primatic' hot water storage cylinder, in place of the conventional indirect storage cylinder, removes the need to install a feed and expansion cistern at high level or in the loft. Figure 7.8 shows a traditional fully vented single feed system having indirect secondary hot water storage supplied by a conventional boiler, and also providing space heating. Figure 7.9 shows three views of the cross-section of the 'Primatic' cylinder: (a) in the process of being filled; (b) when it is completely filled and cold; and (c) when it is hot, with the water in the primary circuit expanded. The cylinder can be fitted with an immersion heater (usually at the top).

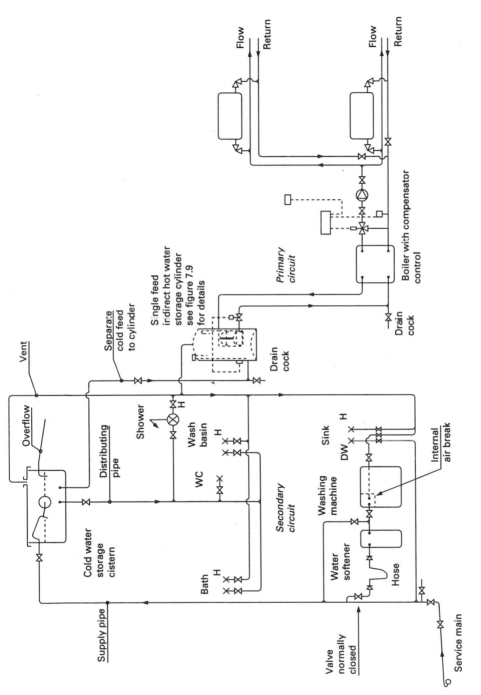

Figure 7.8 Vented single feed system, having indirect secondary hot water storage supplied by a boiler, also providing space heating

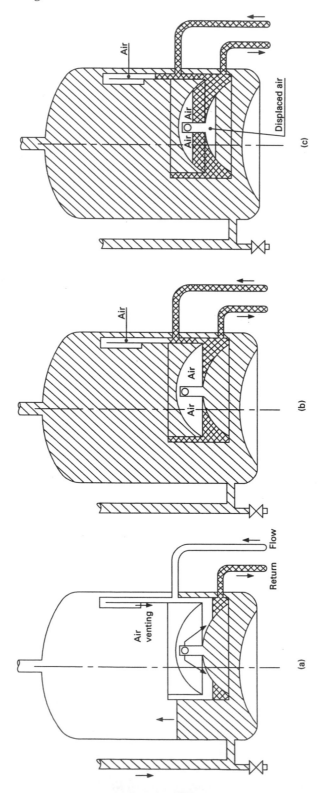

Figure 7.9 Operation of the primatic cylinder: (a) filling, (b) filled and cold and (c) filled and hot with expanded primary

Advantages

(1) to (4) As detailed for the system in subsection 7.5.1.
(5) No feed and expansion cistern is required.
(6) The pipe layout is simpler than that shown in subsection 7.5.1.

Disadvantages

(1) to (3) As detailed for the system in subsection 7.5.1.
(4) Space is required at high level, within the living space, or in the loft, to accommodate the storage cistern
(5) In order for the 'Primatic' unit to operate effectively, there is an upper limit on the permissible volume of the primary circuit.

7.5.3 Method 3

Figure 7.10 shows an unvented system operating at mains pressure having indirect secondary hot water storage heated by a pressurised boiler, and also providing space heating. The hot water is stored at mains pressure in a stainless steel cylinder, or a mild steel cylinder lined with a seamless polythene liner (to prevent corrosion and discoloration of the water). Figure 7.11 shows a cross-section of a pressurised cylinder, together with details of the obligatory safety fittings. Note that the water inside the pressurised steel cylinder can be heated using immersion heater elements, operating on an off-peak electricity supply, instead of using the boiler as shown, and that the space heating system illustrated can be either vented, or unvented.

Packaged units are available but should not be installed unless they have been approved by the BBA, and the installers of such equipment must themselves be certificated by the same body. (The installation of unvented hot water systems having a capacity of 15 litres or less may be carried out by those who have not been so certificated.)

Prior to the installation of unvented hot water storage systems of a capacity greater than 15 litres, the building control department of the local authority must be informed, as it has a responsibility to ensure that all safety factors are taken into consideration. Note also that an ongoing schedule of maintenance, of the pressure and temperature relief valves and other fittings, must be initiated by the client and be carried out diligently.

Advantages

(1) No storage tanks are required, thus saving space.
(2) The hot water storage facility makes high draw-off rates possible, simultaneously from more than one outlet.
(3) The high head available for pipe sizing permits small diameters and high volume flow rates, and is good for showering.
(4) A simpler pipe layout than that shown in subsection 7.5.1 is possible.

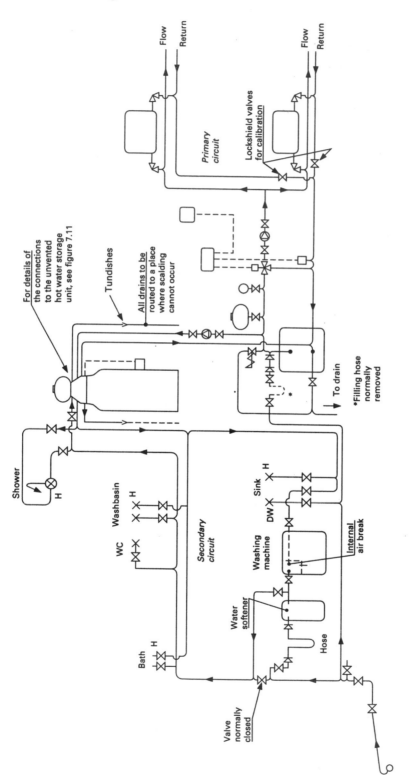

Figure 7.10 Unvented system operating at mains pressure, having indirect secondary hot water storage supplied by a pressurised boiler, also providing space heating

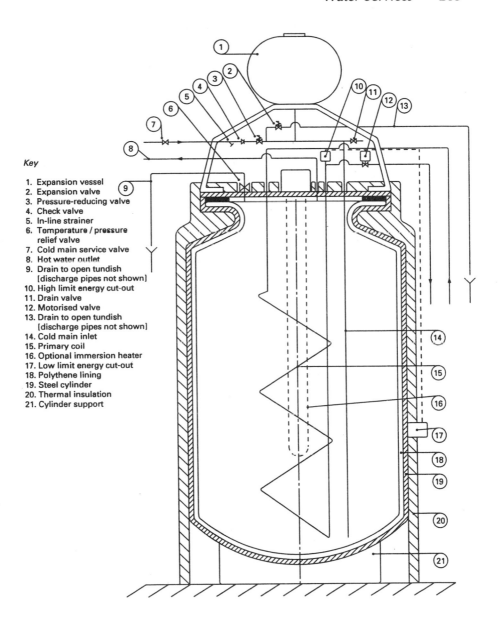

Key

1. Expansion vessel
2. Expansion valve
3. Pressure-reducing valve
4. Check valve
5. In-line strainer
6. Temperature / pressure relief valve
7. Cold main service valve
8. Hot water outlet
9. Drain to open tundish [discharge pipes not shown]
10. High limit energy cut-out
11. Drain valve
12. Motorised valve
13. Drain to open tundish [discharge pipes not shown]
14. Cold main inlet
15. Primary coil
16. Optional immersion heater
17. Low limit energy cut-out
18. Polythene lining
19. Steel cylinder
20. Thermal insulation
21. Cylinder support

Figure 7.11   Steel storage cylinder, with controls for use with unvented hot water

(5) The flow temperature of the water used for space heating can be higher, making smaller radiators than those for the system in subsection 7.5.1 possible (but see disadvantage (4)).

Disadvantages

(1) The continued operation of the system depends on the continuity of supply to the mains.
(2) Whenever pressure vessels are included in a system, it is imperative for safety reasons to ensure the continued integrity of the plant by the instigation of an ongoing schedule of maintenance; in this case the amount of maintenance required is increased from that of the system in subsection 7.5.1.
(3) Persons of a nervous disposition may be concerned at the prospect of a relatively large pressure vessel being on the premises.
(4) Emitters operating at higher temperatures may need to be guarded to prevent injury to small children and the infirm.
(5) Installers must be certificated by the BBA.

7.5.4 Method 4

Figure 7.12 shows an unvented system operating at mains pressure having indirect secondary hot water supplied by a pressurised combination boiler, and also providing space heating, applied to a small dwelling. The combination boiler can provide a prioritised 'instantaneous' supply of hot water for washing purposes, at the expense of a supply of hot water for space heating. The operation of a combination boiler is described in the following subsection, while figure 7.13 shows a simplified layout of the 'Vaillant' combination boiler.

Operation of the boiler in space heating mode (see figure 7.13)

The gas cock 11 and the pilot cock 10 are opened. The solenoid valve 9 is held open using the override button, and the pilot flame is ignited by means of the electrodes 2. Valve 9 is kept open manually until a suitable voltage has become established across the ends of thermocouple 3, enabling the solenoid of valve 9 to become energised. When this occurs, the override button on valve 9 may be released.

The space heating flow and return valves 7 and 14 are opened and the circulator 19 is started, producing a pressure differential across the venturi 20, which opens valve 21. This movement operates a microswitch which energises the solenoid of valve 22, allowing gas to flow to the main burners, where it is ignited by the pilot flame.

Water is recirculated through the heat exchanger and ports 1 and 2 of the thermostatically operated valve 5, until a temperature of 55°C is reached. When this occurs, port 3 opens and port 2 closes, allowing the heated

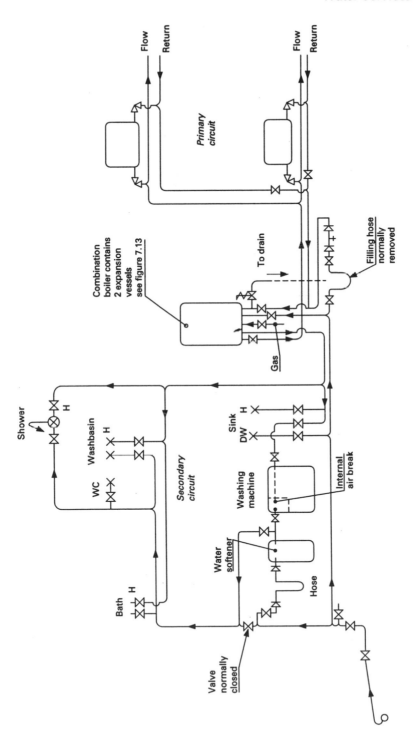

Figure 7.12 Unvented system operating at mains pressure, having indirect secondary hot water supplied by a combination boiler, also providing space heating

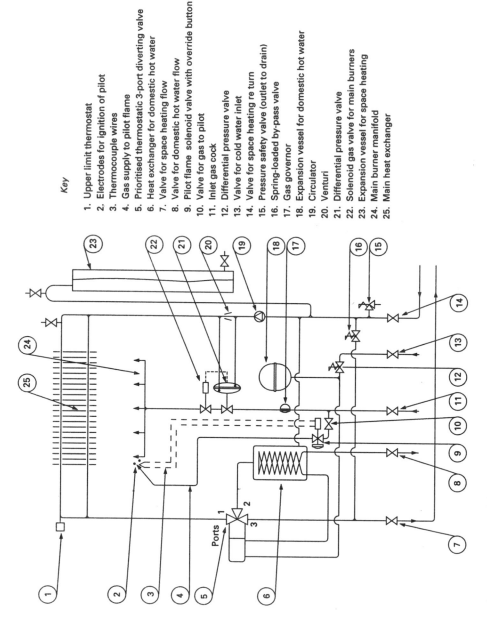

Key

1. Upper limit thermostat
2. Electrodes for ignition of pilot
3. Thermocouple wires
4. Gas supply to pilot flame
5. Prioritised thermostatic 3-port diverting valve
6. Heat exchanger for domestic hot water
7. Valve for space heating flow
8. Valve for domestic hot water flow
9. Pilot flame solenoid valve with override button
10. Valve for gas to pilot
11. Inlet gas cock
12. Differential pressure valve
13. Valve for cold water inlet
14. Valve for space heating re turn
15. Pressure safety valve (outlet to drain)
16. Spring-loaded by-pass valve
17. Gas governor
18. Expansion vessel for domestic hot water
19. Circulator
20. Venturi
21. Differential pressure valve
22. Solenoid gas valve for main burners
23. Expansion vessel for space heating
24. Main burner manifold
25. Main heat exchanger

Figure 7.13 Simplified layout of the 'Vaillant' combination boiler

water to flow around the space heating circuit. The opening of port 3 operates a microswitch, which activates a temperature control circuit, making it possible for the flow temperature selector (not shown) to operate at any setting between 60°C and 90°C.

The appliance is equipped with a built-in by-pass controlled by a spring-loaded valve 16, which opens whenever an unusually high pressure difference occurs across the flow and return of the space heating system. This ensures that there is always a satisfactory volume flow rate of water passing through the heat exchanger, thus avoiding 'kettling'.

Operation of the boiler when supplying prioritised domestic hot water see figure 7.13)

When any hot tap is opened, the pressure on the downstream side of the spring-loaded diaphragm valve 12 falls, allowing the mains pressure acting on the other side of the diaphragm to deflect it upwards, lifting the piston of the valve. This action establishes flow to the tap and operates a microswitch which activates the circulator 19 and the solenoid of the gas valve 22, permitting gas to flow to the main burners.

The cold water now circulating around the thermostatic element of valve 5 closes port 3 and diverts the hot water from the space heating circuit, through ports 1 and 2, to the heat exchanger 6, thus heating the cold water contained in the helically coiled copper tube.

When the demand for hot water ceases, valve 12 closes and the hot water flowing through ports 1 and 2 heats up the thermostatic element of valve 5, closing port 2 and reopening port 3, thus making it possible for the flow to the space heating circuit to be reinstated.

Advantages

(1) No storage tanks are required, thus saving space.
(2) The high head available for pipe sizing permits small diameters and high volume flow rates, and is good for showering.
(3) A simpler pipe layout than that shown in subsections 7.5.1 and 7.5.2 is possible.
(4) There are no thermal storage losses.
(5) The flow temperature of the water used for space heating can be higher, making smaller radiators than those for the system in subsection 7.5.1 possible (but see disadvantage (5)).
(6) Installers do not have to be certificated by the BBA.

Disadvantages

(1) The time to draw a bath can be protracted, particularly when other hot taps are being used.
(2) Prioritisation of output can be unsatisfactory, particularly when starting from cold.

(3) Continued operation of the system depends on the continuity of supply to the water main.

(4) Increased maintenance is required when compared with that for the systems detailed in subsections 7.5.1 and 7.5.2.

(5) Radiators operating at high temperature may need to be guarded to prevent injury to small children and the infirm.

7.5.5 Method 5

Figure 7.14 shows an unvented system, operating at mains pressure having indirect secondary hot water supplied by a conventional boiler, connected to a vented space heating system via a thermal store. Such an installation achieves the benefits of mains pressure hot water and traditional fully vented space heating, without the need to install a storage cylinder operating at mains pressure. Instead, a 'thermal store' is supplied, which consists of a vented cylinder, containing water directly in contact with the boiler. A feed and expansion cistern can be integral with the unit (as shown), or it may be fitted in an alternative position, to suit the positioning of the emitters. The inside of the cylinder is fitted with a finned coil, which is connected to the mains water supply pipe. The output of hot water from this pipe can continue for about 4 minutes at a rate typically three or four times that of the volume flow rate from a combination boiler, and this is more than ample for drawing a bath. Readers may have noticed that the arrangement just described is the reverse of an indirect cylinder.

In this method, pressure from the main is restricted to the inside of the pipe coil, and the strain energy stored in the pipe is very much less than that stored inside the pressurised cylinder used in the system detailed in subsection 7.5.2 which, it will be recalled, has to be made of steel in order to withstand the stresses involved.

The concept of the use of a 'thermal store' is not new to industrial process engineers, who use 'steam accumulators' (or thermal stores) to provide a store of available energy. The accumulator, operating at a low pressure, accepts exhaust steam taken from a previous process operating at a higher pressure, and the energy stored in the accumulator may then be used for further process work at a lower pressure.

In this method, the thermal store evens out the fluctuating demands of the heating and hot water circuits, reducing the frequency with which the boiler is fired. This reduces the total energy losses from the flue, which has to be heated up at each firing, with the result that fuel is saved. As heat is instantly available from the thermal store, it is possible to make a 'delayed start' to the daily heating programme. In the smaller installations it is also possible to reduce the margin allowed on the boiler for accelerating power, resulting in a smaller boiler and less capital cost. Information on tests carried out by the engineers at Watson House Re-

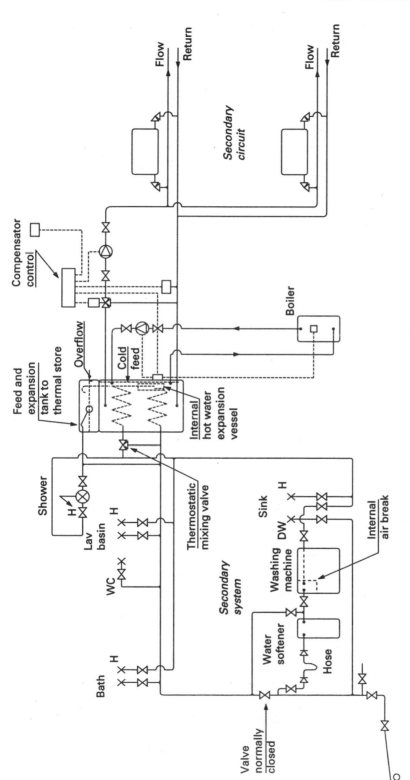

Figure 7.14 Unvented system operating at mains pressure, having indirect secondary hot water supplied by a conventional boiler, connected to a vented space heating system, via a thermal store

search Station, Peterborough Road, London, SW6 3HN, on a selection of area and regional offices, is given in [7.6].

Advantages

(1) The thermal store makes high draw-off rates possible, simultaneously from more than one outlet.
(2) The high head available for pipe sizing permits small diameters and high volume flow rates, and is good for showering.
(3) A simpler pipe layout than that shown in subsection 7.5.1, but comparable to that shown in subsection 7.5.2, is possible.
(4) The frequency of boiler firing is reduced, thus saving fuel.
(5) A smaller boiler than those for the systems detailed in subsections 7.5.1 and 7.5.2 are required.
(6) An 'instant' emitter warm-up operates, because of the thermal store, so reducing the time to reach room operating temperature.
(7) No pressure vessel is needed, therefore no pressure and temperature relief valves are required, unlike the system detailed in subsection 7.5.2.
(8) Installers do not have to be certificated by the BBA.

Disadvantages

(1) Heat losses from the thermal store are greater than all other methods, but are typically no more than 5% of the heat demand over a heating season [7.6]; by judicious siting, the heat loss can provide useful heat to the airing cupboard (as in the systems detailed in subsections 7.5.1 and 7.5.2), together with background heat to a room.
(2) A feed and expansion cistern is required (usually fitted integral with the thermal store and taking up no loft space).

7.6   Provision of domestic hot water in commercial and industrial systems

7.6.1   Use of calorifiers

In such installations, hot water is often heated and stored in cylinders that are very much larger than those used in dwellings. They are referred to as 'calorifiers' and can be mounted horizontally, or vertically (if space is limited), with the contents being heated by low-, medium-, or high-pressure hot water, or by steam. The heating elements can take the form of 'U'-shaped (hairpin) tubes, helically coiled tubes, or an annular surface. The tubes are connected to 'headers', which can be removed to gain access to the shell of the calorifier for maintenance purposes (see figure 7.15).

Whenever water containing dissolved salts is heated, scale is precipi-

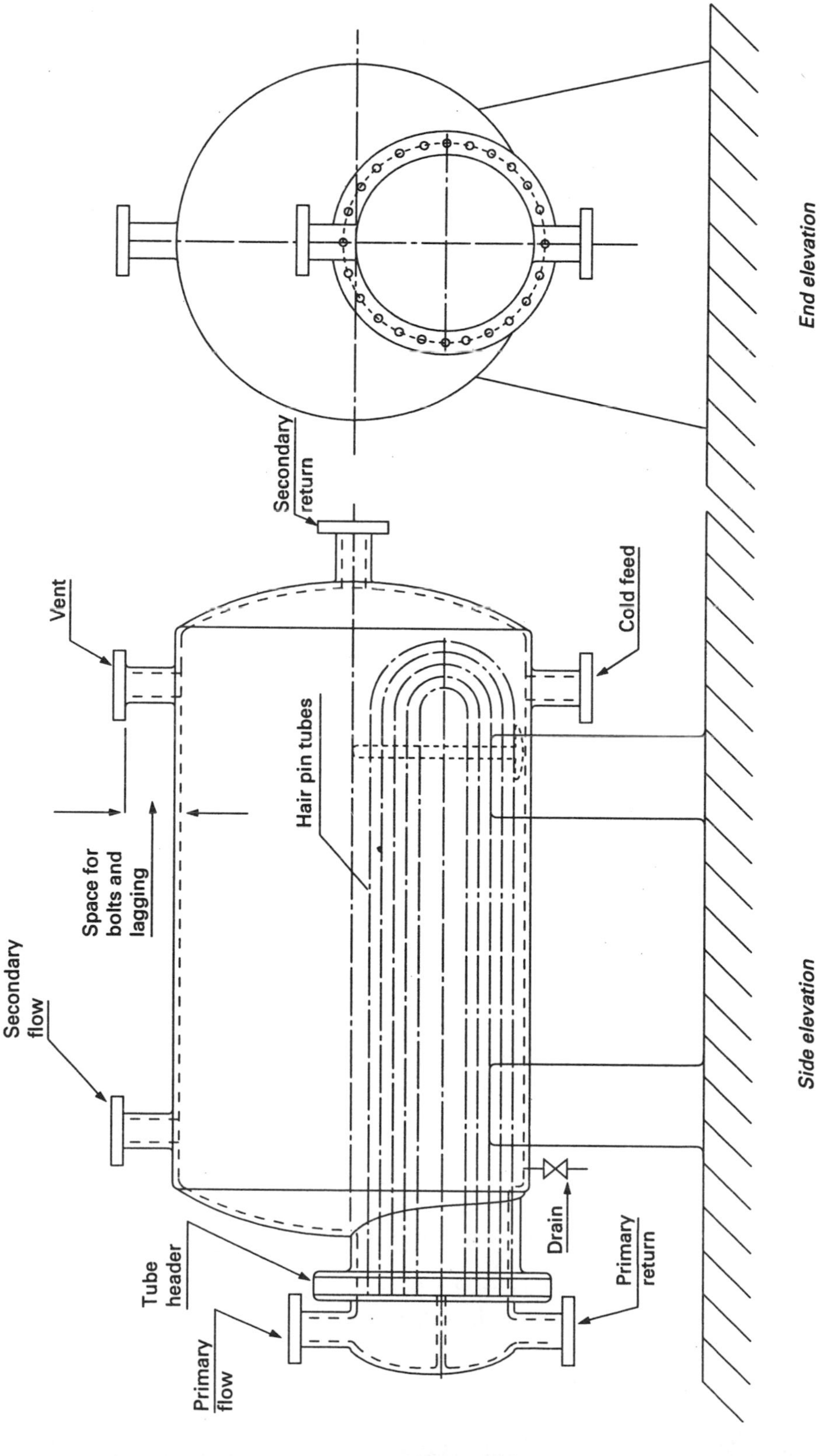

End elevation

Side elevation

Vent

Secondary return

Cold feed

Space for bolts and lagging

Hair pin tubes

Secondary flow

Tube header

Primary flow

Drain

Primary return

Figure 7.15 Horizontal water to water calorifier

tated on to the surface of the heating elements, from which, after re-peated expansion and contraction it is subsequently dislodged, falling to the lowest part of the unit from where it must be removed periodically.

Galvanic action will occur when different metals in close proximity are surrounded by water, or when metals having a high level of electro-chemical activity are upstream of those metals that are lower in the scale. Most water systems use copper pipes and, when this is the case, it becomes logical to fabricate the calorifier from the same material. Some medium to large copper vessels have a rod of aluminium fixed internally and in good electrical contact, so that the aluminium becomes an anode and the copper acts as a cathode. In this way, any 'stray' electrical currents are effectively concentrated at the anode, which becomes sacrificial and may be renewed from time to time, more cheaply and certainly more conveniently than the copper calorifier.

For large constructions likely to be under mains pressure, the cost of the copper becomes prohibitive so that any alternative cost-effective sol-ution will always be considered. It was at one time common practice to fabricate calorifiers using steel that had been galvanised after manufac-ture, but because of the seemingly random effects of galvanic action causing pitting, this practice has now become less frequent. Instead, for those parts of the calorifier that are in contact with the secondary water, some manufacturers are now using steel coated with an enamel glaze (glass lining). This makes the shell effectively corrosion resistant as well as sim-plifying the cleaning out of the precipitated scale.

Conventional calorifiers, which are mounted vertically, have less space available for the heating elements than those mounted horizontally; so that, for the same thermal conditions and equal storage volumes, the re-covery rate for vertical calorifiers is usually lower. In an attempt to save on the thermal losses associated with the storage of large volumes of hot water, the present tendency in calorifier design is to provide for a high recovery rate with a minimum storage volume, and these units are re-ferred to as 'high output semi-storage calorifiers'. Such calorifiers are potentially space saving, have lower thermal losses and are equipped with a circulator for temperature averaging to combat the possible growth of *Legionella* bacteria inside the cooler parts of the calorifier shell. This aspect is discussed in Chapter 2, subsection 2.19.6.

In one development with which the author was involved, it was found to be convenient to use two packaged boiler and calorifier sets, with each boiler having a calorifier mounted horizontally above it in the same casing. Figure 7.16 gives an annotated cross-section of the arrangement. The shell of the calorifier was made of steel and contained the primary water, which was circulated to and from the steel shell boiler by means of a pump, having a cast iron casing.

The secondary water was separated from the primary water by means of a large-diameter stainless steel cylinder, which was mounted inside

Key

1. Forced draught oil or gas burner
2. Hinged door with sight glass
3. Path of combustion gases
4. Primary flow to calorifier and heating
5. Tapping points for instruments
6. Cleaning door in stainless steel shell
7. Pocket for thermostat
8. Insulated jacket
9. Domestic hot water secondary flow
10. Primary calorifier vent
11. Domestic hot water secondary return
12. Domestic hot water cold feed
13. Pump for primary flow to calorifier
14. Primary boiler vent
15. Flow to space heating
16. Primary return from calorifier
17. Return from space heating
18. Outlet to flue
19. Pressure relief door
20. Drain

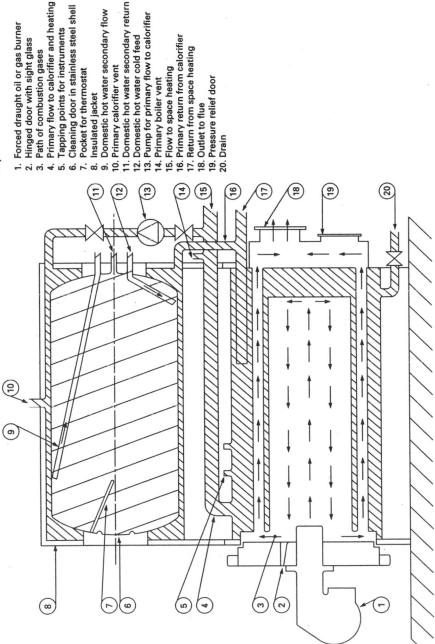

Figure 7.16 Combination boiler and calorifier *(after Hoval Farrar Ltd, Nottinghamshire, UK)*

the shell of the calorifier. This had a cleaning door fitted to one end and the heated water was then circulated around the system by means of a second pump made of non-ferrous materials. (If a pump made of cast iron had been used, the oxygen present in the water would have combined with the iron to form iron oxide (rust), which would have caused discoloration of the domestic hot water.)

7.6.2   Review of 'point of use' heaters

Not all commercial or industrial installations require large quantities of hot water and, in many cases, domestic hot water is needed at positions that are remote from any central source of heat supply. In such cases, it is obviously uneconomic, both in running and in capital costs, to circulate heated water from a calorifier to every distant outlet. In addition, many commercial buildings are occupied by more than one tenant, each of whom (it must be assumed) has a strong aversion to paying for the hot water used by other tenants.

Point of use water heaters operate using gas or electricity as their source of energy and in the case of electricity, depending on the type of heater, the supply may be either on- or off-peak. The water heaters may be broadly classified as being either of the 'instantaneous' type, or of the storage type.

7.6.3   Instantaneous outlet-controlled gas-fired water heaters

The term 'instantaneous' is used to describe those gas water heaters that are connected directly to the cold water main and produce heated water, available instantaneously at a valved outlet. The heaters operate at a pressure above atmospheric and may therefore be described as being of the 'pressure type'. Appliances are available to supply single or multi-point outlets, and this type of heater appeals to those who are prepared to pay for the use of energy only at the time when it is needed, so saving on the capital cost and high thermal losses associated with an unlagged (or indifferently lagged) hot water storage cylinder, boiler and pipes.

Modern multi-point appliances are room sealed, taking the air required for combustion from outside the building and discharging the products of combustion back to the outside air, via a balanced flue. It has to be noted, **and it cannot be stressed too highly**, that it is essential for the flueing arrangements to be kept in perfect working order at all times, to avoid consequences that too often in the past have proved to be lethal to the users.

The maximum volume flow rate of hot water expected from such a heater is of the order of 0.13 litre/s and, assuming that a bath of 120 litres requires approximately half this volume in hot water, it will take about 8 minutes to draw a bath. Clearly, the production of hot water can be described as instantaneous, but the time required to draw a bath cannot.

The principle of operation is similar to that of the combination boiler described in the previous section. The unit has one stage of a two-stage gas valve kept open by the continued operation of a pilot flame. The water, which is outlet controlled, flows through a venturi and produces a pressure differential across a diaphragm, which opens the second stage of the gas valve, allowing the main burners to ignite.

The products of combustion are then allowed to pass across a heat exchanger, consisting of tubes and fins, which absorb the energy as the water is passing through the appliance.

The unit has virtually no storage capacity and does not require an open vent. When the flow of water stops, the supply of gas to the main burner is cut off and any further expansion of the water remaining in the appliance due to the effect of residual heat is accommodated by the cold supply pipe, which should not include a non-return valve, or a valve having a loose jumper.

7.6.4 Gas-fired circulators

A circulator takes the form of a small boiler, having inlet and outlet connections that are plumbed into the side of an adjacent hot water storage cylinder. The combined unit, which may be direct, or indirect, is small enough to be fitted under a draining board, or in a cupboard; figure 7.17 gives the details.

When the unit is in a confined space, sufficient fixed ventilation must be provided to ensure that the gas burner has a plentiful supply of fresh air for combustion purposes and the flueing arrangements must comply with the local Building Regulations and the manufacturer's recommendations.

When the circulator is connected to a direct cylinder, it is possible to fit an economy valve, which routes the cool water returning to the appliance either from near the top of the cylinder, or from the bottom. In this way it is possible to heat either the water in the top part of the cylinder, or the whole of the contents.

7.6.5 Direct gas-fired vented storage water heaters

In the UK, these appliances tend to be used in commercial or industrial premises and they vary in capacity from 100 litres with a recovery rate of about 150 litre/hour to 2700 litres with a recovery rate in the region of 4500 litre/hour; figure 7.18(a) shows the cross-section of a typical heater.

The units are suitable for use in areas where the water is soft; if they are used in hard water areas, then the scale formed from the precipitation of dissolved salts must be removed on a regular basis. Those parts of the heater that are in contact with water are made from steel that has been fused with a corrosion resistant enamel/glass lining.

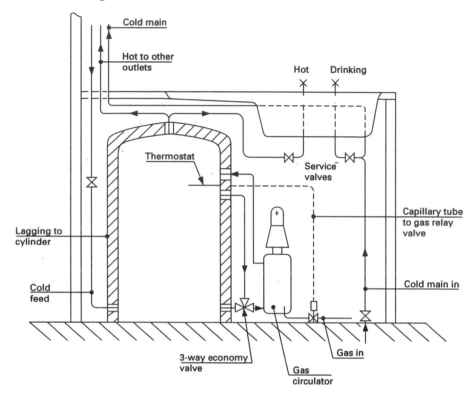

Figure 7.17 Gas circulator and cylinder under draining board

7.6.6 Direct gas-fired unvented storage water heaters

The recent changes in the water bye-laws have made it possible, provided that certain precautions are taken, to connect the cold water main
directly into unvented water heating apparatus; figure 7.18(b) shows a
typical arrangement. The result is to provide a self-contained water heating system which may be sited in any convenient position without the
need to provide a cold water storage cistern.

7.6.7 Instantaneous inlet-controlled electric water heaters

These heaters have been developed for use as single hot water outlets,
either providing a spray, angled directly downwards from the front face
of the control cabinet, or a single spray outlet from a swivel arm, which
can reach across two adjacent washbasins. Instantaneous electric water
heaters are physically very much smaller than instantaneous gas heaters
and do not, of course, require a flue. This type of heater may be classified further as being of the 'non-pressure type; this refers to the fact that
at all times the water inside the heater is subjected only to atmospheric
pressure.

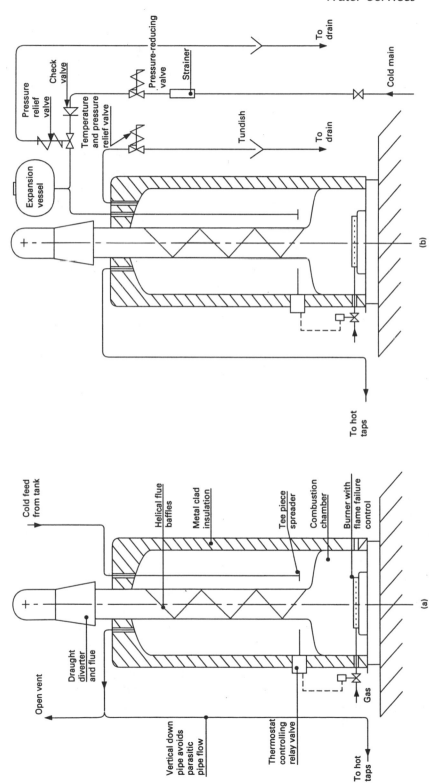

Figure 7.18 (a) Direct gas-fired vented storage water heater. (b) Direct gas-fired unvented storage water heater

The angled spray version is ideally suited for use in public places. It is operated by means of a pushbutton switch, which opens a solenoid-controlled water valve at the inlet and connects the 3 kW heating element to the power supply for a period of 25 seconds. The water flow rate is then adjusted automatically, to give a preset outlet temperature, making the operation of the appliance independent of normal variations in mains pressure.

When the solenoid valve is closed, the expansion of the contents of the heater due to the effect of residual heat takes place in the direction of the outlet pipe, which must remain open at all times. The water is heated inside a copper cylinder, which is fitted with a pressure relief valve and a drain plug, the latter for use in case freezing conditions should occur. A thermal cut-out operates if the water temperature reaches 54°C.

The swivel outlet versions, having capacities of 3 kW and 6 kW, are suitable for use in domestic and commercial premises, and both the period and temperature of operation are user controlled.

7.6.8 *Instantaneous inlet-controlled electric showers*

These appliances are also of the non-pressure type and for a fixed power input, the outlet temperature of the water is controlled by opening or closing an inlet valve, so that a high volume flow rate results in a small rise in temperature, and a low volume flow rate produces a larger rise in temperature. Some shower units provide an option to vary the power input in increments of about 2 kW, from a 4 kW setting for summer use, to a 9.5 kW setting for use in the winter. In some models, further seasonal adjustment, for differing inlet water temperatures, is available at the shower head, by removing, or inserting, plain or serrated flow rings. Figure 7.19(a) gives some internal details of a typical shower control unit.

A printed circuit board incorporating dedicated microprocessor chips is sometimes referred to as an 'application specific integrated circuit' (ASIC) and the incorporation of such circuits into electric shower appliances has made it possible to offer a series of control features that operate automatically. For instance, on switching off the heater, the solenoid valve will remain open for about 8 seconds, allowing residual heat to be removed, so that subsequent immediate use of the shower will not result in a discharge of very hot water on to an unsuspecting user. Similar circuitry can also provide visual and aural indications of the operational state of the appliance. Touch-sensitive switches can be used to change the power input to the heat exchanger and to adjust the volume flow rate, using a motorised inlet valve. A pressure-sensitive switch cuts off the power to the heating elements and signals a low-pressure condition at the water inlet, which could be due to other appliances being used simultaneously with the shower unit.

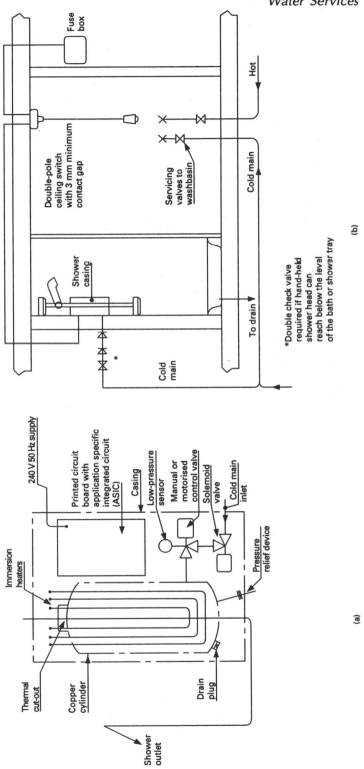

Figure 7.19 (a) Internal details of an electric shower unit. (b) Installation details of electric shower unit

At the outlet from the shell of the heat exchanger, a self-resetting safety cut-out switch is provided which, when the unit is in operational mode, limits the output water temperature to a safe preset level. A separate piped outlet from the shell incorporates a pressure relief disc, which will rupture if the internal pressure becomes higher than normal. This could occur if the spray head becomes blocked, if the flexible hose becomes blocked or kinked, or if the water inside the shower has frozen. Under freezing conditions, an attempt to use the shower by switching on the power would produce a very rapid rise in pressure inside the heat exchanger, presenting a potentionally dangerous situation to the user; accordingly, the manu-facturer of this type of shower (Triton Plc. of Nuneaton) has included in its trade literature a statement that 'the shower must not be positioned where it will be subjected to such (freezing) conditions'.

If in the case of a shower already installed it is known that freezing is likely to occur, it is important that the appliance be isolated electrically and the shell of the heat exchanger be drained, so preventing the forma-tion of ice. Subsequent reuse of the shower should only be attempted after cold water has been allowed to refill the heat exchanger, when the unit may then be reconnected to the electrical supply. Figure 7.19(b) gives some installation details.

7.6.9 *Quasi-multi-point instantaneous electric water heater*

At least one manufacturer produces an instantaneous electric water heater for connection to a cold main, which may be used with the outlets either open or closed. A closed outlet means that the water inside the heater is subjected to a pressure that is in excess of atmospheric pressure; in this case, the appliance can be described as being of the 'pressure type'.

The internal copper cylinder is constructed to withstand a working pressure of 7 bar and can be drained in case freezing should occur. The heating elements are connected to the supply by means of a reed switch, which is activated by the water flowing through the appliance. In case of mal-function, a safety cut-out switch is preset to operate at 88°C. The opera-tion of the unit is similar to that of the instantaneous electric shower unit previously described, except that a preset pressure relief valve (which is mandatory) is set to operate at a pressure of 10 bar, discharging via an open tundish to a drain sited where any outflow will be free of frost and will be noticed. The appliance may be connected to both a single wash-basin and a spray tap/shower. However it is not intended that both out-lets shall be used at the same time.

7.6.10 *Inlet-controlled electric storage water heaters*

These appliances may be described as being of the 'non-pressure type'; they are intended to be fitted close to the point of use and, provided that the storage capacity is less than 15 litres, they may be connected to the cold water main, see figure 7.20(a). The swivel outlet, which must always remain open, can be used to supply either one of two adjacent wash-basins, or a sink.

By opening the inlet valve, cold water enters the cistern from the bottom, displacing the heated water, which is then discharged from the open supply pipe at the point of use. On heating up, the water expands and the volume of expansion is displaced from the cistern via the open outlet. During this process, the outlet 'drips' and it is natural for the user to try and prevent this, by closing the inlet valve more tightly. Clearly this action can have no effect on the rate of drip but, in practice, it is found that the washer on the seat of the inlet valve has to be renewed more often than those of other valves.

An interesting development by IMI Santon Ltd of Newport, Gwent, UK, has been to fit a combined shut-off and diverting valve, in place of the inlet control valve, which gives the option of mixing the incoming cold water with the outgoing hot water, providing a tempered outflow. In this particular model, the storage vessel is made from heat-stabilised poly-propylene, instead of the more usual copper.

The 3 kW immersion heater can raise 2 litres of cold water (enough for handwashing) through 50°C in about 2.5 minutes. The units are insulated and finished to a high standard, using metal-clad polyurethane foam, which minimises the frequency of 'topping up' required by the thermostatically controlled immersion heater. A back-up 'safety' thermal cut-out is wired in series with the thermostat, in case of malfunction.

A similar appliance is produced for fitting underneath a sink, where it may be connected to the cold main, via a vented tap or vented kitchen mixer unit, by using an inlet valve having an extended spindle, see figures 7.20(b) and (c).

7.6.11 Outlet-controlled electric storage water heaters

Spray taps require a smaller volume flow rate of water than conventional taps and, as a general rule, it is found that a power input of 1 kW is sufficient to heat the water used by one spray tap.

In an installation consisting of six spray taps, it is possible to fit a 6 kW heater having a capacity of 15 litres, which will prove to be satisfactory in most washrooms. Figure 7.21(a) gives some installation details of the 'pressure type' heater specified. Like the unit in the previous section, the outlet water temperature is controlled by an adjustable thermostat, wired in series with an adjustable upper limit thermal cut-out switch, which provides additional safety.

In some installations where the demand for hot water is heavy, it is an obvious advantage to store the water at a high temperature and to deliver it at a lower temperature, so making the best use of the storage capacity available. It is known that the dissolved salts in water begin to precipitate

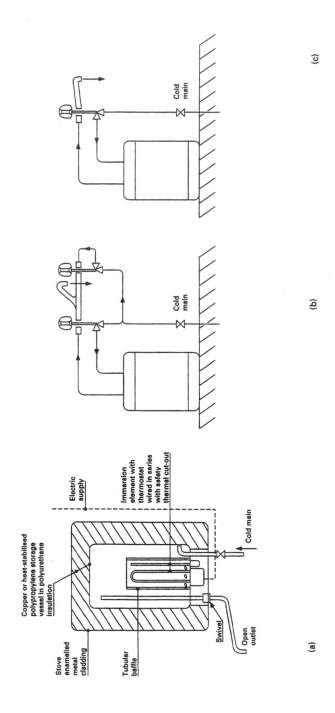

Figure 7.20 (a) Small electric oversink inlet-controlled non-pressure type storage water heater. (b) Small electric undersink inlet-controlled non-pressure type storage water heater with mixer tap assembly. (c) Small electric undersink inlet-controlled non-pressure type storage water heater with single tap assembly

(a)

(b)

(c)

Copper or heat-stabilised polypropylene storage vessel in polyurethane insulation

Stove enamelled metal cladding

Tubular baffle

Electric supply

Immersion element with thermostat wired in series with safety thermal cut-out

Cold main

Swivel

Open outlet

Cold main

Cold main

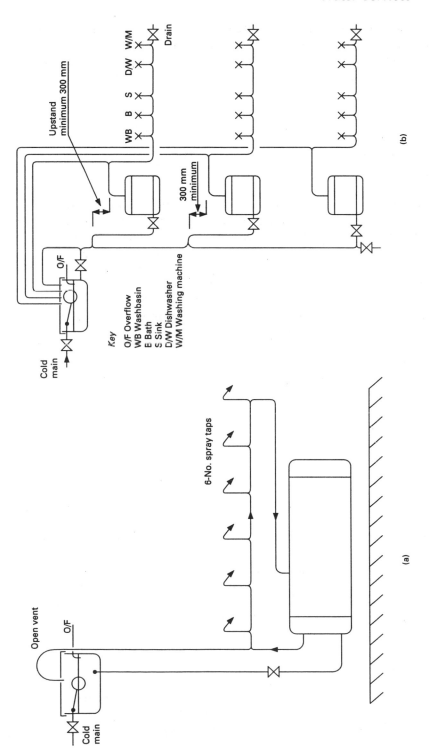

Figure 7.21 (a) 15 litre, 6 kW electric undersink outlet-controlled pressure type storage water heater. (b) Method of connecting storage heaters to avoid flow of hot water from one floor to another for a cistern-fed vented system

from the solution at temperatures that vary according to the hardness of the water and, in an attempt to help cut down on maintenance, IMI Santon Ltd has made some general recommendations in its trade literature for suitable storage temperatures, which are broadly linked to the hardness of water. These are reproduced below:

Recommended temperatures for thermostat settings
Soft (or softened) water	82°C
Medium hard water	71°C
Hard water	66°C

In its range of small storage water heaters, the company produces a range of appliances of the pressure type, which include the facility to store water at a higher temperature and to deliver it to the taps at an adjustable lower value. The units range from 8 litres capacity with a 3 kW element to 15 litres capacity with a 6 kW element, and when the mains pressure is less than 6 bar, they may be connected directly to the mains. When the mains pressure is greater than 6 bar a standard approved connection kit is available, consisting of the following units connected in order from the mains to the appliance:

- in-line strainer
- pressure-reducing valve
- cold water draw-off tee
- non-return valve
- expansion vessel
- expansion relief valve.

These small storage units, and units having capacities up to 455 litres for domestic use and up to 3000 litres for industrial use, are available from manufacturers and are notable for the high standard of metal-clad thermal insulation provided. They can be obtained in circular, elliptical or rectangular form in all sizes, some being small enough to fit underneath a draining board (udb), or in a cupboard, while others may be small in diameter and tall, to provide for good stratification of the hot and cold water when the unit is connected to an off-peak supply.

The immersion heaters can be placed so that the system can provide just enough hot water for occasional hand washing (the 'economy' mode of operation), or for the full capacity of the cylinder to be available for more extensive use.

Such units may be used with advantage in multi-storey office buildings, provided the static pressure on the inside of every cylinder is kept below the specified operating pressure for the type of unit being used. Figure 7.21(b) shows the method of connecting a number of cistern-fed and vented hot water storage cylinders so as to prevent users on a lower floor of a block of flats from obtaining their hot water at the expense of those living above them. Figure 7.22(a) gives details of the connections required for a cistern-fed unvented storage water heater, and figures 7.22(b) and (c) of the connections required for a supply pipe (mains)-fed and unvented system, when the pressure in the cold main is greater than 6 bar and when it is smaller than 6 bar. Bye-laws 90 and 91 of the *Water Supply Bye-laws Guide* give the relevant details.

7.6.12 *Outlet-controlled combination electric storage water heaters*

Such appliances consist of a hot water storage vessel surmounted by an internally plumbed feed and expansion cistern, all enclosed in a single stove enamelled metal casing containing polyurethane foam insulation. The unit provides a self-contained hot water supply, which only requires connection to a cold main and an electrical supply. The head of water available from such a unit depends on the vertical distance between the free surface of the water in the feed and expansion cistern and the highest outlet tap, and it is clear that when the outlets are on the same floor as the unit, the available head is limited. In a multi-storey development, in order to overcome this drawback, it may be possible to position the unit on the floor immediately above the hot water outlets being served.

7.7 Methods of distribution of water supplies to high and low rise buildings

In urban areas the pressure in the water main is normally sufficient to supply water outlets satisfactorily, up to and including the sixth floor of a development, and if in such cases it is required to supply to higher levels, then the water must be pumped. The pumps used are of the multi-stage centrifugal type which, because of the stalling characteristic of the impeller vanes, will operate safely when directed into a closed hydraulic circuit; whereas a positive displacement pump will generate pressures that are high enough to stall the pump motor.

7.7.1 *System for high rise building, volume controlled, using constant-speed pumps*

The pressure of water behind any tap used for drinking or washing purposes must be limited to about 3 bar so that, when it is opened, the flow of water is not so vigorous that both hands and feet are wetted simultaneously. The pressure of 3 bar is equivalent to a head of water of about 30 m, or 10 storeys in a block of flats, at which vertical intervals, 'break' tanks are positioned (see figure 7.23). Note that in this case the 'lift' of the pumps extends to 60 m (or 20 storeys), before other pump sets are installed to deal with the next 20 storeys. The number of storeys contained in a 'lift' must obviously depend on the building and the type of pump sets that are available.

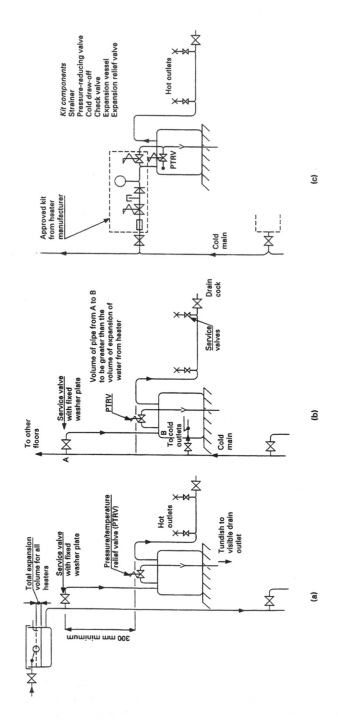

Figure 7.22 (a) Cistern-fed unvented water heater. (b) Supply-pipe-fed unvented water heater, pressure < 6 bar. (c) Supply-pipe-fed unvented water heater, pressure > 6 bar

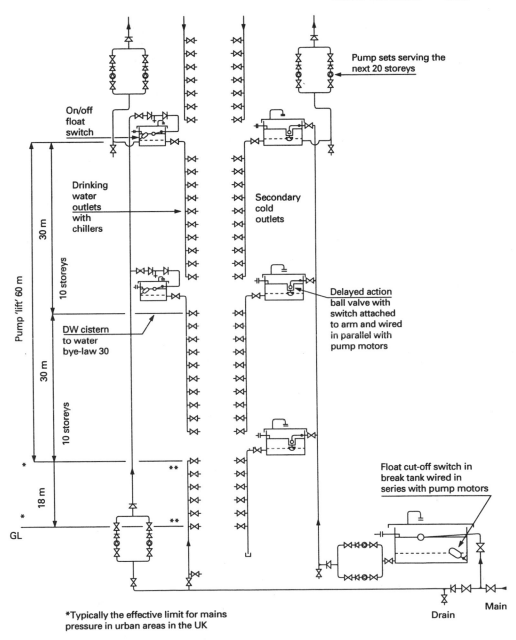

On/off float switch

Drinking water outlets with chillers

Pump 'lift' 60 m

30 m

10 storeys

DW cistern to water bye-law 30

30 m

10 storeys

18 m

GL

*

*

**

**

Pump sets serving the next 20 storeys

Secondary cold outlets

Delayed action ball valve with switch attached to arm and wired in parallel with pump motors

Float cut-off switch in break tank wired in series with pump motors

Main

Drain

*Typically the effective limit for mains pressure in urban areas in the UK

**Stop valves and double check valves with an upstand may be required at each mains pressure outlet

Figure 7.23 Boosted water services for a high rise building, volume-controlled using constant speed pumps

The system is said to be 'volume controlled' because the change in the volume of water in the system is used to activate the motor of the duty pump. In the case illustrated in figure 7.23, the level of water in the secondary cold break tanks is allowed to fall to the positions indicated by the lower broken lines, before any one of the switches attached to the delayed action ball valves energise the constant-speed duty pump motors. In this way, the number of starts are minimised, resulting in less wear on the contacts of the electrical control gear.

The break tanks used to store drinking water have a relatively small capacity of between 5 and 7 litres per dwelling and must be manufactured and fitted in accordance with water bye-law 30. They should be sited in cool positions and be insulated against heat and frost. Each tank must be proof against dust and all insects, and to this end, both the vent and the overflow pipes incorporate filters. The access cover must be sealed and screwed down. The drinking water break tanks shown in figure 7.23 are fitted with float switches, operating in the same way as the switches attached to the arms of the delayed action ball valves fitted to the secondary cold water break tanks.

It is common practice to apportion the cold water storage capacity required for the building so that two-thirds of it is contained in the lower break tank, and the remainder is contained in the smaller break tanks (and pipework) situated in the rest of the building.

7.7.2   System for low rise building, volume controlled, using constant-speed pumps

For buildings that do not require a repeated number of pumped 'lifts', but are still tall enough to require the water services to be boosted, an alternative method of supply that is mainly volume controlled, using constant-speed pumps, is shown in figure 7.24.

Water for drinking and other purposes is pumped via a break tank (situated at low level) to the roof, where it is taken to the delayed action ball valve of a cold water storage tank, via an enlarged part of the pumped main, referred to as a drinking water header. The volume of this section of the main is determined allowing for a capacity of between 5 and 7 litres per dwelling.

The drinking water is supplied to those outlets that are above the limit that mains pressure can reach by means of the standing head of water provided by the drinking water header. This is made possible by the fall of the float of the automatic air vent, allowing air into the system whenever water is being used. This process continues until pipe line float switch 1 falls, energising the pump. The pump will then continue to operate until pipe line pressure switch 2 opens. Figure 7.24 shows the switching functions, with the float switch in the break tank able to move to the open position to switch off the pump, so guarding against the possible

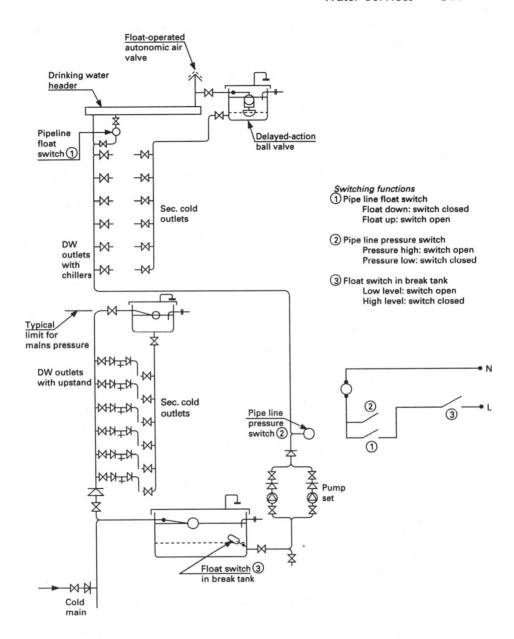

Figure 7.24 Boosted water services for a low rise building, volume-controlled using constant speed pumps

shortage of water in the low level break tank, and allowing the impeller of the pump to run 'dry'.

7.7.3 System for low rise building, volume controlled

This layout, using a pneumatic accumulator drawing from a basement break tank, with constant speed pumps and roof tank(s), is shown in figure 7.25. It is suitable for an office block or a block of flats, and involves the use of a cylindrical pressure vessel containing air and water. The cylinder acts as an accumulator, keeping the system under pressure and providing the energy required to supply each outlet with a suitable volume flow rate of water. The pumps become active only when the level of water in the cylinder requires topping up.

The air in the cylinder is absorbed by the water, making it occasionally necessary for it too to be 'topped up', and this is accomplished by using a small air compressor. The cylinder, with mountings, air compressor, water pumps, valves and associated controls, is assembled into a package referred to as a 'cold water booster set'. The sets are free-standing and, among the various options available, the steel accumulator cylinder may be mounted vertically, or horizontally. Also, in order to reduce the volume required for the accumulator, the manufacturer (Messrs. G.C. Pillinger and Co. (Engineers) Ltd of Croydon, Surrey) claims that increasing the number of multi-stage pumps used will result in a saving in both space and capital cost.

Reference to figure 7.25 will show that the first six floors of the building are served with drinking water outlets, which are taken directly from the cold main, while the upper six floors are supplied with drinking water outlets taken directly from the pressurised distribution pipe. This pipe continues to rise and feeds the delayed action ball valve, fitted to the cold water storage tank sited on the roof, from which the secondary cold outlets for all twelve storeys are taken. Users of the secondary cold outlets on the lower six floors will be inconvenienced with a surfeit of pressure (head of water), which may be alleviated by fitting a pressure-reducing valve in the position shown, by fitting a break tank, or by fitting orifice plates to the take-off points at each floor.

7.7.4 As 7.7.3 but connected via a by-pass to the main

This arrangement is suitable for a block of flats and is shown in figure 7.26. Using a booster set having pumps running at constant speed, the secondary cold water is pumped from a break tank sited in the basement to a storage tank (or tanks) on the roof of the building and distributed to each flat as shown. In this case, the drinking water is taken directly from the cold main using a constant-speed booster set and an air/water accumulator, which pressurises the distribution pipes serving each

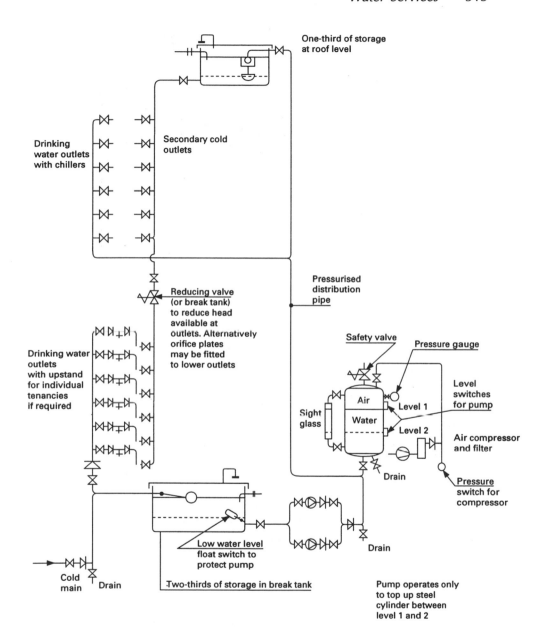

Figure 7.25 Boosted water services for a low rise building, volume-controlled using a pneumatic accumulator

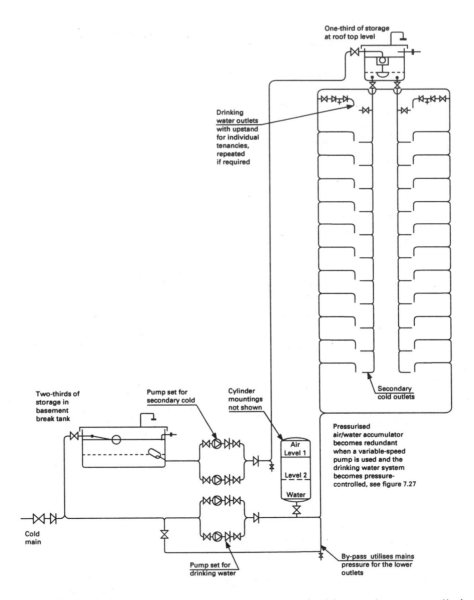

One-third of storage
at roof top level

Drinking
water outlets
with upstand
for individual
tenancies,
repeated
if required

Two-thirds of
storage in
basement
break tank

Pump set for
secondary cold

Cylinder
mountings
not shown

Secondary
cold outlets

Air
Level 1

Level 2

Water

Pressurised
air/water accumulator
becomes redundant
when a variable-speed
pump is used and the
drinking water system
becomes pressure-
controlled, see figure 7.27

Cold
main

Pump set for
drinking water

By-pass utilises mains
pressure for the lower
outlets

Figure 7.26 Boosted water services for a low rise building, volume-controlled
using a pneumatic accumulator connected to the cold main

flat. Again, in order to avoid the effects of excessive head at the lower
outlets, they should be fitted with flow restrictors. The provision of the
by-pass to the drinking water booster set enables mains pressure to be
utilised in serving the lower outlets.

7.7.5 Use of variable-speed pumps in a pressure-controlled system

The constant ideal pressure is associated with different speeds of the impeller, so that by providing a suitable speed control system, it should be possible to produce a booster set that gives a (near) constant-pressure output, which would be equivalent to the output from the traditional air/water accumulator, so making the latter redundant. As with all electronic control systems, it is possible to introduce many other variations in the way the booster set can operate; these include the automatic changeover from one duty pump to another and, depending on the protocol of the existing building management system (bms), all the relevant operational functions may be monitored and controlled at a remote station, if required.

7.7.6 Provision of domestic hot water and space heating to a low rise block of flats

Figure 7.27 shows one way of achieving the above, when the bulk of the energy converting equipment is centralised.

The domestic hot water is provided for each 'lift' by means of a calorifier as shown. Both calorifiers are heated using an unvented domestic hot water boiler, having a circulating pump fitted in the return pipe. Readers will recognise the system as being eminently suited to operation using thermosiphonic flow. However, the circulator is included so that smaller pipe sizes may be used, consistent with obtaining an enhanced output from the heat exchangers in the calorifiers, as a result of employing a pumped flow. For the first lift, the head of water supported by the boiler is about 21 m, so that a cast iron boiler would be suitable (see Chapter 4, subsection 4.4.1).

The upper cold water storage tank is supplied with water from a boosting unit, which is not shown on the figure. Water from each tank is taken via a cold feed pipe to the secondary side of the appropriate calorifier, where it is heated, accumulating in the top of the vessel. This hot water is then pumped around the secondary circuit, to which all the hot water outlets are connected, and the circulator is sized to make good the thermal loss from this run of pipe (which must be lagged). Some of the older installations incorporate a header which, during periods of heavy usage, helps to ensure that the upper outlets do not run dry. However, such headers may be omitted provided the cold feed to the calorifier is sized generously. Also note that, whenever a hot outlet is opened on the ring main, the water will be drawn from two directions, making it possible to specify a relatively smaller diameter for the pipe used in the secondary circuit, compared with when the flow is unidirectional.

The water for space heating purposes is heated in an unvented boiler and taken by vertical flow and return pipes to valved connections in each flat. The size of the expansion vessel depends on the volumetric contents

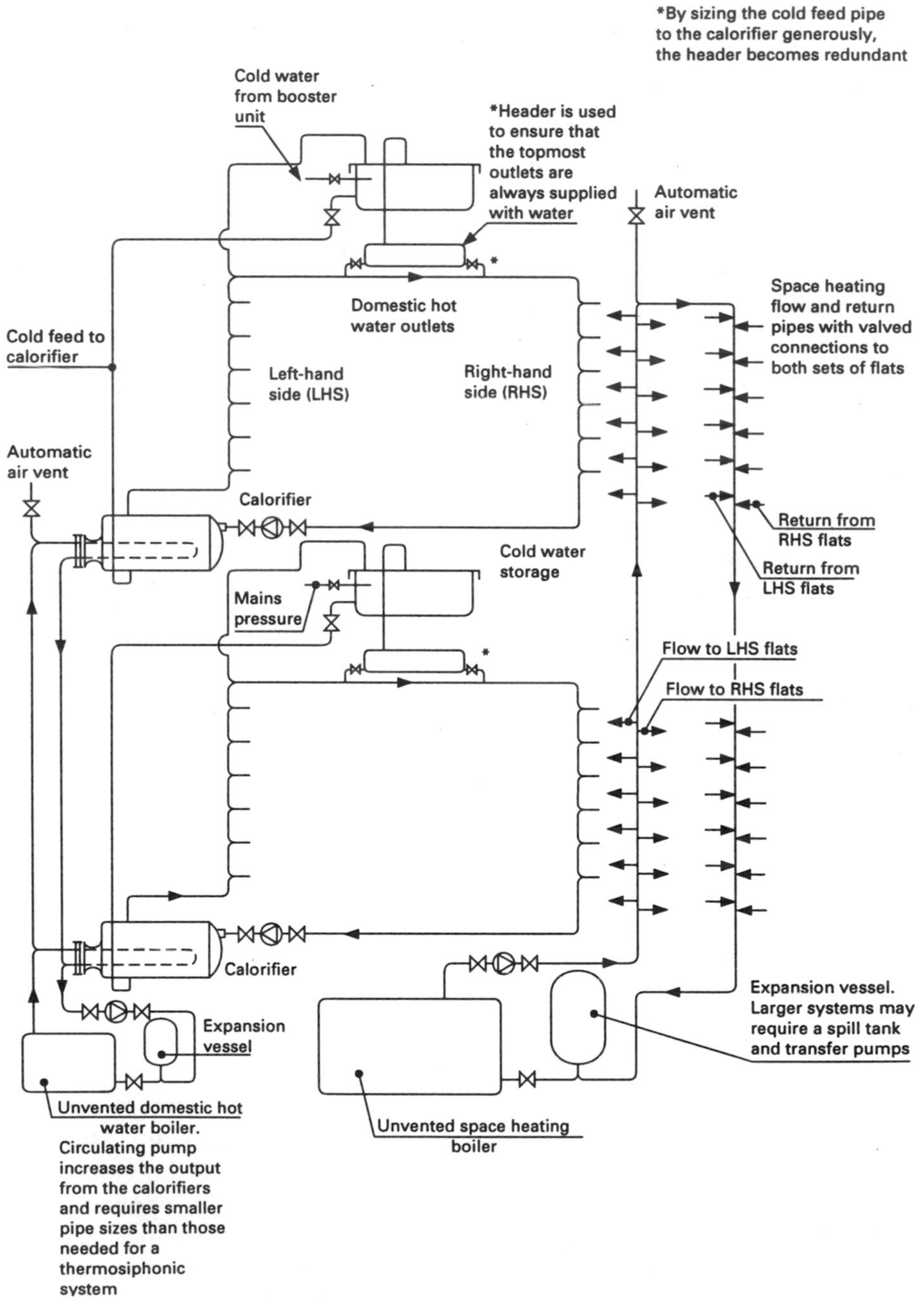

*By sizing the cold feed pipe
to the calorifier generously,
the header becomes redundant

Cold water
from booster
unit

*Header is used
to ensure that
the topmost
outlets are
always supplied
with water

Automatic
air vent

Domestic hot
water outlets

Space heating
flow and return
pipes with valved
connections to
both sets of flats

Cold feed to
calorifier

Left-hand
side (LHS)

Right-hand
side (RHS)

Automatic
air vent

Calorifier

Cold water
storage

Return from
RHS flats

Return from
LHS flats

Mains
pressure

Flow to LHS flats

Flow to RHS flats

Calorifier

Expansion
vessel

Expansion vessel.
Larger systems may
require a spill tank
and transfer pumps

Unvented domestic hot
water boiler.
Circulating pump
increases the output
from the calorifiers
and requires smaller
pipe sizes than those
needed for a
thermosiphonic
system

Unvented space heating
boiler

Figure 7.27   Provision of centralised domestic hot water and space heating for
a low rise block of flats

of the system and this must be ascertained. For very large systems, it will be necessary to specify that a spill tank with transfer pumps is used in conjunction with the simple expansion vessel.

As with the domestic hot water circuit, the system is well suited to being operated using thermosiphonic flow, however in the interests of obtaining a quicker thermal response and in being able to use smaller pipe sizes, the circulator is included. Note that the pump does not have to 'lift' the water from the bottom of the system to the top, as the system is filled with water before the pump is switched on. Figure 7.27 shows a building having 12 storeys and a roof top tank, so that the overall height of the building is about 39 m, which is about the limit for a conventional cast iron boiler, and in this case the use of a steel boiler is to be preferred.

A more cost-effective and fuel-efficient way of providing domestic hot water and space heating for the system shown in figure 7.27 is to omit the separate boiler used for providing domestic hot water and to connect the domestic hot water risers indirectly to the space heating boiler. Every space heating boiler operates in an 'on/off' cycle, and it is during the 'off' period that boiler power is available to support the domestic hot water requirements, which are usually a small fraction of the space heating load. It is advantageous for the space heating load to be supplied from two or more smaller boilers connected in parallel, so that during the summer period, when the space heating facility is not required, or on occasions when only minimal space heating is needed, it becomes possible to operate with fewer boilers. It also helps if the heat exchangers contained in the calorifiers have a high heat transfer rate, so that the output of the space heating boiler(s) can be accommodated. Such a system will obviously reduce the 'downtime' of the boilers, producing fewer starts, with fewer attendant flue losses.

7.8 Technology of pipe flow

For a pipe which does not flow full, the pressure acting at each of its ends is the same and it behaves as an open channel. For a pipe which does flow full, the contents are subjected to pressures which are different at each end and the analysis required to deal with this is different from the previous case. In the sections following, the analysis refers to pipes which are flowing full, and such a study is required in order that the pipes contained in a system may be sized to carry a working substance at a volume flow rate commensurate with the design function of that system.

7.8.1 Definition of a fluid

A fluid is that substance that deforms continuously under shear stress, however small.

The use of the word 'fluid' incorporates all liquids and gases. Liquids are regarded as being incompressible and gases as readily compressible. The theory and practice of pipe sizing apply mainly to incompressible fluids (liquids), but may also be applied to compressible fluids (gases) provided that, during the process, the density of the gas remains reasonably constant and there is no change in phase.

It is beyond the remit of this book to explain in detail all the developments involved in the study of the flow of fluids in pipes, however a brief review of some of the more important early developments that were precursory to our present day knowledge follows. Students requiring to study the subject of fluid mechanics will find a full account in [7.7] and [7.8].

7.8.2 Review of earlier findings

The Hagen–Poiseuille law, formulated in the 1880s, showed that for small-scale apparatus, when fluid of a given viscosity flows in a given pipe, the head loss due to friction (h_f) is proportional to the mean velocity of flow (u).

Subsequently, Darcy showed that for large-scale apparatus, when fluid of a given viscosity flows in a given pipe, the head loss due to friction (h_f) is proportional to the square of the mean velocity of flow (u^2). His equation linking these parameters also included a 'friction factor' (f).

In 1886, Reynolds showed that for all fluids there were two kinds of flow, laminar and turbulent, separated by a region of critical flow. He also showed that for any given flow rate, by evaluating the ratio of inertia force to viscous force, the type of flow could be identified. This ratio is referred to as Reynolds number and it is given by:

$$Re = \frac{ud\rho}{\mu}$$

where Re = Reynolds number (dimensionless)
   u = mean velocity of the fluid (m/s)
   d = internal diameter of the pipe (m)
   ρ = density of the fluid (kg/m^3)
   μ = absolute viscosity of the fluid (Pa s).
When Re < 2100 laminar flow exists
When Re > 2500 turbulent flow exists
When Re < 2500 > 2100 the flow regime is in the critical zone.

It should be noted that these values of Reynolds number (2100 and 2500) may vary, depending on whether or not the system is subjected to external vibrations at the time of testing.

7.8.3 Laminar flow

Laminar (or streamline) flow occurs when the fluid flowing in a pipe moves as if it were constrained within concentric tubes along its length or, if the fluid is contained in a duct, it moves as if it were constrained in parallel layers along its length.

For laminar flow in a pipe, the head loss due to friction is dependent only to a small extent on the roughness of the boundary surfaces, but it is affected greatly by the viscosity of the fluid, with shear stresses appearing across adjacent concentric tubes; it can be shown that, for laminar flow of a fluid, of given viscosity flowing in a given pipe, the head loss due to friction (h_f) is proportional to the mean velocity of flow (u).

Laminar flow can occur whenever fluid movement is brought about by the change in density of rising and falling columns of fluid, as a result of differences in temperature; this phenomenon is referred to as thermosiphon flow. One example of this occurs when water circulates between a boiler and a storage cistern, without the use of a pump. Another example occurs inside a radiator that is emitting heat (which may or may not be on a pumped circuit). However, not all examples of thermosiphonic flow exhibit the regime of laminar flow exclusively, and values of Reynolds number greater than 3000 are frequently found when thermosiphonic flow conditions are operating.

Another example (also involving changes in density) occurs in the filtration process using reverse osmosis, where with the application of a small pressure difference, liquid of a high density passes through a semipermeable membrane into a fluid having a lower density. Laminar flow may also be found to exist within oil films used for lubrication purposes in bearings. Further examples of laminar flow occur where oil, natural gas, water and other fluids flow through porous ground strata, and also where fluids seep underneath large man-made structures.

7.8.4 Turbulent flow

Turbulent flow occurs in a pipe when the fluid moves in random eddy motions along its length. The regime of turbulent flow may follow on from that of laminar flow, owing to the fluid encountering protuberances forming part of the walls of the pipe, which are large enough and occur frequently enough to disturb the smooth longitudinal progression of the concentric (telescopic) tubes associated with laminar flow.

For turbulent flow in a pipe, the head loss due to friction is dependent on the relative roughness of the boundary surfaces and also the inertia of the fluid. The relative roughness of a pipe is typified by the ratio k/d, where k is taken to be the mean height of the protuberances along the wall of the pipe and d is the diameter. Inertia force is typified by the product of the mass of a particle of fluid and its acceleration. It can be shown that for turbulent flow of a fluid having a given viscosity and flowing in a given pipe, the head loss due to friction (h_f) is proportional to the square of the mean velocity (u^2).

For fluid having a constant viscosity flowing in a pipe under constant head, the results of all three investigators may be illustrated graphically by plotting the velocity of flow against the head loss due to friction/unit length.

In the domain of building and other engineering services, the mode of turbulent flow is the most common and it is to be found occasionally in systems using thermosiphon flow and always in any system using pumps, circulators, or fans.

7.8.5 Later experimental work

Subsequently, Colebrook and White, who carried out their tests at the National Physical Laboratory, UK, developed a formula that could be applied to commercial pipes, which linked f, k, d and Re for turbulent flow (when Re > 3000) as follows:

$$f^{-0.5} = -4\log_{10}[k/3.7d + 1.255/\text{Re}f^{0.5}]$$

Moody, an American engineer, produced a graphical solution in the form of a diagram, applicable to commercial pipes, known as the 'Moody diagram'; this appears in the 1988 edition of the CIBSE Guide.

In building services, systems in different buildings operate under conditions that are broadly similar; for instance, in temperate climates, the supply of cold water may be taken as being at a temperature of 10°C and the supply of hot water may be taken as being at 75°C. The pipes may be made of copper or plastic, of black mild steel, or of galvanised steel. The engineers who have compiled the CIBSE Guide have, for the convenience of their members, produced flow tables for each of these temperatures, giving the head loss/unit length of pipe, against the mass flow rate to be expected, for pipes of different diameters, made of different materials. The equations used for the exercise are derivatives of the Hagen–Poiseuille and Darcy laws and the Colebrook–White formula, and the values computed may therefore be used with confidence.

As an additional service, the CIBSE Guide also contains tables of correction factors (multipliers), which may be applied to a particular flow table to obtain values for water at 150°C flowing in steel pipes, or for water flowing in pipes that are encrusted with scale or rust. The CIBSE Guide also contains tables that cover the flow of air in ducts and pipes, the flow of saturated steam in steel tubes, and the flow of natural gas in steel and copper pipes.

7.9 Pipe sizing exercise for thermosiphonic flow

Pipe sizing is an exercise in combining an analytical approach with the exercise of judgement, and it furnishes a good example of the philosophy required in the solution of many of the problems to be found in engineering in general and building services in particular.

Pipes are manufactured in discrete but nominal sizes and, while staying within the tolerances specified by the appropriate British Standard, the diameters vary slightly from one production unit to another.

In practice, the required mass flow rate for a particular section of pipework may be such that the actual diameter of pipe required lies in between the nominal sizes that are manufactured, and it follows that the practical result will at times not be that which the design calculations demand. Most of the imbalance between the design flow rate and the actual flow rate may subsequently be removed with the use of regulating valves, and this process is referred to as one of 'calibration', or 'balancing'. The search for extreme accuracy in pipe sizing is not warranted and it is hoped that readers will become aware of the truth of this statement as they work through the examples given in the sections that follow.

There are many different ways of sizing pipes, and each method can be shown to be tried and trusted by its advocates. Some methods even require a prior knowledge of the pipe size, and the number and type of pipe fittings which it is proposed will be used. With hindsight born of the experience of others and his own, the author has found this approach to be unnecessary and excessively time-consuming.

As the water flows through the fittings, the head losses may be allowed for by assuming that the run of pipe is longer than it is in reality, and a new 'equivalent' length, $L_e = 1.25 \times L$, obtained. It should be noted that some designers use a multiplier of 1.3.

Most systems are larger than those used as examples in this text and, in such systems, there is more scope for indulging in cost-saving exercises aimed at using smaller-diameter pipes, or pipes made of a cheaper material. Clearly, when carried to extremes, this approach can be overdone, resulting in a system that operates indifferently. The author has always found that the best way to produce a realistically priced system of which the designer can be proud is to produce detailed drawings of the proposed scheme, in which all the equipment is specified. There should never be any areas that are left for a tenderer to 'estimate', as whenever such estimates are made, the overall cost becomes disproportionately high, making many previous economies in design nugatory.

Example 1

The hot water installation shown in figure 7.28 consists of a boiler serving an indirect cylinder sited in a single-storey building. The system has to

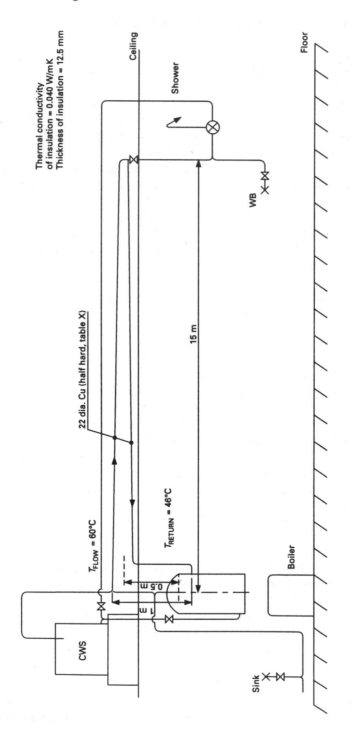

Figure 7.28 Thermosiphon or gravity flow, preventing 'dead leg'

Table 7.1 Circulating pressures for gravity flow hot water systems

Flow temperature (°C)	Circulating pressure (Pa) per metre height for the following temperature differences (flow-return) (°C)						
	10	*12*	*14*	*16*	*18*	*20*	*22*
60	47.4	56.1	64.6	72.8	80.7	88.4	95.7
65	50.3	59.7	68.8	77.6	86.2	94.6	103
70	53.1	63.1	72.8	82.3	91.5	101	109
75	55.9	66.4	76.7	86.8	96.6	106	116
80	58.5	69.6	80.5	91.1	102	112	122

Data drawn from CIBSE Guide C, *Reference data* (1986), by permission of the Chartered Institution of Building Service Engineers.

supply a shower that is remote from the source of heat, resulting in a 15 m 'dead leg', which contravenes the local water bye-laws and would be inconvenient to the user. It is proposed to install a heating loop operating under thermosiphonic conditions, consisting of a 22 mm diameter copper tube (half hard, table X), which is to be insulated to a thickness of 12.5 mm with insulation having a thermal conductivity of 0.040 W/m K.

Assuming flow and return temperatures of 60°C and 46°C, check that the mass flow rate needed to sustain thermosiphon flow will be possible. Evaluate Reynolds number and categorise the flow regime. At a temperature of 53°C, water has a density ρ of 986.7 kg/m^3 and an absolute viscosity μ of 0.521 × 10^{-3} Pa s. The internal diameter of the tube is 20.22 mm.

Solution

From table 7.1, taken from the CIBSE Guide, a flow temperature of 60°C and a temperature drop of 14 K will produce a circulating pressure CP = 64.6 Pa/m of height.

From figure 7.28, the height of the system is taken as 0.5 m (in this case a vertical distance taken between the centre line of the flow and return connections at the cylinder and that of the heating loop).

The circulating pressure available is 64.6 × 0.5 Pa, or 32.3 Pa.

The approximate length of run of pipework L = 2 × 15 + 1 m, or 31 m.

The equivalent length, allowing for the head loss as water flows through the fittings may be estimated as L_e = 1.25 × 31 m, or 38.75 m

The expected pressure loss/m run (Δp_l) is given by Δp_l = 32.3/38.75 pa/m run, or 0.83 Pa/m run.

Reference to table 7.2, taken from the CIBSE Guide, shows that the nearest value of Δp_l that is viable is 0.8 Pa/m run.

From table 7.3, taken from the CIBSE Guide, a nominal pipe diameter of 20 mm covered with insulation 12.5 mm thick and having a thermal conductivity of 0.040 W/m K will emit 0.31 W/m °C temperature difference

Table 7.2 Flow of water at 75°C in copper pipes (table X)

Δp_l	v	15 mm M	l_e	22 mm M	l_e	28 mm M	l_e	35 mm M	l_e	42 mm M	l_e	v	Δp_l
0.1						0.003	0.2	0.007	0.4	0.016*	0.8		0.1
0.2						0.006	0.3	0.014*	0.8	0.023	0.9		0.2
0.3				0.003	0.2	0.009	0.5	0.017	0.7	0.028	0.9		0.3
0.4				0.004	0.2	0.012	0.6	0.020	0.7	0.033†	0.9		0.4
0.5				0.005	0.3	0.015*	0.8	0.023	0.7	0.037	0.9		0.5
0.6				0.006	0.3	0.014	0.6	0.025	0.7	0.041	1.0		0.6
0.7				0.007	0.4	0.015	0.6	0.027	0.7	0.045	1.0		0.7
0.8				0.008	0.4	0.017	0.6	0.029†	0.7	0.049	1.0		0.8
0.9				0.010	0.5	0.018	0.6	0.031	0.8	0.052	1.0		0.9
1.0				0.011*	0.6	0.018	0.6	0.032	0.8	0.056	1.0	0.05	1.0
1.5		0.003	0.2	0.012	0.5	0.022†	0.6	0.041	0.8	0.070	1.1		1.5
2.0		0.004	0.2	0.014	0.5	0.027	0.6	0.049	0.9	0.083	1.2		2.0
2.5		0.005	0.3	0.015	0.5	0.030	0.7	0.056	0.9	0.095	1.2		2.5
3.0		0.006	0.3	0.017	0.5	0.034	0.7	0.062	0.9	0.105	1.2		3.0
3.5	0.05	0.008*	0.4	0.018†	0.5	0.037	0.7	0.068	1.0	0.115	1.3	0.10	3.5

M = mass flow rate (kg/s); l_e = equivalent length of pipe (ζ = 1) (m); Δp_l = pressure loss per unit length (Pa/m); v = velocity (m/s).

*Re = 2000; †Re = 3000.

Data drawn from CIBSE Guide C, *Reference data* (1986), by permission of the Chartered Institution of Building Services Engineers.

Table 7.3 Heat emission or absorption from insulated pipes

Nominal pipe size (mm)	Heat emission or absorption from insulated pipework (W/m run per °C temperature difference) for insulation thermal conductivity of 0.040 W/m K at thickness of insulation (mm):				
	12.5	19	25	38	50
15	0.27	0.22	0.19	0.16	0.14
20	0.31	0.25	0.22	0.18	0.16
25	0.36	0.29	0.25	0.20	0.18
32	0.43	0.34	0.29	0.23	0.20

Note: The pipe sizes are to BS 1387 and BS 3600. It is assumed that the outside surface of the insulation has been painted, is in still air at 20°C and h_{so} = 10 W/m² K.

Data drawn from CIBSE Guide C, *Reference data* (1986), by permission of the Chartered Institution of Building Services Engineers.

between the mean surface temperature of the pipe and that of the ambient air.

Hence heat loss rate from the pipe is given by:

$$\{[(60 + 46)/2] - 20\} \times 0.31 \times 31 \text{ W} \quad \text{or} \quad 317 \text{ W}$$

The mass flow rate of water M kg/s needed to sustain the power output P kW from the lagged pipework with a temperature drop of ΔT K may be obtained from:

$$P = M \times c_p \times \Delta T \text{ kW}$$

where c_p = the specific heat capacity of water at a temperature of 70°C, taken as 4.18 kJ/kg K. Transposing:

$$M = P/c_p \times \Delta T \text{ kg/s}$$

or $M = 317/4.18 \times 10^3 \times 14$ kg/s or $M = 0.0054$ kg/s.

Entering table 7.2 at a value of $\Delta p_l = 0.8$ Pa/m run, it will be seen that a 22 mm diameter copper pipe will carry a mass flow rate of water of 0.008 kg/s at a temperature of 75°C.

As 0.008 kg/s > 0.0054 kg/s and the value of Δp_l has been slightly understated, it is assumed that the pipe size will be satisfactory, even though the mean temperature of the water circulating is 53°C and not 75°C as quoted in the table. At this point, it is necessary to remind the reader that the thermal conditions governing the operation of the system will always be variable, and that both the flow and return temperatures will vary continuously from those assumed.

For Reynolds number Re:

$$Re = u \times d \times \rho/\mu$$

The velocity u (m/s) may be obtained from:

$$M = a \times u \times \rho$$

Transposing:

$$u = M/a \times \rho$$

The area of the tube is given by:

$$a = \pi \times (20.22/10^3)^2/4 \text{ m}^2 \quad \text{or} \quad a = 0.00032 \text{ m}^2$$

Then $u = 0.008/0.00032 \times 986.7$ m/s or $u = 0.025$ m/s.

This value for the velocity may also be obtained by interpolation from table 7.2, namely:

$u = 0.05/2 = 0.025$ m/s

Substituting for Re:

Re $= (0.025 \times 20.22/10^3) \times 986.7/0.000521$ or Re $= 957$

which indicates that the flow is laminar.

7.10 Pump and pipe sizing exercises for low-pressure hot water space heating systems

Example 2

The installation shown in figure 7.29(a) consists of a single pipe pumped ring main, constructed of medium grade black steel pipes. The estimated fabric and air losses for the building are 14 kW and seven emitters, each having a capacity of 2 kW, are to be installed. In order to avoid the sound of water flowing in the pipes, the velocity must not be greater than 1.5 m/s.

Determine suitable sizes for the pipes, pump, boiler and expansion tank, when the temperature drop around the system is 15 K.

Solution

It is necessary to allow for a heat emission rate from the (as yet unsized), pipe. A common assumption is that the emission rate from the pipe is $0.33 \times$ (the heat loss rate from the emitters).

Hence estimated pipe emission rate $= 0.33 \times 14$ kW or 4.6 kW, and the total emission rate for the system $= 1.33 \times 14$ kW or 18.6 kW

Using the equation $P = M \times c_p \times \Delta T$, transposing for M and inserting values, we have:

$M = 18.6/4.18 \times 15$ kg/s or 0.297 kg/s

At this point it should be noted that by assuming a greater value for the temperature drop around the system, the calculated mass flow rate can be made less, resulting in a smaller pump and smaller pipe sizes.

In the case of systems using single pipe flow, the distribution of heat around the system is likely to be uneven, as the emitters closest to the boiler will receive water at a higher temperature than those connected further downstream. If this effect cannot be corrected by throttling the valves supplying the upstream emitters, then larger emitters will have to

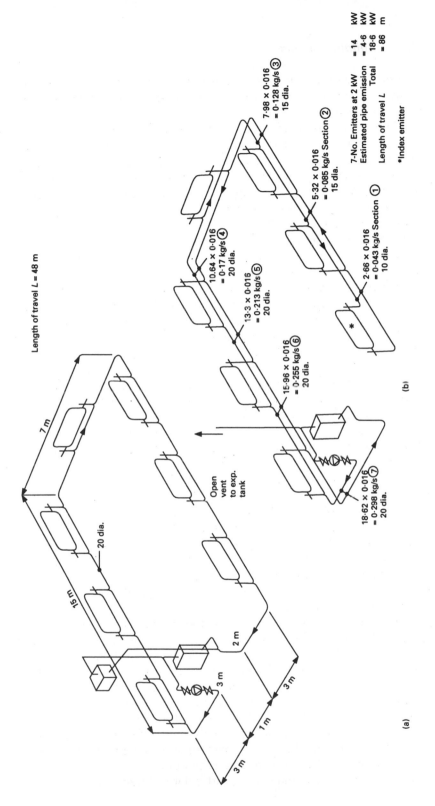

Length of travel L = 48 m

7 m

Open
vent
to exp.
tank

20 dia.

15 m

3 m

2 m

3 m

1 m

3 m

(a)

7·98 × 0·016
= 0·128 kg/s ③
15 dia.

5·32 × 0·016
= 0·085 kg/s Section ②
15 dia.

7-No. Emitters at 2 kW   = 14   kW
Estimated pipe emission   = 4·6   kW
                    Total   = 18·6   kW
Length of travel L   = 86   m

*Index emitter

10·64 × 0·016
= 0·17 kg/s ④
20 dia.

13·3 × 0·016
= 0·213 kg/s ⑤
20 dia.

2·66 × 0·016
= 0·043 kg/s Section ①
10 dia.

15·96 × 0·016
= 0·255 kg/s ⑥
20 dia.

*

(b)

18·62 × 0·016
= 0·298 kg/s ⑦
20 dia.

Figure 7.29 (a) Single pipe flow. (b) Double pipe flow

327

be specified at the downstream positions. In the UK, the maximum value for the temperature drop around any low-pressure heating system does not normally exceed 17 K.

Reference to pump manufactures' catalogues indicate that a pump is available that will deliver a mass flow rate of water of 0.3 kg/s when running at 2200 rev/min, against a pressure of 32.5 kPa. By operating a switch, the same pump offers two other speeds, one higher and one lower than that quoted, giving a choice of characteristic curve, to suit other conditions in other systems.

Length of travel L is 48 m.

The equivalent length $L_e = 1.25 \times 48$ m, or 60 m.

For the system, the pressure drop available/m run of pipe is given by $\Delta p_l = 32.5 \times 10^3/60$ Pa/m run, or 542 Pa/m run.

Reference to table 7.4, taken from the CIBSE Guide, shows that the nearest value of Δp_l which is viable is 540 Pa/m run, and at this rate of pressure loss a 20 mm diameter pipe will carry a mass flow rate of 0.338 kg/s with a velocity (estimated from the table) of approximately 0.9 m/s.

As 0.338 kg/s > 0.297 kg/s, this pipe size will be satisfactory for the purposes of mass flow rate and as 0.9 m/s < 1.5 m/s there will be no problems associated with noise coming from the pipes.

A check from table 7.5, taken from the CIBSE Guide, will show that for an ambient temperature of 20°C and a mean flow temperature of 75°C (a temperature difference of 55 K), the rate of heat emission from the flow pipe/m run is 75 W/m run.

Therefore the expected heat emission rate from the flow pipe is given by 48×75 W, or 3600 W.

3600 W < 4600 W so that the original emission allowance was slightly high. The boiler has to provide 14 + 3.6 kW, or 17.6 kW to keep the system operating steadily under the design conditions chosen. Assuming that 15% of this value is needed for ease of warming up, then the boiler should be capable of providing heat at the rate of 17.6×1.15 kW, or 20.2 kW, and a boiler having a capacity of 20 kW would in practice be satisfactory.

The reader will have noticed that the original value for air and fabric losses, 14 kW, has been supplemented by an extra 3.6 kW input from the emission of the pipes, and it could be argued that this extra input should be removed from the system, by deleting, say, two emitters, each having an output of 2 kW. In this way the cost of the emitters and the labour in fitting them are saved, the boiler may be made smaller and the downtime of the boiler would be reduced, making the annual amount of flue losses less and increasing the thermal efficiency of the system. It must also be noted that the impact of the system on the environment will be diminished by a measurable amount.

In a competitive market, and particularly when dealing with a larger system, the savings to be made in this way must be taken into account and may necessitate a recalculation of the pipe sizes. With smaller schemes, such as the one given in this example, the gains made in this way may be ignored, resulting in a system that will have spare capacity, for use whenever the outside temperature falls below the system design temperature. There may also be other overriding factors to be considered, such as the need to provide a quick warm up, owing to intermittent usage of the building, or it may be that the client is planning a small additional construction phase in the near future, which is to be heated using the existing boiler.

The size of boiler is related to both the total volume of water contained in the system and its subsequent expansion when heated. Table 7.6, taken from the CIBSE Guide, gives the approximate volume of expansion expected for low-pressure hot water heating systems comprising emitters (radiators). For a 20 kW boiler, the volume of expansion is given as 18 litres. As a general rule this means that an expansion tank will have a nominal capacity ($L \times B \times H$) equal to three times the volume of expansion. Hence the expansion tank will have a nominal capacity of 54 litres. The top third of the tank will be taken up by the overflow and ball valve connections, the middle third by the water of expansion, and the lower third by the outflow connection fitted 50 mm above the base of the tank.

Example 3

The installation shown in figure 7.29(b) consists of a double pipe flow and return low-pressure hot water system, intended to provide space heating to the same building as that used in 7.29(a). Calculate and specify suitable pipe sizes for the scheme using the same basic data as that already given in example 2. Also recommend a suitable boiler rating.

Solution

As before, it is necessary to assume a total heat emission rate from the (as yet unsized) pipes. Let this be 0.33 × (the heat loss from the emitters). Hence the pipe emission rate = 0.33 × 14 kW or 4.6 kW which, in this case, may be assumed to be distributed evenly around the system at the rate of 4.6/7 kW/emitter, or 0.66 kW/emitter, making the new output of each emitter 2.66 kW.

The total emission rate from the system is given by 1.33 × 14 kW, or 18.6 kW. Using the equation $P = M \times c_p \times \Delta T$, transposing for M and inserting values for an emitter having an output of 1 kW will facilitate subsequent calculations for emitters having sizes that are multiples of 1 kW. Hence $M = 1/4.18 \times 15$ kg/s, or 0.016 kg/s for 1 kW of emitter.

Reference to figure 7.29(b) shows an index emitter. This is the emitter at the end of the circuit, and it will incur the highest head loss due to

Table 7.4 Flow of water at 75°C in black steel pipes

Δp_l	v	10 mm M	l_e	15 mm M	l_e	20 mm M	l_e	25 mm M	l_e	32 mm M	l_e	40 mm M	l_e	50 mm M	l_e	v	Δp_l
4.0	0.05	0.006	0.3	0.011	0.3	0.023	0.5	0.043	0.7	0.092	1.1	0.140	1.3	0.264	1.8		4.0
4.5		0.007	0.4	0.011	0.3	0.024	0.5	0.046	0.7	0.099	1.1	0.149	1.3	0.282	1.9		4.5
5.0		0.007*	0.4	0.012	0.3	0.026	0.5	0.049	0.7	0.105	1.1	0.158	1.4	0.299	1.9		5.0
5.5		0.006	0.3	0.012	0.3	0.027	0.5	0.052	0.7	0.110	1.1	0.167	1.4	0.315	1.9	0.15	5.5
6.0		0.007	0.3	0.013	0.3	0.029	0.5	0.055	0.8	0.116	1.1	0.175	1.4	0.331	1.9		6.0
6.5		0.007	0.3	0.014	0.3	0.030	0.5	0.057	0.8	0.121	1.1	0.183	1.4	0.346	2.0		6.5
7.0		0.007	0.3	0.014†	0.4	0.032	0.5	0.060	0.8	0.127	1.1	0.191	1.4	0.361	2.0		7.0
7.5		0.008	0.3	0.015	0.4	0.033	0.6	0.062	0.8	0.131	1.2	0.198	1.4	0.375	2.0		7.5
8.0		0.008	0.3	0.015	0.4	0.034	0.6	0.064	0.8	0.136	1.2	0.206	1.4	0.388	2.0		8.0
8.5		0.008	0.3	0.016	0.4	0.035	0.6	0.066	0.8	0.141	1.2	0.213	1.4	0.401	2.0		8.5
9.0		0.008	0.3	0.016	0.4	0.036	0.6	0.069	0.8	0.146	1.2	0.220	1.5	0.414	2.0		9.0
9.5		0.008	0.3	0.017	0.4	0.037	0.6	0.071	0.8	0.150	1.2	0.226	1.5	0.427	2.0		9.5
10.0		0.009	0.3	0.017	0.4	0.039	0.6	0.073	0.8	0.154	1.2	0.233	1.5	0.439	2.0		10.0
12.5		0.010	0.3	0.020	0.4	0.044	0.6	0.082	0.8	0.175	1.2	0.263	1.5	0.496	2.1		12.5
15.0		0.011†	0.3	0.022	0.4	0.049	0.6	0.091	0.8	0.193	1.2	0.291	1.5	0.548	2.1		15.0
17.5		0.012	0.3	0.024	0.4	0.053	0.6	0.099	0.9	0.210	1.3	0.317	1.6	0.596	2.2		17.5
20.0		0.012	0.3	0.026	0.4	0.057	0.6	0.107	0.9	0.226	1.3	0.341	1.6	0.641	2.2		20.0
22.5		0.013	0.3	0.027	0.4	0.061	0.6	0.114	0.9	0.242	1.3	0.363	1.6	0.683	2.2		22.5
25.0		0.014	0.3	0.029	0.4	0.065	0.6	0.121	0.9	0.256	1.3	0.385	1.6	0.723	2.2		25.0
27.5		0.015	0.3	0.031	0.4	0.068	0.6	0.128	0.9	0.270	1.3	0.405	1.6	0.761	2.2		27.5
30.0		0.016	0.3	0.032	0.4	0.071	0.7	0.134	0.9	0.283	1.3	0.425	1.6	0.798	2.2		30.0
32.5		0.016	0.3	0.034	0.4	0.075	0.7	0.140	0.9	0.295	1.3	0.444	1.6	0.833	2.3	0.30	32.5
35.0		0.017	0.3	0.035	0.4	0.078	0.7	0.146	0.9	0.307	1.3	0.462	1.7	0.867	2.3		35.0
37.5		0.018	0.3	0.036	0.4	0.081	0.7	0.151	0.9	0.319	1.4	0.479	1.7	0.899	2.3		37.5
40.0	0.15	0.018	0.3	0.038	0.4	0.084	0.7	0.157	0.9	0.330	1.4	0.496	1.7	0.931	2.3		40.0

M	Δp_l	v	Δp_l	v	Δp_l	v	Δp_l	v	Δp_l	v	Δp_l	v	Δp_l	v	M
100.0	0.031	0.3	0.062	0.5	0.138	0.7	0.258	1.0	0.540	1.5	0.810	1.8	1.52	2.4	100.0
120.0	0.034	0.3	0.069	0.5	0.152	0.7	0.284	1.0	0.595	1.5	0.893	1.8	1.67	2.4	120.0
140.0	0.037	0.3	0.075	0.5	0.165	0.8	0.308	1.0	0.646	1.5	0.968	1.8	1.81	2.5	140.0
160.0	0.040	0.4	0.081	0.5	0.178	0.8	0.331	1.0	0.693	1.5	1.04	1.8	1.94	2.5	160.0
180.0	0.042	0.4	0.086	0.5	0.189	0.8	0.353	1.0	0.738	1.5	1.11	1.8	2.06	2.5	180.0
200.0	0.045	0.4	0.091	0.5	0.200	0.8	0.373	1.1	0.780	1.5	1.17	1.9	2.18	2.5	200.0
220.0	0.047	0.4	0.096	0.5	0.211	0.8	0.392	1.1	0.820	1.5	1.28	1.9	2.29	2.5	220.0
240.0	0.050	0.4	0.100	0.5	0.221	0.8	0.411	1.1	0.858	1.5	1.29	1.9	2.40	2.5	240.0
260.0	0.052	0.4	0.105	0.5	0.230	0.8	0.428	1.1	0.895	1.5	1.34	1.9	2.50	2.5	260.0
280.0	0.054	0.4	0.109	0.5	0.239	0.8	0.445	1.1	0.931	1.5	1.39	1.9	2.60	2.6	280.0
300.0	0.056	0.4	0.113	0.5	0.248	0.8	0.462	1.1	0.965	1.5	1.44	1.9	2.69	2.6	300.0
320.0	0.058	0.4	0.117	0.5	0.257	0.8	0.478	1.1	0.998	1.6	1.49	1.9	2.78	2.6	320.0
340.0	0.060	0.4	0.121	0.5	0.265	0.8	0.493	1.1	1.03	1.6	1.54	1.9	2.87	2.6	340.0
360.0	0.062	0.4	0.125	0.5	0.273	0.8	0.508	1.1	1.06	1.6	1.59	1.9	2.96	2.6	360.0
380.0	0.064	0.4	0.128	0.5	0.281	0.8	0.523	1.1	1.09	1.6	1.63	1.9	3.04	2.6	380.0
400.0	0.065	0.4	0.132	0.5	0.289	0.8	0.537	1.1	1.12	1.6	1.68	1.9	3.12	2.6	400.0
420.0	0.067	0.4	0.135	0.5	0.297	0.8	0.551	1.1	1.15	1.6	1.72	1.9	3.20	2.6	420.0
440.0	0.069	0.4	0.139	0.5	0.304	0.8	0.564	1.1	1.18	1.6	1.76	1.9	3.28	2.6	440.0
460.0	0.070	0.4	0.142	0.5	0.311	0.8	0.578	1.1	1.21	1.6	1.80	1.9	3.36	2.6	460.0
480.0	0.072	0.4	0.145	0.5	0.318	0.8	0.591	1.1	1.23	1.6	1.84	1.9	3.43	2.6	480.0
500.0	0.074	0.4	0.148	0.5	0.325	0.8	0.603	1.1	1.25	1.6	1.88	1.9	3.51	2.6	500.0
520.0	0.075	0.4	0.151	0.5	0.332	0.8	0.616	1.1	1.29	1.6	1.92	1.9	3.58	2.6	520.0
540.0	0.077	0.4	0.154	0.6	0.338	0.8	0.628	1.1	1.31	1.6	1.96	1.9	3.65	2.6	540.0
560.0	0.078	0.4	0.157	0.6	0.345	0.8	0.640	1.1	1.34	1.6	2.00	1.9	3.72	2.6	560.0
580.0	0.080	0.4	0.160	0.6	0.351	0.8	0.652	1.1	1.36	1.6	2.03	1.9	3.78	2.6	580.0

Column-group markers: 0.30, 0.50, 1.0, 1.5.

M = mass flow rate (kg/s); l_e = equivalent length of pipe (ζ = 1) (m); Δp_l = pressure loss per unit length (Pa/m); v = velocity (m/s).
*Re = 2000; †Re = 3000.

Data drawn from CIBSE Guide C, *Reference data* (1986), by permission of the Chartered Institution of Building Services Engineers.

Table 7.5 Heat emission from single horizontal steel pipes ε = 0.95 freely exposed in surroundings at 20°C

Temp. diff. between surface and surroundings (°C)	Heat emission (W/m) for Pipe nominal size* (mm):			
	15	20	25	32
45	48	59	71	87
50	55	67	81	99
55	62	75	92	112
60	69	84	102	125

*These pipe sizes are to BS 1387: 1967 and BS 3600: 1973.

Data drawn from CIBSE Guide C, *Reference data* (1986), by permission of the Chartered Institution of Building Services Engineers.

Table 7.6 Approximate sizes of feed and expansion cisterns for low-pressure hot water heating systems

Boiler or water heater rating (kW)	Cistern size (litre)	Ball-valve size (mm)	Cold-feed size (mm)	Open-vent size (mm)	Over-flow size (mm)
15	18	15	20	25	25
22	18	15	20	25	32
30	36	15	20	25	32
45	36	15	20	25	32

Notes:
1. Cistern sizes are actual.
2. Cistern sizes are based on radiator-heating systems and are approximate only.
3. The ball-valve sizes apply to installations where an adequate mains water pressure is available at the ball valve.

Data drawn from CIBSE Guide C, *Reference data* (1986), by permission of the Chartered Institution of Building Services Engineers.

friction. It follows that the other emitters will have proportionally lower friction losses and, by arranging for the index emitter to be served, the remaining units will automatically be supplied.

The index emitter served by the pipes in section 1 of the diagram will require a mass flow rate of 2.66 × 0.016 kg/s, or 0.043 kg/s, and the pipes serving section 2 will require a mass flow rate of 5.32 × 0.016 kg/s, or 0.085 kg/s. The mass flow rates required by the pipes serving the remaining sections have been calculated and their values are given on the diagram. Reference to pump manufacturers' catalogues indicates that a pump is available that will deliver a mass flow rate of water of 0.3 kg/s when running at 2200 rev/min, against a pressure of 32.5 kPa.

Length of travel L is 86 m.

The equivalent length L_e = 1.25 × 86 m, or 107.5 m.

The pressure drop available/m run of pipe is given by Δp_1 = 32.5 × 10^3/107.5 Pa/m run, or 302 Pa/m run.

Reference to table 7.4 shows that the nearest value of Δp_1 that is viable is 300 Pa/m run, and it is this value of pressure loss/metre run that will be used to size the pipes.

The thinking behind the selection of pipe sizes is set out in detail in the following sections. As a designer becomes more practised in the task, it will be found that he can determine the relevant size of pipe without setting down all the details indicated below.

Section 1

Required mass flow rate = 0.043 kg/s.

Choosing a 10 mm diameter pipe, 0.056 kg/s > 0.043 kg/s, leaving a net extra mass flow rate of 0.056 − 0.043 kg/s = +0.013 kg/s.

The velocity corresponding to a mass flow rate of 0.043 kg/s (estimated from table 7.4 is approximately 0.35 m/s, and as 0.35 m/s < 1.5 m/s, this is satisfactory.

Section 2

Required mass flow rate = 0.085 kg/s.

Choosing a 15 mm diameter pipe, 0.113 kg/s > 0.085 kg/s, leaving a net extra mass flow rate of 0.113 − 0.085 + 0.013 kg/s = +0.041 kg/s.

The velocity corresponding to a mass flow rate of 0.085 kg/s (estimated from table 7.4) is approximately 0.4 m/s, and as 0.4 m/s < 1.5 m/s, this is satisfactory.

Section 3

Required mass flow rate = 0.128 kg/s.

Choosing a 15 mm diameter pipe, 0.113 kg/s < 0.128 kg/s, leaving a net extra mass flow rate of 0.113 − 0.128 + 0.041 kg/s = +0.026 kg/s.

The velocity corresponding to a mass flow rate of 0.128 kg/s (estimated from table 7.4) is approximately 0.6 m/s, and as 0.6 m/s < 1.5 m/s, this is satisfactory.

Section 4

Required mass flow rate = 0.17 kg/s.

Choosing a 20 mm diameter pipe, 0.248 kg/s > 0.17 kg/s, leaving a net extra mass flow rate of 0.248 − 0.17 + 0.026 kg/s = 0.104 kg/s.

The velocity corresponding to a mass flow rate of 0.17 kg/s (estimated from table 7.4) is approximately 0.5 m/s, and as 0.5 m/s < 1.5 m/s, this is satisfactory.

Section 5

Required mass flow rate = 0.213 kg/s.
Choosing a 20 mm diameter pipe, 0.248 kg/s > 0.213 kg/s, leaving a net extra mass flow rate of 0.248 − 0.213 + 0.104 kg/s = 0.139 kg/s.
The velocity corresponding to a mass flow rate of 0.213 kg/s (estimated from table 7.4) is approximately 0.6 m/s, and as 0.6 m/s < 1.5 m/s, this is satisfactory.

Section 6

Required mass flow rate = 0.255 kg/s.
Choosing a 20 mm diameter pipe, 0.248 kg/s < 0.255 kg/s, leaving a net extra mass flow rate of 0.248 − 0.255 + 0.139 kg/s = 0.132 kg/s.
The velocity (estimated from table 7.4) is approximately 0.6 m/s, and as 0.6 m/s < 1.5 m/s, this is satisfactory.

Section 7

Required mass flow rate = 0.298 kg/s.
Choosing a 20 mm diameter pipe, 0.248 kg/s < 0.298 kg/s, leaving a net extra mass flow rate of 0.248 − 0.298 + 0.132 kg/s = 0.082 kg/s.
The velocity corresponding to a mass flow rate of 0.298 kg/s (estimated from table 7.4) is approximately 0.75 m/s, and as 0.75 m/s < 1.5 m/s, this is satisfactory.

Now that the pipes have been sized, it is possible and prudent to calculate the emission rate to be expected from them and from tables this comes to approximately 6 kW, which is 1.4 kW above the estimated figure of 4.6 kW.
Section 7 indicates a possible surplus mass flow rate of 0.082 kg/s which is equivalent to a power supply of 0.082/0.016 kW, or 5.13 kW, and it may be concluded that the pipework of the system will cope easily with the expected design conditions.
The boiler has to provide 14 + 6 kW or 20 kW to keep the system operating steadily under the design conditions chosen. Assuming that 15% of this value is needed for ease of warming up, then the boiler should be capable of providing heat at the rate of 20 × 1.15 kW, or 23 kW, and a boiler having a capacity of 23 kW would in practice be satisfactory.
It is interesting to compare this figure with the rating for the boiler quoted in example 2, in which the same building is heated using a single pipe flow circuit. The difference between the two systems occurs because of the extra emission rate available from the greater surface area of pipework used in the double pipe flow and return system, and as such it highlights the need for a possible reduction in the sizes of the emitters. In a scheme of this size, the difference in boiler size is small

and, in line with the comments given at the end of example 2, could be ignored.

7.11 Pipe sizing exercises for domestic hot and cold water supply systems

7.11.1 Pipe sizing when the pattern of use is known

This situation will occur where a large number of washbasins, or wash fountains, are intended to serve the needs of a group of manual workers coming off shift at the same time; or where a team of sportsmen all require the instantaneous use of showering facilities; or where a number of pieces of apparatus requiring a known consumption rate of water are to be used simultaneously.

In these cases, the systems must be designed to deal with the expected instantaneous use of the appliances and the total volume flow rate for all the outlets must be supplied by the branch pipe. Satisfactory volume flow rates for fittings attached to various domestic appliances are given in table 7.7, taken from the CIBSE Guide.

7.11.2 Pipe sizing when the pattern of use is not known

This situation must be regarded as the norm. It arises whenever a system of pipework is to be designed to serve domestic and sanitary appliances being used in both large and small buildings.

In order to design a system involving hot or cold water supplies, when the pattern of use is not known, information is required on the following three items:

Item 1 – The number and type of fittings. These can be ascertained from the design drawings.
Item 2 – The volume flow rate required for each fitting. These can be obtained from table 7.7.
Item 3 – An indication of the cumulative effect of a number of fittings on the theoretical volume flow rate.

Obviously, not all the fittings connected to the system will be in use at the same time and it can be expected that, as the number is increased, an even smaller proportion of fittings will be in operation. Traditionally, information of this kind is presented as shown in figure 7.30 when, with the aid of table 7.7, the theoretical volume flow rate may be determined for any branch pipe serving any number of fittings.

The expected simultaneous demand may be obtained by evaluating the product of the theoretical volume flow rate and the diversity factor, obtained from the graph. The use of items 1 to 3 in the manner indicated

Table 7.7 Volume flow rates required at outlets

Sanitary appliance	Flow rate (litre/s)
Basin (spray)	0.05
Basin (tap)	0.15
Bath (private)	0.3
Bath (public)	0.6
Flushing cistern	0.1
Shower (nozzle)	0.15
Shower (100 mm rose)	0.4
Sink (15 mm tap)	0.2
Sink (20 mm tap)	0.3
Wash fountain	0.4

Reproduced from CIBSE Guide B, *Reference data* (1986), by permission of the Chartered Institution of Building Services Engineers.

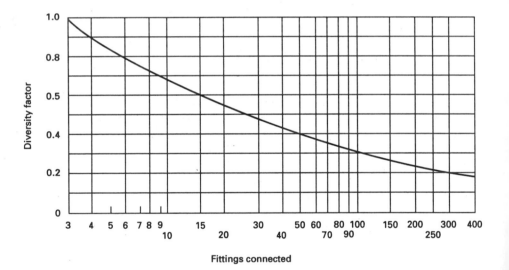

Figure 7.30 Diversity factor against number of fittings connected

has been found to produce systems that were workable, but also slightly oversized. The bibliography on design methods is extensive and the CIBSE Guide has developed a design method applicable to the design of systems where the pattern of use is not known. The resulting pipe sizes tend to be smaller than those obtained from the earlier design methods, and it is to be hoped that this method might become recognised as the standard approach to the solution of design problems associated with pipe sizing in this category.

In order to operate the CIBSE method, additional information is required, which is detailed under item 4.

Item 4

This involves an estimate of the period and frequency of operation of each fitting connected to the system, which depends on the type of building and its proposed pattern of use. Pertinent to this, it is advantageous to evaluate a ratio, referred to as the usage ratio *p*, defined as:

p= the average time a demand is imposed on a fitting for each
occasion of use/the average time between occasions of use

The determination of values for both items 3 and 4 requires a knowledge of the theory of probability and of the methods involved in finding solutions to the equations. In sections B6 and B4 of the CIBSE Guide, the relevant theory and the solutions to the equations have been set out, and in the presentation of these results, the Guide has produced a theoretical set of 'demand units' for a value of *p* = 0.1, based on the operation of a cold tap fitted to a washbasin. The tap is assumed to have a volume flow rate of 0.15 litre/s and to be used at intervals of 300 s. Under these circumstances, the theoretical demand unit is deemed to have a value of unity and, by solving the equations, correspondingly higher and lower values for theoretical demand units have been obtained for other appliances, operating with different volume flow rates, over other appropriate intervals of time.

The solutions to the equations contain fractions of a decimal which, for subsequent ease of computation, have been eliminated by multiplying the theoretical values by 10 and then rounding up the remaining fractions to give whole numbers. The resulting figures are referred to as 'practical demand units' and they are reproduced from the Guide in table 7.8. They are associated with various appliances being operated in three different modes, one of which assumes 'private' use, another 'public' use, and the last 'congested' use. The information given in table 7.9, also taken from the Guide, links the practical demand units with the expected simultaneous demand for water in litre/s.

When using the demand units to facilitate pipe sizing, the information given in table 7.8 is used to allocate the appropriate demand units to the fittings served by the branch pipe under consideration. These demand units are then summed for that section of pipe, and the corresponding simultaneous demands may be ascertained by referring to table 7.9.

It is singularly unfortunate that the operation of some fittings does not conform to the mathematical model used in the calculations. For instance, the mode of operation of a shower fitting and a spray tap requires a continuous flow of water for the duration of their use. It is also true that

Table 7.8 Practical demand units

Fitting	Type of application		
	Congested	Public	Private
Basin	10	5	3
Bath	47	25	12
Sink	43	22	11
Urinal cistern	—	—	—
WC (13.5 litre)	35	15	8
WC (9 litre)	22	10	5

Reproduced from CIBSE Guide B, *Reference data* (1986), by permission of the Chartered Institution of Building Services Engineers.

the pattern of operation of dishwashers and washing machines is programmable and peculiar to each model, and that such patterns of use cannot be known during the early stages of a development. These discontinuities may be overcome by carrying an assumed volume flow rate for these appliances into and through the calculation as a separate quantity.

The next part of the pipe sizing exercise involves ensuring that the sum of the head losses incurred as the water flows through the system is never greater than the initial head of water available. These losses can conveniently be grouped into two categories:

(1) Head losses at entry to and exit from the system, together with losses due to bends and changes in cross-section of the pipes. These can be dealt with as in the previous examples, by using a multiplier of 1.25 (some designers use a value of 1.3), applied to the measured length L of the pipe run, to give an equivalent length L_e.
(2) Head loss due to friction as the water flows through the pipes. This can be dealt with by first converting the available head of water H (m) into units of pressure p (Pa) (use $p = H\rho g$, where $\rho = 10^3$ kg/m^3 and $g = 9.81$ m/s^2), and then obtaining a value for the ratio $\Delta p_l = p/L_e$ Pa/m for the index fitting.

As a guide, the index fitting can be taken as being that fitting on the top floor that has the longest pipe run. If there is any doubt as to which of several fittings qualifies as the index fitting, then it is always possible to evaluate Δp_l for each case, when the lowest value of Δp_l indicates that the index fitting lies on that circuit.

Using this value of Δp_l and the flow tables (published in full in the CIBSE Guide), the pipe sizes for the index fitting may then be determined. It is also possible to use the same value of Δp_l to size the rest of

Table 7.9 Simultaneous demand

Demand units	Design demand (litre/s)																			
	0	*50*	*100*	*150*	*200*	*250*	*300*	*350*	*400*	*450*	*500*	*550*	*600*	*650*	*700*	*750*	*800*	*850*	*900*	*950*
0	0.0	0.3	0.5	0.6	0.8	0.9	1.0	1.2	1.3	1.4	1.5	1.6	1.7	1.9	2.0	2.1	2.2	2.3	2.4	2.5
1000	2.6	2.7	2.8	2.9	3.0	3.1	3.2	3.3	3.4	3.5	3.6	3.7	3.8	3.9	4.0	4.1	4.2	4.3	4.4	4.5
2000	4.6	4.7	4.8	4.9	5.0	5.1	5.1	5.2	5.3	5.4	5.5	5.6	5.7	5.8	5.9	6.0	6.1	6.2	6.3	6.4
3000	6.4	6.5	6.6	6.7	6.8	6.9	7.0	7.1	7.2	7.3	7.4	7.4	7.5	7.6	7.7	7.8	7.9	8.0	8.1	8.2

Reproduced from CIBSE Guide B, *Reference data* (1986), by permission of the Chartered Institution of Building Services Engineers.

the system, confident in the belief that if the volume flow rate to the index fitting is satisfactory, then it is an *a priori* case that fittings less difficult to reach will be assured of a volume flow rate greater than the design value.

When dealing with multi-storey buildings, the lower floors usually have a surfeit of head available and, to avoid causing the users of the appliances embarrassment, from unexpected splashing whenever a tap is opened, it is necessary to restrict the volume flow rate to the appliance. It follows that the pipework serving the lower outlets may be sized less generously, by using a value for Δp_l that is appropriate to that outlet.

It is hoped that the mechanics of the pipe sizing process and the elegance of the method, suggested by the CIBSE Guide, will become apparent to the reader after working through the following example.

Example 4

The fittings, requiring a supply of cold water in a proposed block of ten flats each having three bedrooms, are to be arranged as shown in figure 7.31(a). The pipework, connected to the cistern comprising the down feeds, is to be constructed of galvanised steel, and the pipework connected to the down feeds and the appliances is to be of copper. Determine:

(a) suitable pipe sizes for the inflow and outflow at the cistern;
(b) a suitable capacity for the cistern to give protection against a 24 hour shutdown of the mains.

Assume that the head of water available from the galvanised steel supply pipe at street level is 21 m and that the total head loss across the ball valve is 2.0 m.

Solution

It is suggested that a photocopy of figure 7.31(a) be made and that the boxes shown on the diagram be completed while working through the problem. The demand units relating to each fitting can be obtained from table 7.8 (type of application, private) and these have been marked against each fitting as shown in figure 7.31(b) at first floor level. Demand units are not applicable for three of the appliances (the shower, washing machine and dishwasher), and the volume flow rates assumed for each of these units have also been added to the sketch. The WC on the fourth floor becomes the index fitting.

Boxes 1 to 6

The value of the demand unit for each box may be ascertained and totalled cumulatively from box 1 to box 6. The corresponding design demand

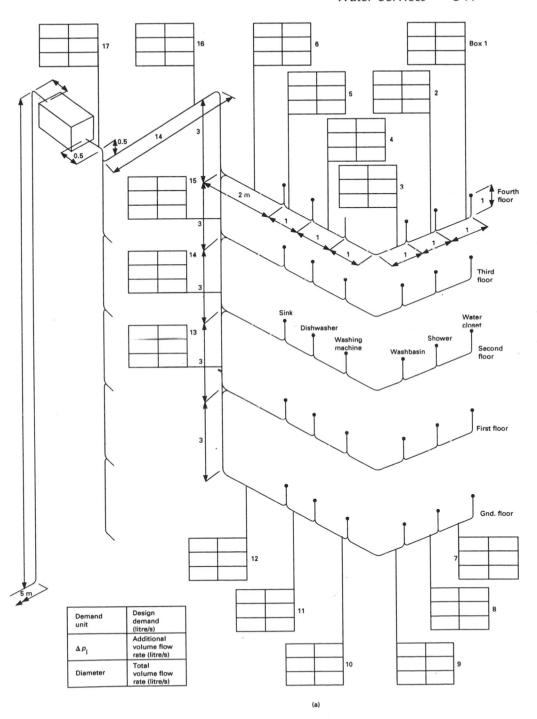

Figure 7.31   (a) Diagram for example 4

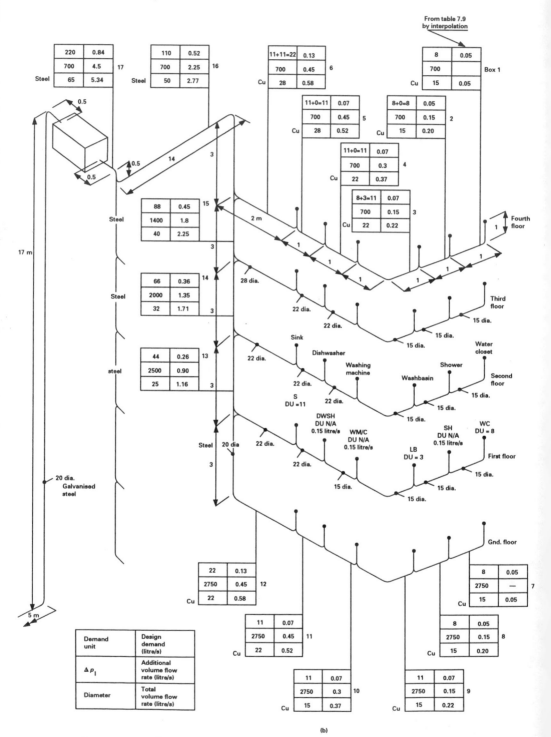

Figure 7.31 (b) Solution to example 4

expected, for each demand unit, may then be obtained (by interpolation) from table 7.9 and added to the boxes. Note that the pipe associated with box 2 has to carry a volume flow rate of 0.15 litre/s in addition to the design demand of 0.05 litre/s, making a total volume flow rate of 0.2 litre/s.

The pipe associated with box 4 has to carry a volume flow rate of 0.15 + 0.15 = 0.3 litre/s, in addition to the design demand of 0.07 litre/s, making a total volume flow rate of 0.37 litre/s, see figure 7.31(b).

Boxes 7 to 17

The demand units and the total volume flow rates may be compiled in a similar manner for all the remaining boxes, as shown in figure 7.31(b).

Determination of pressure loss/unit length for boxes 1 to 6 and 16 and 17

The head available from the tank outlet to the index fitting, H, is 2.5 m.

As $p = H\rho g$, then $p = 2.5 \times 10^3 \times 9.81$ Pa, or 24 525 Pa.

Length of travel from tank outlet to the index fitting, L, is 27 m

The equivalent length, L_e, is given by $L_e = 1.25 \times 27$ m, or 33.75 m.

As $\Delta p_l = p/L_e$ Pa/m run, then $\Delta p_l = 24\,525/33.75$ Pa/m run, or 727 Pa/m run.

Using the flow table 7.10, taken from the CIBSE Guide, for copper tube, table X for water at 10°C, the pipework contained in the flats on the fourth floor may now be sized. The nearest viable value for $\Delta p_l = 700$ Pa/m run.

Box 1

Required volume flow rate = 0.05 litre/s.

A 12 mm diameter copper tube would easily carry the required volume flow rate but, as it is considered uneconomic by plumbers to carry fittings of this size, it is rarely used for this type of work and a 15 mm diameter copper tube is selected.

Choosing a 15 mm diameter pipe, 0.110 litre/s > 0.05 litre/s, leaving a net extra volume flow rate of 0.110 − 0.05 litre/s = +0.06 litre/s.

The velocity corresponding to a volume flow rate of 0.110 litre/s (estimated from table 7.10) is approximately 0.6 m/s and, as 0.6 m/s < 1.5 m/s, this is satisfactory.

Box 2

Required volume flow rate = 0.2 litre/s.

Choosing a 15 mm diameter pipe, 0.110 litre/s < 0.2 litre/s, leaving a net extra volume flow rate of 0.110 − 0.2 + 0.06 litre/s = −0.03 litre/s. (This small deficit may be neglected.)

Table 7.10 Flow of water at 10°C in copper pipes (table X)

Δp_l	v	12 mm M	l_e	15 mm M	l_e	22 mm M	l_e	28 mm M	l_e	35 mm M	l_e	42 mm M	l_e	v	Δp_l
130		0.023	0.2	0.041†	0.3	0.123	0.6	0.251	0.8	0.456	1.1	0.772	1.5		130
140		0.024	0.2	0.043	0.3	0.128	0.6	0.262	0.8	0.475	1.2	0.805	1.5		140
150		0.025	0.2	0.045	0.3	0.134	0.6	0.272	0.8	0.494	1.2	0.838	1.5		150
160		0.026	0.2	0.047	0.3	0.139	0.6	0.283	0.9	0.513	1.2	0.869	1.6		160
170		0.027	0.2	0.048	0.3	0.144	0.6	0.293	0.9	0.531	1.2	0.899	1.6		170
180	0.3	0.027	0.2	0.050	0.3	0.149	0.6	0.302	0.9	0.549	1.2	0.929	1.6		180
190		0.028	0.2	0.052	0.3	0.153	0.6	0.312	0.9	0.566	1.2	0.958	1.6		190
200		0.029	0.2	0.053	0.3	0.158	0.6	0.321	0.9	0.583	1.2	0.986	1.6		200
225		0.031	0.2	0.057	0.3	0.169	0.6	0.344	0.9	0.623	1.2	1.05	1.6		225
250		0.032†	0.2	0.061	0.3	0.180	0.6	0.365	0.9	0.662	1.3	1.12	1.6		250
275		0.034	0.2	0.064	0.4	0.190	0.6	0.385	0.9	0.698	1.3	1.18	1.7	1.0	275
300		0.036	0.3	0.067	0.4	0.200	0.6	0.405	0.9	0.734	1.3	1.24	1.7		300
325		0.037	0.3	0.071	0.4	0.209	0.7	0.424	1.0	0.768	1.3	1.30	1.7		325
350		0.039	0.3	0.074	0.4	0.218	0.7	0.442	1.0	0.801	1.3	1.35	1.7		350
375		0.041	0.3	0.077	0.4	0.227	0.7	0.460	1.0	0.833	1.3	1.41	1.7		375
400		0.042	0.3	0.080	0.4	0.236	0.7	0.477	1.0	0.864	1.3	1.46	1.8		400
425		0.044	0.3	0.083	0.4	0.244	0.7	0.494	1.0	0.894	1.3	1.51	1.8		425
450	0.5	0.045	0.3	0.085	0.4	0.252	0.7	0.511	1.0	0.924	1.4	1.56	1.8		450
475		0.047	0.3	0.088	0.4	0.260	0.7	0.527	1.0	0.952	1.4	1.61	1.8		475
500		0.048	0.3	0.091	0.4	0.268	0.7	0.542	1.0	0.980	1.4	1.66	1.8		500
550		0.051	0.3	0.096	0.4	0.283	0.7	0.572	1.0	1.04	1.4	1.75	1.8		550
600		0.054	0.3	0.101	0.4	0.297	0.7	0.601	1.0	1.09	1.4	1.83	1.8	1.5	600
650		0.056	0.3	0.106	0.4	0.311	0.7	0.629	1.0	1.14	1.4	1.92	1.9		650
700		0.059	0.3	0.110	0.4	0.324	0.7	0.656	1.1	1.19	1.4	2.00	1.9		700
750		0.061	0.3	0.115	0.4	0.337	0.7	0.682	1.1	1.23	1.4	2.08	1.9		750
800		0.063	0.3	0.119	0.4	0.350	0.7	0.708	1.1	1.28	1.5	2.16	1.9		800
850		0.066	0.3	0.123	0.4	0.362	0.8	0.732	1.1	1.32	1.5	2.23	1.9		850

Table 7.10 (continued)

Δp_l	v	12 mm M	12 mm l_e	15 mm M	15 mm l_e	22 mm M	22 mm l_e	28 mm M	28 mm l_e	35 mm M	35 mm l_e	42 mm M	42 mm l_e	v	Δp_l
900		0.068	0.3	0.127	0.4	0.374	0.8	0.757	1.1	1.37	1.5	2.30	1.9		900
950		0.070	0.3	0.131	0.4	0.386	0.8	0.780	1.1	1.41	1.5	2.37	1.9		950
1000		0.072	0.3	0.135	0.4	0.398	0.8	0.803	1.1	1.45	1.5	2.44	2.0	2.0	1000
1100		0.076	0.3	0.143	0.4	0.420	0.8	0.847	1.1	1.53	1.5	2.58	2.0		1100
1200		0.080	0.3	0.150	0.4	0.441	0.8	0.890	1.1	1.61	1.5	2.71	2.0		1200
1300		0.084	0.3	0.157	0.5	0.461	0.8	0.931	1.1	1.68	1.6	2.83	2.0		1300
1400		0.088	0.3	0.164	0.5	0.481	0.8	0.971	1.2	1.75	1.6	2.95	2.0		1400
1500	1.0	0.091	0.3	0.171	0.5	0.500	0.8	1.00	1.2	1.82	1.6	3.06	2.1		1500
1600		0.095	0.3	0.177	0.5	0.519	0.8	1.05	1.2	1.89	1.6	3.18	2.1		1600
1700		0.098	0.3	0.184	0.5	0.537	0.8	1.08	1.2	1.95	1.6	3.29	2.1		1700
1800		0.101	0.3	0.190	0.5	0.555	0.8	1.12	1.2	2.02	1.6	3.39	2.1		1800
1900		0.104	0.3	0.196	0.5	0.572	0.8	1.15	1.2	2.08	1.6	3.50	2.1		1900
2000		0.108	0.3	0.201	0.5	0.589	0.8	1.19	1.2	2.14	1.6	3.60	2.1	3.0	2000
2250		0.115	0.4	0.215	0.5	0.629	0.9	1.27	1.2	2.28	1.7	3.84	2.2		2250
2500		0.122	0.4	0.229	0.5	0.668	0.9	1.35	1.2	2.42	1.7	4.07	2.2		2500
2750		0.129	0.4	0.242	0.5	0.705	0.9	1.42	1.3	2.55	1.7	4.29	2.2		2750
3000	1.5	0.136	0.4	0.254	0.5	0.740	0.9	1.49	1.3	2.68	1.7	4.51	2.2		3000
3250		0.142	0.4	0.266	0.5	0.774	0.9	1.56	1.3	2.80	1.7	4.71	2.2		3250
3500		0.148	0.4	0.277	0.5	0.807	0.9	1.62	1.3	2.92	1.7	4.91	2.3	4.0	3500
3750		0.154	0.4	0.288	0.5	0.839	0.9	1.69	1.3	3.03	1.8	5.10	2.3		3750

M = mass flow rate (kg/s); l_e = equivalent length of pipe ($\zeta = 1$) (m); Δp_l = pressure loss per unit length (Pa/m); v = velocity (m/s).

*Re = 2000; †Re = 3000.

Data drawn from CIBSE Guide C, *Reference data* (1986), by permission of the Chartered Institution of Building Services Engineers.

The velocity corresponding to a volume flow rate of 0.110 litre/s (estimated from table 7.10) is approximately 0.6 m/s and, as 0.6 m/s < 1.5 m/s, this is satisfactory.

Box 3

Required volume flow rate = 0.22 litre/s.

Choosing a 22 mm diameter pipe, 0.324 litre/s > 0.22 litre/s, leaving a net extra volume flow rate of 0.324 − 0.22 − 0.03 litre/s = +0.074 litre/s.

The velocity corresponding to a volume flow rate of 0.324 litre/s (estimated from table 7.10) is approximately 1.0 m/s and, as 1.0 m/s < 1.5 m/s, this is satisfactory.

Boxes 4, 5 and 6

Similar working may be applied to each of these boxes to determine the pipe sizes, and their values may be inserted in the appropriate spaces, as indicated in figure 7.31(b)

Determination of pressure loss/unit length for the ground floor

The head available from the tank outlet to the furthest fitting on the ground floor is 14.5 m and $p = 14.5 \times 10^3 \times 9.81$ Pa, or 142 245 Pa.

The pipe travel from the tank outlet to the furthest fitting, L, is 39 m, and $L_e = 1.25 \times 39$ m, or 48.75 m. $\Delta p_l = 142\,245/48.75$ Pa/m run, or 2918 Pa/m run.

The nearest viable value for Δp_l from table 7.10 is 2750 Pa/m run, and this value has been used to size the pipes for the ground floor.

Value of pressure loss/unit length for the first floor

Δp_l is 2507 Pa/m run and a viable value for Δp_l is 2500 Pa/m run, which has been used to size the pipes for the first floor.

Value of pressure loss/unit length for the second floor

Δp_l for the second floor is 2022 Pa/m run and a viable value for Δp_l is 2000 Pa/m run, which has been used to size the pipes for the second floor.

Value of pressure loss/unit length for the third floor

Δp_l for the third floor is 1439 Pa/m run and a viable value for Δp_l is 1400 Pa/m run, which has been used to size the pipes for the third floor.

The down feed pipes may now be sized by referring to table 7.11, taken from the CIBSE Guide, when the pipe sizes shown on figure 7.31(b) may be obtained.

7.11.3 Size of cistern

The CIBSE Guide (table B4.2) suggests that, for dwellings, houses and flats (up to 4 bedrooms), an allowance of 120 litre/bedroom is satisfactory to cover an interruption in supply of 24 hours' duration. Therefore the actual capacity of the cistern is given by $10 \times 3 \times 120$ litres, or 3600 litres.

7.11.4 Size of galvanised steel supply pipe

Head available to overcome friction in the pipe, $H = 21 - 17 - 2$ m, or 2 m; and $p = 2 \times 10^3 \times 9.81$ Pa, or 19 620 Pa. Also $L = 17 + 5 + 0.5$ m, or 22.5 m, and the equivalent length, $l_e = 1.25 \times 22.5$ m, or 28.125 m. Hence $\Delta p_l = 19\,620/28.125$ Pa/m run, or 698 Pa/m run (say 700 Pa/m run).

It is possible that the head of water available in the cold main will fall below 21 m for part of the 24 hour cycle. If this occurs for 20 hours of the day and all of the capacity has been used, then this leaves just 4 hours for the contents of the cistern to be made up.

Under these conditions a suitable make-up rate is given by $3600/4 \times 3600$ litre/s, or 0.25 litre/s. From table 7.11, a 20 mm diameter galvanised steel pipe will be satisfactory. Note that where the galvanised steel pipes join the copper pipes, it is essential to interpose gunmetal fittings in order to limit possible electrolytic action occurring in the system.

7.12 Rating and capacity of hot water storage calorifiers

7.12.1 Calorifier sizing when the pattern of use is known

The exercise of choosing a calorifier requires prior knowledge of the amount of hot water required, the frequency of demand and the possible recovery rate. For an industrial development, details of these three factors may be available, or may be ascertained fairly easily with a little investigative work.

Example 5

The information given in table A in figure 7.32 has been compiled by the production manager of a small industrial unit, using information obtained from the quoted hot water consumption rates of his equipment, combined with the time of operation of the processes to be used in the plant. Determine:

(a) a suitable storage capacity for a calorifier;
(b) a suitable recovery rate;
(c) an estimate of the power required.

Table 7.11 Flow of water at 10°C in galvanised steel pipes

Δp₁	v	10 mm M	10 mm lₑ	15 mm M	15 mm lₑ	20 mm M	20 mm lₑ	25 mm M	25 mm lₑ	32 mm M	32 mm lₑ	40 mm M	40 mm lₑ	v	Δp₁
400		0.037	0.20	0.081	0.31	0.194	0.49	0.366	0.68	0.814	1.0	1.25	1.3	1.0	400
450		0.039	0.20	0.086	0.31	0.207	0.49	0.390	0.68	0.866	1.0	1.33	1.3		450
500		0.041	0.20	0.091	0.31	0.219	0.50	0.412	0.69	0.915	1.0	1.41	1.3		500
550		0.044	0.21	0.096	0.31	0.230	0.50	0.433	0.69	0.962	1.0	1.48	1.3		550
600		0.046	0.21	0.100	0.32	0.241	0.50	0.453	0.69	1.01	1.0	1.55	1.3		600
700		0.050	0.21	0.109	0.32	0.262	0.51	0.492	0.70	1.09	1.0	1.68	1.3		700
800		0.053	0.21	0.117	0.32	0.281	0.51	0.527	0.70	1.17	1.1	1.80	1.3		800
900		0.057	0.21	0.124	0.32	0.299	0.51	0.561	0.71	1.24	1.1	1.91	1.3		900
1000		0.060	0.22	0.130	0.33	0.316	0.51	0.592	0.71	1.31	1.1	2.02	1.3		1000
1250		0.068	0.22	0.148	0.33	0.355	0.52	0.665	0.72	1.47	1.1	2.26	1.3		1250
1500		0.075	0.22	0.163	0.33	0.390	0.52	0.732	0.72	1.62	1.1	2.49	1.3	2.0	1500
1750		0.081	0.22	0.177	0.34	0.423	0.53	0.792	0.73	1.75	1.1	2.69	1.3		1750
2000		0.087	0.23	0.190	0.34	0.453	0.53	0.849	0.73	1.88	1.1	2.88	1.4		2000
2250	1.0	0.092	0.23	0.202	0.34	0.482	0.53	0.902	0.73	2.00	1.1	3.06	1.4		2250
2500		0.098	0.23	0.213	0.34	0.509	0.53	0.952	0.74	2.11	1.1	3.23	1.4		2500
2750		0.103	0.23	0.224	0.34	0.534	0.54	1.00	0.74	2.21	1.1	3.39	1.4		2750
3000		0.108	0.23	0.234	0.34	0.559	0.54	1.05	0.74	2.31	1.1	3.54	1.4		3000
3500		0.117	0.23	0.254	0.35	0.605	0.54	1.13	0.74	2.50	1.1	3.84	1.4	3.0	3500

Table 7.11 (continued)

Δp_l	v	50 mm M	l_e	65 mm M	l_e	80 mm M	l_e	100 mm M	l_e	125 mm M	l_e	150 mm M	l_e	v	Δp_l
175		1.54	1.7	3.20	2.4	5.00	3.0	10.2	4.3	18.4	5.8	29.8	7.3		175
200		1.66	1.7	3.44	2.5	5.37	3.1	11.0	4.4	19.7	5.8	31.9	7.4		200
225		1.76	1.7	3.66	2.5	5.79	3.1	11.6	4.4	21.0	5.8	33.9	7.4		225
250		1.86	1.7	3.86	2.5	6.03	3.1	12.3	4.4	22.2	5.9	35.8	7.4		250
275	1.0	1.96	1.7	4.06	2.5	6.34	3.1	12.9	4.4	23.3	5.9	37.6	7.4	2.0	275
300		2.05	1.7	4.25	2.5	6.63	3.1	13.5	4.4	24.3	5.9	39.3	7.4		300
350		2.22	1.8	4.60	2.5	7.18	3.1	14.6	4.4	26.3	5.9	42.5	7.5		350
400		2.38	1.8	4.93	2.5	7.69	3.2	15.7	4.5	28.2	5.9	45.6	7.5		400
450		2.53	1.8	5.24	2.5	8.17	3.2	16.6	4.5	30.0	6.0	48.4	7.5		450
500		2.68	1.8	5.54	2.6	8.63	3.2	17.6	4.5	31.6	6.0	51.1	7.5		500
550		2.81	1.8	5.82	2.6	9.06	3.2	18.5	4.5	33.2	6.0	53.6	7.6		550
600		2.94	1.8	6.08	2.6	9.48	3.2	19.3	4.5	34.7	6.0	56.0	7.6	3.0	600
700		3.19	1.8	6.58	2.6	10.3	3.2	20.9	4.5	37.6	6.0	60.6	7.6		700
800		3.41	1.8	7.05	2.6	11.0	3.2	22.4	4.5	40.2	6.0	64.9	7.6		800
900		3.63	1.8	7.49	2.6	11.7	3.2	23.7	4.6	42.7	6.0	68.9	7.6		900
1000	2.0	3.83	1.8	7.91	2.6	12.3	3.2	25.1	4.6	45.0	6.1	72.7	7.6	4.0	1000

M = mass flow rate (kg/s); l_e = equivalent length of pipe ($\zeta = 1$) (m); Δp_l = pressure loss per unit length (Pa/m); v = velocity (m/s).

*Data drawn from CIBSE Guide C, *Reference data* (1986), by permission of the Chartered Institution of Building Services Engineers.

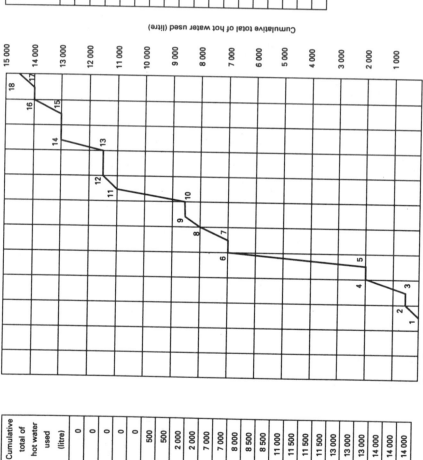

Table B

Position on graph	Storage before change (litre)	Adjustments Draw-off (litre)	Adjustments Recovery (litre)	Storage after change (litre)
1	5000	-0	+0	5000
2	5000	-500	+500	5000
3	5000	-0	+0	5000
4	5000	-1500	+1000	4500
5	4500	-0	+500	5000
6	5000	-5000	+1000	1000
7	1000	-0	+1000	2000
8	2000	-1000	+1000	2000
9	2000	-500	+1000	2500
10	2500	-0	+1000	3500
11	3500	-2500	+1000	2000
12	2000	-500	+1000	2500
13	2500	-0	+2000	4500
14	4500	-1500	+1000	4000
15	4000	-0	+1000	5000
16	5000	-1000	+1000	5000
17	5000	-0	+0	5000
18	5000	-500	+500	5000

Cumulative total of hot water used (litre)

Hour of day

Table A

Hour of the day	Cumulative total of hot water used (litre)
01	0
02	0
03	0
04	0
05	0
06	500
07	500
08	2 000
09	2 000
10	7 000
11	7 000
12	8 000
13	8 500
14	8 500
15	11 000
16	11 500
17	11 500
18	11 500
19	13 000
20	13 000
21	13 000
22	14 000
23	14 000
24	14 000

Figure 7.32 Calorifier sizing when the pattern of use is known

Solution

The graph shown on figure 7.32 may be plotted from the information given in table A. The greatest draw-off rate corresponds to the steepest slope, and this occurs between points 5 and 6 on the graph and has a value of 5000 litre/h.

As a first approximation, it has been found to be generally satisfactory to make the storage capacity of the calorifier equal to the value of this quantity, that is 5000 litres.

At point 6 on the graph, the volume of hot water contained in the calorifier has its lowest value and it is logical to base the recovery rate on the next 5 hour period from points 6 to 11, when 4000 litres of hot water must be made up. Under these conditions the recovery rate required is given by 4000/5 = 800 litre/h.

For ease of working, a recovery rate of 1000 litre/h could be assumed and this figure may then be used to compile table B, which gives an indication of the volume of hot water available from storage at each of the numbered points on the graph.

An inspection of the figures in the last column will reveal that, at point 6, the lowest volume available is 1000 litres. It is sometimes suggested that the volume of hot water remaining in the calorifier should not be less than 20% of the storage volume and, in this case, (20/100) × 5000 = 1000 litres, which is equal to the volume of hot water remaining at point 6.

After a heavy draw-off period, such as that which occurs between 5 and 6, it is likely that the temperature stratification within the storage calorifier will have been destroyed and, in an attempt to minimise this effect, it is usual to increase the storage volume by 25%, making the proposed new storage capacity equal to 1.25 × 5000 litres, or 6250 litres.

In this instance, the arrangement of having a storage capacity of 6250 litres with a recovery rate of 1000 litre/h will be satisfactory.

At this stage, the magnitude of the storage capacity and the value of the recovery rate may have to be varied to suit the actual conditions on the site, and this may be facilitated by compiling a new table, in the style of table B, until a satisfactory practical solution is obtained.

The power P (kW) required to support the recovery rate, may be calculated from the equation $P = m \times C_p \times \Delta T$

Here m = 1000/3600 kg/s, C_p = 4.2 kJ/kg K and ΔT = (65 − 10) K. Hence P = (1000/3600) × 4.2 × 55 kW, or 64 kW.

If the boiler output is dedicated only to the calorifier, an allowance must be made for any casing losses from the calorifier, for emission losses from the connecting pipes and for ease of warming up.

7.12.2 *Calorifier sizing when the pattern of use is not known*

When the pattern of use is not known, the choice of a suitable storage capacity and recovery rate has an infinite number of solutions, and because of this a practical solution to the problem is fraught with difficulties.

In the absence of more detailed information, a suggested first step is to use the empirical approach given in example 5 to decide on the maximum volume of draw-off/h and then make the storage capacity equal to this value, sizing the boiler so that it is capable of heating this quantity over a period of from 2 to 3 h, depending on the assumed pattern of use expected in the building.

In order to arrive at values that are realistic, recourse may be made to published information of the average consumption of hot water at 65°C in litre/person day and the average peak consumption of hot water in litre/h person day, available for different types of accommodation. These figures, together with published information dealing with the capacity of various sanitary appliances, indicate the use of equipment that, in the experience of the author, usually results in an installation of oversized boilers and calorifiers.

The CIBSE Guide has published data, based on the work of many investigators, which provides a solution for those types of building for which information is available. This work is presented in the sections B4-11 to B4-19 of the Guide and covers the supply of energy both on- and off-peak. It deals with the supply of domestic hot water (hot water for services) and hot water for catering purposes. Examples 6 and 7 cover the calculations when using energy on-peak with a conventional boiler and calorifier, and off-peak when using electrical energy and a storage cylinder.

Example 6

Domestic hot water at a temperature of 60°C is to be provided for use in a 100-bed students' hostel from a storage calorifier heated by a boiler. The recovery period is to be 1.5 h. Determine:

(a) a suitable storage capacity for the calorifier, assuming a 'mixing allowance' of 25% of the nominal capacity and a cold feed temperature of 10°C;
(b) a suitable power for the boiler, allowing for a plant warming-up factor of 1.15 and a combined emission loss from the calorifier casing and connecting pipework of 5 kW.

Solution

The information given on figure 7.33, graph (a), has been taken from the CIBSE Guide and is apposite to a students' hostel housing between 80 to 320 persons/day; it applies when the temperature of the stored water is

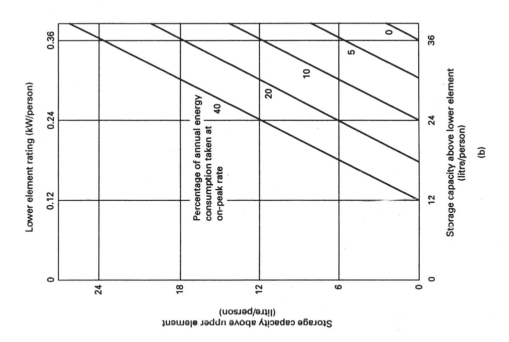

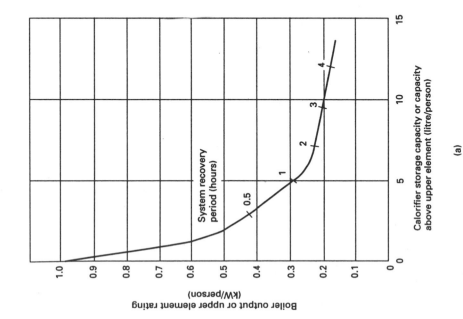

Figure 7.33 Graphs of plant sizing guide for services to a student hostel catering for 80–320 persons/day

at 65°C. The curve is obtained by plotting the required boiler output (or upper element rating), in kW/person, against the storage capacity of the calorifier (or storage capacity above the upper element), in litre/person. Superimposed on the curve are points representing various recovery rates. The axes of the graph each have a dual title because the curve can be used when dealing with systems utilising both on- and off-peak energy. Entering the graph at a recovery rate of 1.5 h, it can be seen that a nominal boiler output of 0.25 kW/person and a nominal calorifier storage capacity of 6 litre/person are required, when the storage temperature is 65°C.

For the boiler

As the storage temperature is to be 60°C, the boiler output will need to be slightly less and must be modified by the ratio (60 − 10)/(65 − 10). Hence the new nominal boiler power required is 0.25 x 50/55 kW/person, or 0.23 kW/person. The total nominal power required then becomes 0.23 × 100 kW, or 23 kW.

The actual boiler power required is (23 + 5) × 1.15 = 32.2 kW, and if a boiler is available having a rating of 32 kW, then this will be satisfactory.

For the storage capacity

The total nominal capacity required is 6 × 100 litres, or 600 litres. The actual storage capacity using a 'mixing allowance' of 25% of the nominal capacity then becomes 1.25 × 600 litres, or 750 litres.

Example 7

The owner of the hostel in example 6 has requested that a study should be made as to the feasibility of heating the domestic hot water using off-peak electrical energy, using a proportion of annual load taken off-peak of 80%. The recovery period is to be 1.5 h. Determine:

(a) the storage capacity and power required for on-peak use, assuming an emission loss from the casing of 1.5 kW;
(b) the storage capacity and power required for off-peak use.

Solution

A typical storage cylinder for such a system would be mounted vertically and be provided with a bank of immersion heaters, some fitted near the top of the cylinder and some near the bottom.

(a) *For the upper part*

Storage volume needed

From figure 7.33, graph (a): nominal storage factor above upper elements
= 6 litre/person. The actual storage above the upper element = 6 × 100
litres, or 600 litres.

Power needed

From figure 7.33, graph (a): nominal rating factor for the upper element
= 0.25 kW/person. The nominal rating factor for the upper element adjusted for the difference in storage temperature is then 0.25 × (60 − 10)/
(65 − 10) kW/person, or 0.23 kW/person. The nominal rating for the
upper element = 0.23 × 100 kW, or 23 kW. The actual rating for the
upper element allowing for the casing emission loss = 23 + 1.5 kW, or
24.5 kW, say 25 kW.

(b) *For the lower part*

Storage volume needed

From figure 7.33, graph (b): enter the graph on the vertical axis at 6 litre/
person and, where the horizontal line intersects the 20% line of percentage annual energy consumption taken at on-peak rate, draw a vertical
line and read off the nominal storage factor above the lower element =
24 litre/person.

Nominal storage above the lower element = 24 × 100 litres, or 2400
litres.

The heat loss from the casing over 16 h = 1.5 × 16 × 3600 kJ, or
86.4 MJ.

80% of this loss, that is 0.8 x 86.4 MJ, or 69.12 MJ, is to be made up
off-peak.

This may be turned into an equivalent mass of water using

$$Q = m \times c_p \times \Delta T$$

where Q = heat to be made up off-peak (kJ)
m = mass of water (kg)
c_p = specific heat capacity for water = 4.2 kJ/kg K
ΔT = temperature difference through which the water is to be
raised (K).

Transposing and inserting values, we have:
$m = 69.12 \times 10^3/4.2 \times (60 - 10)$ kg or $m = 329$ kg
Hence the equivalent volume is 329 litres.

The amended storage above the lower element = 2400 + 329 litres, or
2729 litres.

Power needed

From figure 7.33, graph (b): reading upwards on the vertical line through the previous point of intersection, the nominal rating factor for the lower element = 0.24 kW/person and the nominal rating for the lower element = 0.24 × 100 kW, or 24 kW.

This figure must be increased proportionally to heat the equivalent volume. Thus:

24 kW heats 2400 litres
24/2400 kW heats 1 litre and
24 × 2729/2400 heats 2729 litres
or 27.3 kW heats 2729 litres

Hence make the actual rating for the lower element = 27 kW.

In an attempt to minimise the effect of turbulence during heavy periods of draw-off and depending on the way in which the cold feed is introduced into the storage cylinder, it is customary to add an allowance on the calculated storage capacity of 25%. Thus total volume of storage vessel above the lower element is given by 1.25 × 2729 litres, or 3411 litres, say 3400 litres.

References

7.1. White S.F. and Mays G.D. *Water Supply Bye-laws Guide*, 2nd edition, 1989. Ellis Horwood Ltd, Chichester, for Water Research Centre.
7.2. Department of the Environment. *Backsiphonage*, 1974. HMSO, London.
7.3. BS 6700: 1987 *Design installation, testing and maintenance of services supplying water for domestic use within buildings and their curtilages*. BSI, London.
7.4. Garrett R.H. *Hot and Cold Water Supply*, 1991. BSP Professional Books, Oxford.
7.5. Coe A.L. *Water Supply and Plumbing Practices in Continental Europe*, Hutchinson Benham, London. [Copy available in the Science Reference Library, 25 Southampton Buildings, London WC2 (Tel.: 0171 323 7494)]
7.6. Barnes J.V., Enga T.G. and Wright G.F. The potential for improved energy efficiency in existing office buildings. Presented to the *Conference on the Impact of High Insulation on Buildings and Services Design*. University of Nottingham, March 1987.
7.7. Massey B.S. *Mechanics of Fluids*, 6th edition, 1989. Van Nostrand Reinhold, Wokingham.
7.8. Hamill L. *Understanding Hydraulics*, 1995. Macmillan, London.

8 Fire Safety and Firefighting

8.1 Fire protection code

The Fire Protection Association is the UK's national fire safety organisation, and operating as a division of the Loss Prevention Council, it has produced (May 1992) a code of practice entitled *Fire Protection on Construction Sites* [8.1]. Building contractors are now finding that, when tendering for work, the adoption of this code is more frequently becoming a necessary prerequisite to their being able to obtain insurance against fire risks during the construction period, and can also lead to lower insurance premiums on the completed buildings.

The code gives design and management guidelines, which are aimed at reducing the incidence of fires on site. Some of the topics dealt with include the following:

(1) Scheduling and early implementation of fire components and systems into the building, together with the establishment of escape routes.
(2) Appointment by the main contractor of site fire safety co-ordinators, who will be responsible for all contractors complying with a 'Site Fire Safety Plan'.
(3) Instigation of a 'permit to work' system whenever hot working occurs.
(4) Acquisition of fire certificates (when required), for temporary site accommodation.
(5) Placing of site huts, having regard to the safe distance of any hut from the building under construction.

8.2 Building Fire Plan

The Fire Plan for a building is drawn up by a group of people representing the owner, the occupier, the fire authority, the insurers, the water authority, the design team and possibly a fire consultant. The plan is concerned with finding answers to criteria affecting the methods of fireproofing, the way (or ways) in which a fire will be extinguished, the way in which the occupants will be evacuated from the building, and the way

357

in which the fire brigade is to be summoned. A detailed check list of the topics that have to be addressed is published in section B5-4 of the 1986 CIBSE Guide [8.2], together with a comprehensive bibliography (with dates) on the subjects of Fire Safety Engineering and Fire Fighting Systems.

When the plan is completed, information will be available on firefighting systems and their design parameters for which the building services engineer will be responsible, and because the Fire Plan is considered to be part of a continuing fire safety programme, any alterations to the building and their possible effect on existing systems will become matters for subsequent appraisal by the managers.

Depending on the geographical location, the type of development and the nature of the risk, different authorities have an interest in stipulating the ways in which the fire systems relating to the building should be implemented. In the UK, the District Council deals with Building Regulations, the County Fire Brigade deals with the award of fire certificates and the Health and Safety Executive, or the Environmental Health Officer, deals with those fire risks associated with industrial processes. However, the perceived functions of the various authorities in different regions of the British Isles do vary.

In order to help deal with the specific requirements of these different bodies, CIBSE has published technical memoranda TM 2, 8, 9 and 12, the topics dealt with in these publications being as follows:

TM 2 – Notes on legislation relating to fire and services in buildings
TM 8 – Design notes for ductwork
TM 9 – Notes on non statutory codes and standards relating to fire and services in buildings
TM 12 – Emergency lighting

The specific requirements of the local authorities referred to in this section cover the following four topics.

(1) *Compartmentation and fire separation*

These refer to the size of compartmentations, their fireproofing and the provision of protected escape routes. In the UK, reference should be made to *The Building Regulations 1991*, HMSO, London, or to the latest *Building Regulations* of Scotland, or of Northern Ireland.

Fire-resistant compartments must also be incorporated in the following places:

(1) stores or workplaces containing highly combustible materials and/or flammable liquids;
(2) places used for bulk storage;
(3) plant and equipment rooms;

(4) kitchens;
(5) car parks and garages.

In addition, wherever pipes, cables and ducts pass through fire protected boundaries, the openings around them must be protected to the same degree as the boundary through which they are passing. Ductwork passing through such boundaries must be fabricated from incombustible materials, and both the ductwork and any insulation surrounding it should offer the same resistance to fire as the boundary. The ductwork must also be fitted with either an automatic fire damper operated by a fusible link, or an intumescent-coated honeycomb damper.

(2) *Ventilation and air conditioning systems*

These systems should be operated so that the movement of air in the building is always in a direction away from designated escape routes. Also, smoke detectors must be fitted in any return air duct which, on operation, will either shut down the fans, or discharge the contents of the return air duct to the atmosphere. Fan motor override switches must also be provided and sited in positions that have been agreed with the Fire Service personnel.

(3) *Electrical services*

Any insurance company will testify to the importance of the effect of poor electrical installations on the incidence of fires in buildings, and it is essential that in the UK the current rules and regulations of the Institution of Electrical Engineers (IEE) should be followed in all developments.

(4) *Control of smoke*

The presence of smoke in a building is rightly regarded as being potentially dangerous. It can bring about the suffocation and poisoning of those who inhale it and also bring visibility down to zero, reducing the chances of escape of those caught in a fire. It is obviously important to keep smoke away from protected escape routes, and both active and passive means are employed in this process. Details of the statutory requirements will be found in [8.3] and [8.4]. Design details on the pressurisation of protected escape routes in flats, maisonnettes, offices and shops are given in [8.5]. Information on shopping malls and town centre redevelopment will be found in [8.6], while the treatment of atria will depend on the requirements of both the local Fire Authority and the local authority Building Control Department. Information on the treatment of atria and details of the fire insurer's rules for the fire protection of industrial and commercial buildings will be found in [8.7].

8.3 Smoke control in protected escape routes using pressurisation

The smoke produced by any fire will always move towards a region of low pressure, and it follows that if the pressure in a designated escape route is maintained at a value that is greater than the pressure in the surrounding spaces, the escape route will remain free from smoke.

The difference in pressure required is small and of the order of 50 Pa. Readers should note that higher pressures will increase the amount of force required to open doors, to a point where the difference will become noticeable by an adult; when a building is occupied mainly by children, the floor covering adjacent to the doors should be chosen after taking into account its non-slip properties.

The outer envelope of a building does not provide an airtight enclosure, so that the lateral internal pressure gradient will be affected by the direction and force of the wind. The disposition of pressure will also be affected by the movement of air caused by any ventilation or air conditioning system, and the occurrence of a fire will have the effect of lowering the internal air pressure in the immediate vicinity.

The pressurisation of a designated escape route can be brought about by installing a fan and ductwork system, drawing fresh air into the building, delivering it through grilles and/or registers, placed evenly throughout the escape route, and then allowing it to leave the building.

8.3.1 Importance of removing air from the building

It is essential to the proper functioning of the pressurisation system that the air introduced to the escape route is allowed to leave the building via the external walls of the accommodation space, using exits that are either fortuitous, or have been contrived. This process will ensure that a diminishing pressure gradient is established across the building in the direction of the air flow. If the air was prevented from escaping, the value of the pressure gradient would fall to zero and the passage of smoke would not be controlled. The apparatus needed to remove air should become operational as soon as a fire starts, and it should remain so for the period corresponding to the standard of fire resistance of the structural elements of the building. The Code of Practice [8.5] gives four methods by which air may be removed from the building.

Method 1

This involves the use of gaps inherent in the construction and fitting of window frames. These gaps are referred to as 'crackage' and, for three types of window, the code gives values for the minimum recommended length of crackage, as a function of the net volume flow rate of pressurising air that will be required. In some buildings, natural crackage is very much less than it is in others and, when this is known to be the case, other means of producing a satisfactory exit for the air after it has passed through the occupied space must be used either in whole or in part.

Method 2

This involves the use of vents fitted around the periphery of the space. In order to prevent draughts, they will normally be in the closed position and when the pressurisation system is activated, the vents will open, allowing air to pass to the outside of the building. The vents must be designed to operate with a very small difference in pressure and they should be capable of self-closing, momentarily, against the pressure of a gust of wind. In the case of vents that are automatically controlled, the code makes the proviso that in the event of a fire, only those vents situated on the fire floor should open, and this has the result of increasing the effectiveness of the pressurisation system on the fire floor.

Method 3

This involves the use of ventilation shafts passing vertically through the building and should only be used when methods 1 and 2 are not possible. The preferred construction involves the use of vents on each floor, fitted to the periphery of each vertical shaft. They are normally held in the closed position and when the pressurisation system is activated, only those vents situated on the fire floor are opened. This has the double advantage of preventing smoke from one floor reaching other floors through the vertical shaft and of increasing the effectiveness of the pressurisation system on the fire floor.

Method 4

This involves the use of an extract fan and vertical duct system, constructed to the appropriate standard of fire resistance. As in methods 2 and 3, the extract vents from each floor are normally held in the closed position. When the pressurisation system is activated, only those vents situated on the fire floor are opened. It is important that the mechanical extract fan and motor should be able to operate for a reasonable length of time when handling fumes at temperatures approaching 500°C.

8.3.2 Position of air inlets

An obvious requirement is that the fresh air inlet should not be sited where it is likely to draw in smoke; such a condition is most likely to be met by placing the inlet at or near ground level, and below and at least 5 m (measured horizontally) away from any duct outlets, which in an emergency may be discharging smoke. Note also that the ductwork should not be fitted with automatic dampers designed to close down the system in the event of a fire.

When pressurising a staircase with a corridor and office accommodation adjacent, one possible arrangement for a low rise building having more than one staircase is to use a vertical builders' duct that has been lined and fireproofed, and which delivers fresh air to each floor. An individual fan with stub ducts may then be fitted at each floor level, drawing air from the vertical duct and delivering it, via grilles and/or registers, to the protected spaces. The system should be calibrated so that the pressure in the stairwell is slightly greater than that in the corridor, and the air should then be allowed to escape from the building through the office space by purpose-made vents or via the existing crackage. Note that for buildings having only one staircase, the code recommends that duplicate fans and motors should be provided, and also stipulates ways in which the operation of the system should be linked to a standby generator.

In a multi-storey development it is possible to take advantage of scale and use a single fan (having a duplicate if necessary) located in a plant room. Such a fan supplies fresh air to lined and fireproofed vertical ducts, which then deliver the air to the spaces being pressurised on each floor, via grilles and registers.

8.3.3 Advantages of a pressurisation system

These are as follows:

(1) it will provide a safe escape route for the occupiers;
(2) staircases and lobbies need not be placed on external walls;
(3) provision of smoke shafts for alternative ventilation may not be required;
(4) some 'smoke stop' doors may be omitted from the escape route;
(5) it may be possible to reduce the number of staircases required, based on the density of population;
(6) fortuitous energy losses due to 'natural' methods of ventilation are eliminated.

Two categories of pressurisation are described in [8.5]: single-stage and two-stage. Single-stage pressurisation comes into operation only in the event of a fire. Two-stage pressurisation operates at two pressure levels – the lower pressure is applied continuously and the higher level of pressure is produced only in an emergency. Two-stage pressurisation is preferred but the preference is not exclusive, provided the recommendations made in the code are followed.

8.3.4 Four methods of carrying out the pressurisation of a building

Method 1: Pressurising staircases only

According to the code of practice [8.5], this method should only be used where the horizontal approach from the accommodation to the staircase is minimal and at most is via a simple lobby. A simple lobby is one that has no door opening out, other than a door to a staircase and a door (or doors) leading out to the accommodation. The authors of the code have added an amendment stating that 'during a fire emergency, all protected staircases interconnected by lobbies, corridors or accommodation areas, should be simultaneously pressurised'.

The values of the pressures above atmospheric are expected to be of the order of 50 Pa and 10 Pa for the stairwell and lobby respectively.

Method 2: Pressurising staircases and all, or part of, the horizontal route

This method is to be used for those buildings in which the horizontal approach is not through a simple lobby, but through a lobby having doors to lifts. In this case the pressurisation should be taken into the lobby and possibly into any corridor beyond, and during an emergency all these spaces should be pressurised simultaneously. The code goes on to suggest that the pressurisation of the corridor should be independent of that in the stairwell, that is a separate duct system should be provided for each space and the pressure drop from the staircase to the next space (or spaces) should have decrements of pressure of no more than 5 Pa as the accommodation area is approached.

Method 3: Pressurising lobbies and/or corridors only

This method may be used where there is difficulty in arranging the ductwork necessary for pressurising the staircases. In this case the air required to pressurise the staircases must come from the ductwork that supplies the air to the lobbies or corridors.

Method 4: Pressurising the whole of the building (not recommended)

The operation of this method relies on the opening of vents on the floor where fire breaks out, so allowing the smoke to be forced from the building. Under these conditions the inadvertent opening of a door at the foot of a staircase may cause the stairwell to become filled with smoke, trapping the occupants. Additional 'smoke stop' doors must be fitted to combat the likehood of such an occurrence, and the authors of the code do not recommend this method.

8.3.5 Restrictions in the application of the pressurisation system

The existence of pressurised stairwells in the same building as naturally ventilated stairwells must be avoided. In the event of a fire, it can be expected that the naturally vented area will fill with smoke and under no circumstances should such a stairwell be connected via a corridor and lobby to a pressurised stairwell.

Also, an accommodation area should not be connected to a pressurised stairwell via a ventilated lobby. The lobby will fill with smoke, diluted somewhat by the escape of air from the pressurised stairwell. However the situation is not regarded as being conducive to the safe evacuation of the building and, to avoid such a syndrome, the lobby should be an unventilated one.

8.3.6   Need for integration between an air conditioning system and a two-stage pressurisation system

If the building has a ventilation system or an air conditioning system, it is important that the stale air be removed from the building at points that are remote from the pressurised spaces, and the designers of both systems should be aware of this need.

It is essential that the two systems are compatible. In the event of an emergency, the code suggests that the following events should take place:

(1) any recirculation of air should be discontinued and all exhaust air should be directed to the outside of the building;
(2) any air supply to the accommodation spaces should be discontinued;
(3) the exhaust action may be continued, provided the integrity of the fan and duct system can be assured and provided also that the movement of air is away from the protected escape route.

(In order to comply with point 3 it will be necessary to protect both the exhaust fan and its power supply.)

The change in the mode of operation of the air conditioning system should be triggered by the same transducer (usually a smoke detector) as that used to operate the second stage of the two-stage pressurisation system, and the point is made that the use of a smoke detector fitted into the exhaust duct will not produce the result intended, owing to the dilution in the concentration of smoke in the exhaust air.

8.3.7   System design

The authors of the British Standard [8.5] are to be congratulated on producing a document containing details of a design method (complete with a worked example) that is both comprehensive and understandable. Those readers who need to design, specify, produce, test, or operate and maintain pressurisation systems will find that the information given in the document is essential to their purpose.

8.4   Classification of types of fire hazard

In any development it is the responsibility of the person being insured to confirm, with the insurers, the nature of the hazards involved. To facilitate this process, BS 5306: Part 2: 1990 *Specification for sprinkler systems*

[8.8] has categorised the severity of hazards associated with the various types of occupancy. There are three main types of hazard:

(1) light hazard;
(2) ordinary hazard; and
(3) high hazard.

Both ordinary hazard and high hazard are subclassified into further categories.

8.4.1 Light hazard

This pertains to non-industrial buildings where the amount and combustibility of the contents are low, such as offices, libraries, colleges and prisons. However, for certain parts within these buildings where the area involved exceeds 36 m^2 and which include attics, basements, boiler rooms, kitchens, laundries, storage areas and workrooms, the classification is one of 'ordinary hazard'.

8.4.2 Ordinary hazard groups I to III and group IIIS (special)

The classification of ordinary hazard is subdivided into groups I to III and group IIIS (group I being less hazardous then group IIIS); some typical examples of occupancies that go to make up these sub-groupings are as follows:

Group I   Cement works, jewellery factories, abattoirs, breweries, creameries, restaurants and cafes.

Group II   Selected chemical works, engineering works, brewery bottling factories and confectionery manufacturers.

Group III   Factories manufacturing glass, aircraft, motor vehicles and electronic goods. Also motor garages and car parks (above or below ground).

Groups IIIS   Match factories, distilleries, theatres and film and television studios.

The type of goods and the way in which they are stored present special problems to firefighters, and storage materials have been placed into categories I to IV in order of ascending risk. Some typical examples are as follows:

Category I   Carpets, textiles, fibreboard and groceries.

Category II   Baled waste paper, chipboard and cartons containing alcohols in cans or bottles.

Category III   Bitumen or wax-coated paper, foam plastic and foam rubber products.

Category IV   Foam plastic and foam rubber in sheet or moulded form.

Reference [8.8] shows that the method of storage has been classified into types of storage S1, S2, S3, etc., and is followed by recommendations (for those situations where a sprinkler system is to be used), having regard to the maximum storage height, the positioning of the sprinklers and the density of storage. S1 type of storage refers to free-standing or block stacking, S2 refers to post or box pallets in single rows, and S3 to post or box pallets in multiple rows.

8.4.3 High hazard

This hazard covers commercial and industrial occupancies having abnormal fire loads, and the genre is subclassified into four categories:

(a) process high hazards;
(b) high-piled storage hazards;
(c) potable spirit storage hazards;
(d) oil and flammable liquid hazards.

(a) Process high hazards

Processes using materials mainly of a hazardous nature likely to develop into rapidly and intensely burning fires are themselves subclassified as type 1, 2, 3 or 4, as shown in table 4 of [8.8].

(b) High-piled storage hazards

Goods including packaging, stored so as to be likely to produce exceptionally intense fires with a high rate of heat release, are to be classified as high-hazard, with special recommendations with regard to the height of the sprinkler system as in clause 5.4.3 of [8.8].

(c) Potable spirit storage hazards

Potable spirit not in racked barrels stored to certain specified heights, and potable spirits in racked barrels exceeding the same heights shall be classified according to the details given in clause 5.4.4 of [8.8].

(d) Oil and flammable liquid hazards

This hazard is to be found in occupancies where oil and flammable liquids are stored or used, in such quantities and in such a manner, that ordinary sprinkler protection may not be effective, and clause 5.4.5 of [8.8] suggests consideration of the suitability of a deluge system, fitted with medium- or high-velocity sprayers.

8.5   Sprinkler systems, manual and automatic

One of the first rules in establishing a successful firefighting technique is to be in a position to extinguish the outbreak as soon as it occurs, and it is this truism that resulted in the initial development of sprinkler systems.

Reference [8.9] mentions that around the middle of the 19th century, manually operated sprinkler systems consisting of perforated pipes and roses were to be found in extensive use in the cotton mills of New England, but these systems had several serious disadvantages.

In the latter part of the 19th century, automatic sprinklers were developed.

In the UK, the Loss Prevention Council (LPC) superseded the Fire Offices Committee of London (FOC), and has become the foremost authority for setting standards for the manufacture and testing of equipment and installations associated with water-based fire extinguishing systems; it is currently being used by the majority of insurers in the UK. The LPC worked closely with the British Standards Institution on the revision of BS 5306: Part 2 *Specification for sprinkler systems*, and is also involved with revision work sponsored by the International Standards Organisation (ISO). It also produced the authoritative 'Loss Prevention Council Rules for Automatic Sprinkler Installations' (1994).

The systems resulting from the application of the LPC rules will be suitable for a particular type of occupancy and will deliver water with a stated density (intensity) of coverage, over a specified area, for an allotted time.

Statistics on the operation of sprinkler systems, quoted in [8.10], state that:

55% of fires were extinguished by the operation of two or less sprinkler heads
80% of fires were extinguished by eight or less sprinklers and
90% of fires were extinguished by 18 or less sprinklers.

The record is a good one and savings of about 50% may be made on insurance premiums by installing sprinkler systems.

In order to achieve the density of discharge specified, the pressure of water in the system must conform to particular values at certain key positions in the pipework and, for those who have to design sprinkler systems, these values and the density of discharge required for various occupancies are tabulated in [8.8].

8.5.1   Delineation and selection of water supplies for low rise systems

Important aspects of all sprinkler systems are the source and reliability of the water supplies, and [8.8] (clauses 13 and 17) lists the acceptable sources and gives detailed specifications for each of them. There are three types of supplies, classed as single, superior and duplicate, and [8.8] suggests that, wherever it is practical, superior supplies, or duplicate supplies, should be provided.

(1) Single supplies for use with light and ordinary hazard occupancies

A single supply should consist of one of the following:

(a) a town main complying with clause 17.1.1 of [8.8]; or
(b) a single automatic suction pump, drawing from a source complying with clause 17.4.3.1 of [8.8]; or
(c) a single automatic booster pump, drawing water from a town main (water authority permitting), complying with clause 17.1.1 of [8.8].

(2) Superior supplies for use with high hazard occupancies

A superior supply shall consist of one of the following:

(a) a town main; or
(b) two automatic suction pumps drawing water from a suction tank complying with clause 17.4.3.2 of [8.8]; or
(c) two automatic booster pumps; or
(d) an elevated private reservoir; or
(e) a gravity tank; or
(f) a pressure tank, conditional upon this option applying only to light and/or ordinary hazard group 1 occupancies.

(3) Duplicate supplies for use with high hazard occupancies

The various combinations suitable for duplicate water supplies are given in tabular and diagrammatic form in [8.8]. Typically, these could consist of one supply being taken from a gravity tank and the other from a booster pump drawing from an elevated private reservoir. However, many other combinations are possible.

8.5.2 Delineation and selection of water supplies for high rise systems

The water supply for a high rise system shall be either (a) a gravity tank, or (b) an automatic suction pump arrangement, in which each system is served by either a separate pump or a separate stage of a multi-stage pump.

8.5.3 Arrangement of pipework for a sprinkler system

Figure 8.1(a) details the various types of pipework used to supply sprinkler heads, and figure 8.1(b) shows the various arrangements of pipework.
The number of sprinkler heads allowed on range pipes and fed by distribution pipes depends on:

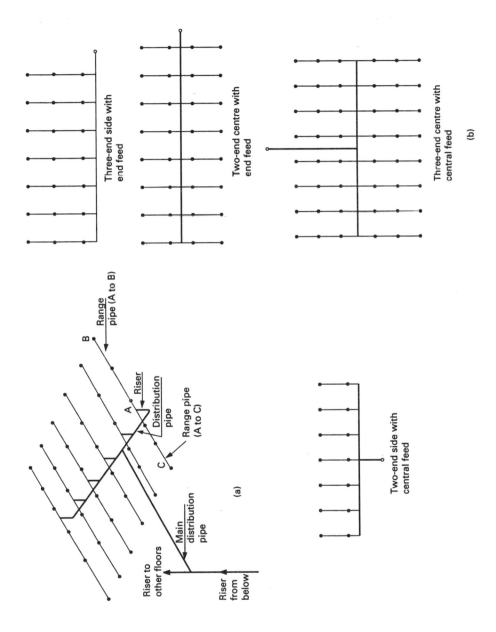

Figure 8.1 (a) Annotation of pipework used in sprinkler systems. (b) Description of pipework arrangements used in sprinkler systems

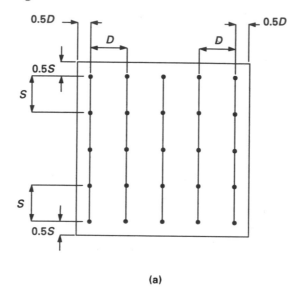

(a)

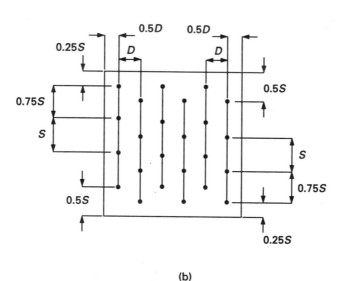

(b)

Figure 8.2 (a) Standard layout of sprinklers in a rectangular matrix. (b) Staggered layout of sprinklers for ordinary hazard systems where $S > 4$ m

(1) the type of hazard;
(2) the pipe arrangement; and
(3) the diameter of both range and distribution pipes.

Figures 8.2(a) and (b) show the ways in which sprinkler heads may be positioned over the area to be protected. The maximum distance between sprinklers, S, on range pipes and between adjacent rows of range pipes, D, and the maximum area of coverage for a sprinkler head ($S \times D$) depend on the type of hazard, and full information on these and other topics is given in clause 26 of [8.8].

Specific information is also given about the placing of sprinklers in film and television studios and in theatres. In the latter case, the sprinkler system becomes a life safety system, as opposed to a system designed mainly to prevent damage to property. For a life safety system, such as might also be installed in covered and enclosed shopping complexes, the additional safeguards required are given in detail in [8.8].

8.6   Types of sprinkler systems

There are five main types of system:

(1) wet pipe;
(2) alternate wet and dry pipe;
(3) dry pipe;
(4) pre-action;
(5) re-cycling.

Three other additional types of system may be combined with (1) and/or (2) to form extensions. These are:

(a) tail-end alternate;
(b) tail-end dry pipe; and
(c) deluge.

(1) Wet pipe system

Generally wet pipe systems are preferred, and as the title suggests, these systems are permanently charged with water and may only be installed in spaces where the temperature will not rise above 70°C and where the premises are not liable to frost damage. Where this is the case, complete self-draining of the system is not important and the sprinkler heads can be fitted either above, or below, the range pipes. When they are fitted below the range pipe, they are referred to as being 'pendent' and because they are not normally self-draining, they are not suitable for use where frost damage can occur.

Included with the control valves fitted to the sprinkler systems is a self-acting alarm valve, which opens when the pressure upstream falls, following the opening of one or more sprinkler heads. This action allows the water under pressure downstream of the valve to flow towards the sprinkler heads, at the same time operating a turbine-driven alarm gong. Pressure gauges complete with shut-off cocks are included in the sprinkler control valve assembly, so that 'running' and 'test' pressures may be observed when the combined test and drain valve is opened, without the need to discharge water from the system.

Reference [8.8] gives recommendations as to the total number of sprinklers that may be connected to an installation, for light hazard, ordinary hazard and high hazard and, depending on the circumstances, these will vary over the range of 500 to 1000. In the case of life safety, 200 sprinklers only may be connected.

When dealing with a high rise building, the difference in height between the lowest and the highest sprinkler in an installation must not be greater than 45 metres. Distribution pipes are to be connected independently to the main rise pipe at the floor being served and no section shall extend to more than one floor, with each section being served by a separate main rise pipe.

(2) Alternate wet and dry pipe system

This arrangement is not recommended for the category of risk covered by high hazard storage. It is used for those developments in which the water in the pipes may freeze as a result of a change in the climate and where the surrounding temperature does not exceed 70°C. The system is operated 'dry' in winter, when the pipes contain compressed air at a pressure of about 2.8 bar (gauge), and 'wet' in the summer, when the pipes are filled with water under pressure.

In some cases this may require a duplication of the automatic alarm valves, which may be connected in parallel, to facilitate an easy changeover at the turn of the seasons. It is known that a greater number of sprinklers open when the system is a dry one than if it were wet, because of the delay that occurs between the loss in air pressure and the discharge of water from the sprinklers.

The system must be self-draining and [8.8] stipulates that, depending on the type of hazard, the number of sprinklers fitted to each installation shall be within the range 125 to 500.

(3) Dry pipe system

This system may only be used when the conditions are such that it is not possible to use a wet pipe system, or an alternate wet and dry type system. Typically, the dry pipe system is used in spaces where the temperature is either kept near to, or below, freezing point, or is at or above 70°C. Such conditions can occur respectively in cold stores and where drying ovens are in use.

In this system, the pipes are filled with air at a pressure that is about one-third to one-half times greater than the maximum water pressure on the downstream side of a special-purpose alarm valve. In the event of a fire, one or more sprinkler heads will open, causing a fall in the pressure of the air contained in the pipes. The special-purpose alarm valve takes the form of a servo-assisted pressure differential valve, which upon a fall in air pressure opens fully in about 20 seconds, allowing the water to actuate a turbine-driven alarm gong and to enter the pipework to which the sprinklers are connected. Attached to the servo section of the differential valve is an 'accelerator', the function of which is to decrease the time required for the fire valve to open after the air pressure has been released.

Because this system is liable to suffer frost damage, it is important that the pipework should be self-draining and, for this reason, the sprinkler heads must normally be fitted above the range pipes. The mass of air contained in the pipes has to be 'topped-up' occasionally, and this is achieved by means of a small air compressor. The requisite number of pressure gauges and a combined test and drain cock are provided. The number of sprinklers that may be fitted is between 125 and 500, and depends on whether or not an accelerator has been fitted to the system.

(4) Pre-action system

These installations differ from the wet or alternate systems in that any action is initiated by an automatic detection system and not by the sprinkler heads, However, the discharge of water into the protected space can only occur when one or more sprinklers open.

There are two types of pre-action system and because their method of control is complicated, they should only be used in those situations where the use of a wet, dry or alternate system is unsuitable.

Type 1: This type of system is used in those cases where the cost of water damage on the protected merchandise is likely to be inordinately high, or unacceptable, as in the case of a library or paper records office. It is likely that the earliest warning of the presence of combustion in the protected space will arise from the action of a smoke detector, and this often gives enough time for the outbreak to be dealt with satisfactorily using portable extinguishers. This obviates any possibly expensive damage that could result from the liberal application of water from sprinklers.

In this case, the arrangement operates as a dry pipe system, charged with air under pressure. When any one of a series of heat or smoke detectors is activated, the visual and audible alarms are operated, giving the occupiers of the building time to attempt to extinguish the fire using portable extinguishers. When a second smoke detector is activated, the air pressure in the pipework is released and the installation control valve is opened, allowing water to flow to the sprinklers. Subsequent opening of one or more sprinkler heads then allows water to be discharged on to the fire.

It should be noted that the control valve cannot be opened solely by a fall in pressure in the pipework, which could be caused by mechanical damage to a sprinkler head, or by the normal operation of a sprinkler, when subjected to heat. The valve can only be opened in response to signals from two smoke or heat detectors, wired one in each of two, adjacent parallel circuits and situated in the affected area. The pitch of these detectors is half the normal recommended value, and additional details about the method of installation of the detection system are given in BS 5839: Part 1.

Type 2: The application of this type of system is preferred to that of type 1, particularly when it is to be used in a high hazard situation and where it is likely that a rapidly developing intense fire will occur.

In this case, the initial activation of a smoke or heat detector operates a visual and audible alarm and releases the air pressure in the pipes; this allows water to flow towards the sprinkler heads, in anticipation of the imminent opening of any one, or several sprinklers. With the pipework primed, the arrangement then operates as a wet pipe system and it can be expected that fewer sprinkler heads will open to deal with a conflagration (see the explanation given earlier in the description of the operation of the alternate wet and dry system – section 8.6 (2)). In the case of malfunction of the detection system, the opening of any sprinkler head will release the air pressure, causing the opening of the control valve and allowing water into the pipework. The system must be self-draining.

For pre-action installations, the maximum numbers of sprinklers that may be fitted to an installation is as follows:

light hazard	500
ordinary hazard	1000
high hazard	1000

(5) Recycling system

The use of this system is generally restricted to situations where it is thought necessary:

(a) to limit water damage after a conflagration;
(b) to prevent inadvertent water damage caused by any accidental mechanical interference with the pipework or sprinklers; and
(c) to be able to carry out work on the installation while ensuring that the system remains in a state of readiness at all times.

In this case, opening of the pre-action valve in response to the operation of any heat detector activates the alarm driven gong and primes the pipework ready for the imminent opening of one or more sprinklers. When this has occurred and the temperature at the place of the conflagration may have dropped, owing to the use by the staff of portable extinguishers, the heat detectors automatically reset to the normal monitoring position.

This has the effect (in some installations) of initiating a hydraulic pressure equalisation process which may typically take 5 minutes to complete and, when this has occurred, the pre-action valve closes. If the fire begins to rekindle, the heat detectors will be reactivated, the pre-action valve will again be opened and water will be directed to the source of the fire.

Other installations use electrical timing units and the use of electrically interlocking circuitry is employed to ensure that the system will not operate solely because of falling air pressure in the pipework. In this way, any accidental damage to either the pipework, or the sprinkler heads will not cause the main valve to open. The re-cycling pre-action installation has the advantage that water damage, after the fire has been extinguished, is kept to a minimum, and the electrical interlocking provided makes it possible to carry out work on the pipework and sprinkler heads without having to close the main stop valve, so that fire monitoring over the remainder of the system is not interrupted. However, operatives working on such an installation must take care that they isolate that part of the system on which they are working from the potentially active main system. The system must be self-draining and the number of sprinklers fitted to such an installation must not exceed 1000.

(6) Tail-end alternate pipe and tail-end dry pipe systems

These systems are essentially of the same form as those described earlier. They may be regarded as an 'add-on' system to a standard sprinkler installation and they are intended to deal with comparatively small areas.

In a development that is mainly heated and served by a conventional wet pipe sprinkler system, it is possible that a small part of the development is either unheated, or is fortuitously overheated and, in order to deal with this, it is permissible to join a tail-end alternate system on to the main wet pipe system.

Another hybrid arrangement, possible in the case of a development that is mainly served by a wet, or alternate system and that contains an isolated space where the temperature is equal to, or higher than 70°C, is to add a tail-end dry system to the main system. Additional and alternative arrangements are prescribed in [8.8].

The number of sprinklers on any tail-end extension must not exceed 100 and, in addition, where more than two tail-end extensions are serviced by one valve set, the total number of sprinklers in the tail-end extensions must not exceed 250.

(7) Deluge system

Some fires, such as those that could occur in paintworks, aircraft hangars and plastic foam and firework factories, can be expected to spread rapidly over a large area, and in these cases it is necessary to discharge water almost instantaneously over the complete area of the conflagration.

This requirement can be met by using a 'deluge' system, consisting of an arrangement of pipework fitted with open sprayers (or projectors), in place of the customary sprinklers. Figure 8.3(a) shows one such system, in which an additional ring of pipework containing air under pressure is run parallel with that containing the open sprayers, and it is fitted with quartzoid bulb heat detectors. If any one bulb fractures, the air pressure is released and the quick opening deluge valve operates, allowing water to flow out through all the permanently open sprayers. An alternative is to use an approved array of heat detectors (see BS 5839: Part 1), any one of which, when activated, will cause the deluge valve to open. Further recommendations with regard to the provision of multiple controls are also shown in figure 8.3(b), and an overriding manual control is described in clause 6.9.2 of [8.8].

8.6.1 Drencher system

The sprinkler systems just described and numbered (1) to (7) are all intended to be used inside buildings. An installation intended for use on the outside of buildings is known as a drencher system, and this is used for those buildings that have a higher than normal fire risk, which may also be situated in areas of dense development. In these cases, water can be discharged through nozzles, over the roof and external openings of a building, to prevent the spread of fire to adjacent structures.

The drencher system is also used backstage in cinemas and theatres, where in the event of a fire, its function is to keep the fire curtain separating the conflagration from the auditorium cool. The system may be controlled either manually or by means of smoke or heat sensors.

In the past, the design of sprinkler systems has tended to be done by fire safety specialists, but increasingly the services design engineer is being called upon to produce such schemes and many engineers will know that such an adjunct to their expertise may usefully become a subject for their continuing professional development (CPD). Both CIBSE at their Mid Career Colleges (Birmingham, Manchester and London) and the Loss Prevention Council (LPC) at Borehamwood (Hertfordshire) run courses on sprinkler system design. The LPC also operate schemes for contractors, who may become initially a 'Registered Installer' and subsequently a 'Certified Installer' when their organisation operates a quality management system modelled on the recommendations given in BS 5750: Part 1.

8.6.2 Provision of a block plan and signs

A block plan of the premises must be displayed close to an entrance (preferably the main entrance), where it can be seen by members of the fire service. The block plan is to show the following data:

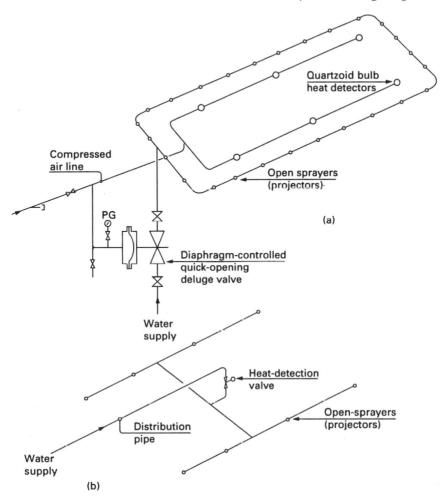

Figure 8.3 (a) Deluge system. (b) Multiple control system

(a) the position and number of sprinkler installations, together with those of the main and subsidiary stop valves;
(b) the height of the highest sprinkler above a datum for each set of installation control valves for each class of hazard and, where the installation is a 'life safety' system, the positions of the zone control valves;
(c) the maximum storage height, together with the position and type of hazard, shaded or coloured, and (if required by the fire authority) the recommended routes to be followed through the building to reach them;
(d) those installations that have been fully hydraulically calculated and, for those that have been pre-calculated, the height of each design

point above the datum and the minimum pressure required to carry out a water supply proving test.

Readers will have noticed the prominent and striking signs (referred to as 'signage' in the trade), such as 'SPRINKLER STOP VALVE INSIDE' and 'SPRINKLER CONTROL VALVE'; these and other signs must be executed and displayed according to the recommendations made in BS 5499: Part 1.

For fully hydraulically calculated pipework, a durable notice shall be fixed to the riser next to each main control valve. This notice is to include technical details pertaining to the installation, and an example of the 'form' of the notice is given in [8.8].

8.6.3 Maintenance of the system

It is obviously important that all installations providing fire safety shall be seen to be as near continuously fully operational as is possible. This means that the user of the building must initiate a programme of checks and arrange to have a regular test, service and maintenance schedule carried out, which includes the keeping of a log book. Such work may be carried out under contract by an accredited sprinkler service organisation.

Spare sprinklers (and spanners) are to be kept on the premises and housed in a cabinet located in an accessible and prominent position, and the numbers of sprinklers held are to conform to a tariff based on the class of hazard as detailed in clause 32.2 of [8.8].

When routine maintenance work is to take place, only the user may give permission (except in an emergency) for the installation or zone to be shut down. Prior to any planned shutdown, the user must instigate a fire check on the premises and where sprinkler installations are shared between two or more occupiers, it is obviously necessary to inform each tenant of the situation, so that they, too, can carry out a fire check and provide extra personnel to combat a possible fire during the period when the system is not operable.

It is to be expected that any monitoring system will occasionally be non-operational during normal working hours and, in order that maximum use may be made of the portable fire extinguishing equipment (if and when required), the supervisory staff should be notified of the times when this will occur and during these periods the area at risk should be patrolled continuously. In some cases it will be necessary to carry out work on an installation outside normal working hours and when this is the case, all fire doors and fire shutters should be closed and sufficient trained personnel must be available to operate portable fire extinguishers if they are needed. It is not always necessary for a complete installation to be shut down and it is usually possible for the section requiring maintenance to be isolated by the use of blank flanges. These must be tagged, and the time of insertion and their position in the system must be logged, to help facilitate their subsequent removal at the earliest opportunity.

In the case of life safety systems, partial or complete shutdown should be avoided; however, when this is not possible, only one zone at a time may be rendered inoperative and the fire authority should be notified in case it may wish to restrict access to the premises during the repair period.

8.6.4 *Refurbishing of the system*

Whenever any part of a sprinkler system is in action, it must only be shut down by order of the fire service, and those sprinkler heads that are subsequently removed should be retained in case other authorities wish to examine them. New heads of the same denomination should be fitted and, in the case of a wet system, checks should be made to ensure that the newly renovated system is fully charged with water.

8.6.5 *Inspection checking and testing of the system*

For the benefit of the user, every installer of a sprinkler system shall provide a suitable programme of inspection and checking. The programme should include details of daily, weekly and quarterly routines that should be followed by the user or the duly accredited agents to ensure all systems are fully operational. In order to avoid false calls, the testing procedures must be agreed by the fire authority, who may be prepared to carry out some of the tests.

For those readers who have responsibility for running the fire services installations in buildings, they will find definitive information about the topics that must be included in all checking and testing programmes in [8.8].

8.7 **Performance of sprinklers**

A sprinkler has two functions:

(1) to detect the presence of a fire; and
(2) to douse it.

The detection of the presence of heat by the two kinds of sprinkler currently in use is brought about by each type utilising the effects of:

(1) a melting alloy; or
(2) an expanding liquid.

Dousing of the fire is taken to mean that the fire is controlled or extinguished and this is achieved by the discharge of water over the affected area in the form of a spray. The pattern of the spray and the size of the water droplets are mainly determined by the orientation and shape of the deflector attached to the sprinkler head.

The melting of an alloy and the prescribed expansion of a liquid take place when a particular temperature is reached, and it is for this reason that sprinklers are sometimes referred to as constant-temperature devices. Note that some fire detectors are designed to be activated in response to a rate of rise of temperature.

8.7.1   Fusible link sprinkler

The solder used for these sprinklers is made from an alloy of bismuth, lead and tin, with traces of cadmium, silver and antimony to help 'fix' the melting point at the required value. It is to be expected that fusible sprinklers taken from the same manufacturing batch using the 'same' solder will, under test, become operational at different temperatures and, in order to produce a standard product that is compatible both within a manufacturing process and between that of other manufacturers, the techniques of quality control must be applied.

Before the constant-temperature change of phase of the fusible alloy can occur, the required enthalpy of fusion must be supplied by the heat from the fire, and every sprinkler that relies on this process will be subject to a delay before the alloy melts. In an attempt to reduce this delay, manufacturers have produced link mechanisms that include an enlarged surface area of copper, so that the heat from the conflagration will be readily absorbed and conducted to the fusible alloy.

There are many ingenious ways in which the fusible link mechanism can be designed, and all manufacturers can be relied upon to produce their own 'special' linkage. Figure 8.4 shows one method, which relies on a captive strut, held in position at a small angle, θ, from the centre line of the head. The angle shown in the figure has been made larger than it is in practice, in order to illustrate the method of operation of the linkage. One end of the strut rests in a small indentation in the sprinkler valve and the other end in a similar slot, formed close to the fulcrum point of a separate curved strip of metal, sometimes referred to as the 'key'. A pair of arms is attached to the other end of the 'key' and partly surrounds the strut. During the assembly of the head, the arms are forced apart by the insertion of an annular bronze cylinder, having a stainless steel ball at each end and containing a fusible alloy. The insertion of the cylinder encloses and restrains the strut.

When the alloy contained by the bronze cylinder melts, the two stainless steel balls are moved inwards by forces derived from the strain energy initially induced in the holding arms, the cylinder falls away and the off-centre strut is no longer restrained in position. The action of the pressure in the pipework is to produce a force along the centre of the

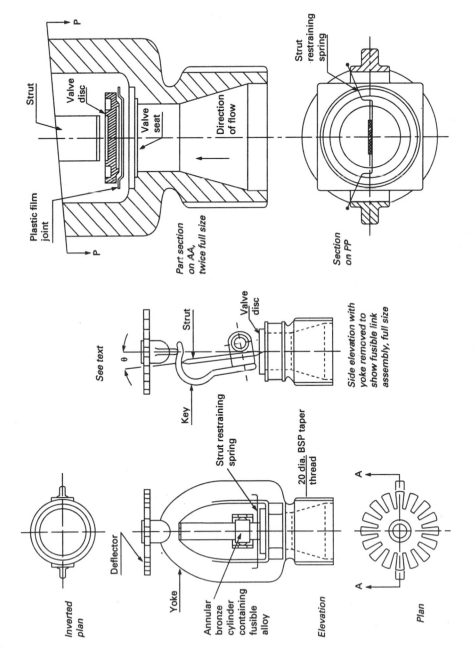

Figure 8.4 Conventional 20 mm diameter fusible link sprinkler

Table 8.1 Temperature and colour codes for sprinklers

Fusible link sprinklers		Glass bulb sprinklers	
Temperature rating (°C)	Colour of yoke arms	Temperature rating (°C)	Colour of liquid in bulb
68/74	Uncoloured	57	Orange
93/100	White	68	Red
141	Blue	79	Yellow
182	Yellow	93	Green
227	Red	141	Blue
		182	Mauve
		204/260	Black

strut that is off-centre to the axis of the head, and it is this force that then applies a bending moment about the fulcrum of the strip of metal forming the 'key', and causing it, also, to fall away from the head. This removes the lower support from the strut, and the valve is released.

The nominal operating temperature of the sprinkler, in degrees Celsius, is stamped on the top flange and on the strut. The finish of the sprinkler may be 'factory or polished brass', chromium plated, black, white, or in some cases 'paint matched'. The range of nominal operating temperatures and colour codes have been agreed internationally and are given in table 8.1.

8.7.2 Glass bulb sprinkler

The operation of this type of sprinkler depends upon the fracturing of a glass bulb containing an alcohol-based liquid and a small air bubble. The continued application of heat to the bulb liquid causes the liquid to expand, until the internal pressure is sufficient to fracture the glass; this has the effect of releasing the valve and allowing the water to be discharged (see figure 8.5).

Variations in the materials used and the design and manufacture of the bulb result in a variation in the nominal temperature of operation of every glass bulb produced. This variation tends to be greater than the variation to be expected from sprinklers of the fusible link type and, as with the fusible link sprinkler, the application of quality control methods to the production process is essential, in order to ensure compatibility of performance of sprinklers obtained from different manufacturers.

Owing to the thermal capacity of the heat-sensitive element, glass bulb sprinklers (in common with fusible link sprinklers) experience a delay before they operate and, in an effort to reduce this delay, manufacturers are now producing 'quick response' glass bulbs that have an outside diameter of 4 mm, which is half that of the earlier conventional bulbs and this has reduced the reaction time by a factor of five.

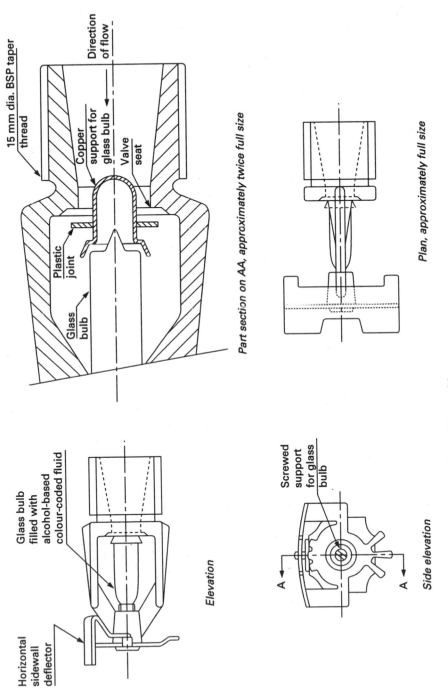

15 mm dia. BSP taper thread

Direction of flow

Copper support for glass bulb

Valve seat

Plastic joint

Glass bulb

Part section on AA, approximately twice full size

Plan, approximately full size

Glass bulb filled with alcohol-based colour-coded fluid

Horizontal sidewall deflector

Elevation

Screwed support for glass bulb

Side elevation

Figure 8.5 Horizontal 15 mm diameter side wall sprinkler

The colour of the alcohol in each sprinkler head corresponds to a particular temperature rating. The code for this has been agreed internationally and the details are given in table 8.1. In the USA it is more common to find that the sprinklers fitted are of the fusible link type, while in Europe the most prevalent type is the glass bulb.

8.7.3 Temperature rating of sprinklers

When selecting a temperature rating for a sprinkler, a general recommendation is that it must not be less than 30°C above the highest expected ambient temperature for the location; in temperate climates this is equivalent to selecting a temperature rating of 68° to 74°C.

Clearly, there will be situations where the ambient temperature will be higher, and clause 25.7 of [8.8] deals precisely with some of the possible variations.

8.7.4 Operation of the sprinkler due to convective heat transfer

Generally it is to be expected that the mode of heat transfer from the fire to the heat-sensitive element of most sprinklers will be mainly convective. Figure 8.6 shows a theoretical model for the transfer of heat by convection from a fire to two flat ceilings, having different heights. In many cases the ceilings and roofs of enclosures may be criss-crossed with beams, and can be sloping, so that it is not always easy to predict the way in which any particular sprinkler will operate.

Spaces that are air conditioned, or heated by means of moving streams of air, usually have air flowing close to the ceiling, and such air currents will tend to disperse the stream of hot gases coming from a fire; the result is that many sprinklers will open somewhat later than is desirable. The aim in the design of a sprinkler system is to contrive that the greatest number of heads open as early in the history of the conflagration as is possible.

The effect of open doors leading to a stairwell and/or a lift shaft from an enclosure where a fire has started is to ensure that the stream of hot gases will effectively be dispersed, so preventing the early operation of any sprinklers fitted in the enclosure. Note that the mounting of sprinkler heads at a lower level, in the mistaken belief that they will operate sooner, is not a valid exercise. Generally, the reaction time for a fusible link sprinkler when placed in a stream of hot gases is shorter than that of a glass bulb type in the same situation.

At this stage it should be noted that automatic roof venting is designed to prevent the spread of smoke due to a local build-up of hot gases, and successful operation of the smoke venting system will militate against the early operation of local sprinklers. Generally where the two systems both form part of the same installation, the temperature rating of the sprinklers should be less than that of the fusible link that operates

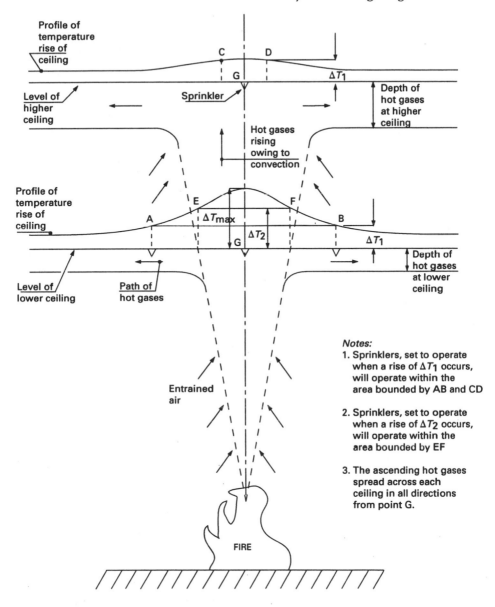

Figure 8.6 Theoretical model for the transfer of heat by convection to sprinklers

the roof vents. However, in those cases where the rapid build-up of smoke may be life threatening and/or where the merchandise is sensitive to smoke, it is obviously important that the roof vents should be the first to open. One solution may be to arrange for one smoke detector to operate a small number of open (dry) sprinkler heads and simultaneously to open the roof vents.

8.7.5 Operation of the sprinkler due to radiative heat transfer

When sprinklers are installed adjacent to ovens and pipes containing hot water or steam, they may be subject to heat transfer in the radiant mode and, when this is the case, the reaction time for a glass bulb sprinkler is shorter than that of a fusible link type sprinkler. Tests have been carried out in the laboratories of the Fire Research Organisation; the reaction times for different types of sprinkler were measured when they were placed in front of a gas-fired radiant panel. One conclusion reached was that the reaction time of both types of sprinkler, when subjected to radiant energy, was longer than when the heat transfer was in the convective mode.

8.7.6 Determination of the orifice factor K

Sprinklers have three nominal thread sizes, 10, 15 and 20 mm, and these correspond to nominal orifice sizes of 10, 15 and 20 mm respectively. For an orifice, the volume flow rate Q (litre/min) is proportional to the pressure at entry to the sprinkler shank P (bar), raised to the power 0.5. Hence:

Q is proportional to $P^{0.5}$

In order to change this relationship into an equation, a constant of proportionality is needed. Hence:

$$Q = k \times P^{0.5}$$

where k is referred to as the k-factor and has units of litre min^{-1} $bar^{-0.5}$.

The value of k for a particular orifice may be determined experimentally by collecting and weighing the amount of water flowing through the sprinkler over a measured time, while keeping the inlet pressure at a known constant value. The values of k are expected to vary between certain prescribed limits and, for each orifice of 10, 15 and, 20 mm diameter, the respective mean values are 57, 80 and 115 litre min^{-1} $bar^{-0.5}$.

8.8 Types of sprinkler

There are many different types of sprinkler and some of these with their uses are described in this section.

Conventional

These may be installed in the upright position where exposed pipework is employed, or pendent where there are finished ceilings, or where the space above the piping is limited. Approximately 40% of the water is discharged upward and the remainder downward, giving a spherical pattern of discharge. They may be used for all hazard classes, with nominal orifice diameters ranging from 10 to 20 mm in accordance with the recommendations made in table 66 of [8.8].

Spray

This is a sprinkler that gives a downward paraboloid pattern of discharge. It may be used for light hazard with a 10 mm diameter orifice, ordinary hazard with a 15 mm diameter orifice, and for high hazard as ceiling or roof sprinkler, having a 15 or 20 mm diameter orifice.

Ceiling or flush

This is a pendent fitting for use in false or finished ceilings; part of the sprinkler is above the ceiling, while the heat-sensitive part is below the lower surface of the ceiling. It may be used for light hazard with a 10 mm diameter orifice, or for ordinary hazard with a 15 mm diameter orifice.

Recessed

All or part of the sprinkler is above the lower surface of the ceiling; it is mounted on the centre line of a recessed inverted cylinder, complete with an external decorative flange (sometimes referred to as an escutcheon), which is adjustable in a vertical direction. It is unsuitable for life safety, and is to be used only in light hazard areas with a 10 mm diameter orifice, or for ordinary hazard with a 15 mm diameter orifice.

Concealed

This is for use with a false ceiling where aesthetic considerations are important; a recessed, spring-loaded cylindrical casing may be positioned over the sprinkler. It is 'locked on' to the yoke by means of plate springs, which are stiff enough to allow the casing to move in a vertical direction through a total displacement of 12 mm, against the force of an enclosed coil spring; this ensures that any small movement of the ceiling is followed by the decorative cover plate. This is fixed to the lower flange of the casing using copper wire lugs and a low melting point alloy, so that heat from a fire causes it to fall away, in readiness for the operation of the sprinkler head, mounted on the centre line of the canister. It is unsuitable for life safety, and is to be used only in light hazard areas with an orifice diameter of 10 mm, or in ordinary hazard areas with a 15 mm diameter orifice.

Horizontal sidewall

This sprinkler gives a half-paraboloid pattern of water, discharged outwards. It is generally used in hotels and hospitals where the pipework can be concealed in corridors, cupboards or service areas. It may be used in light hazard areas with a 10 mm diameter orifice, or for ordinary hazard with a 15 mm diameter orifice.

Intermediate level

This is used in warehouses in multiple-level storage systems. A shield covers the top of the sprinkler, preventing water emitted from other sprinklers (mounted above) from coming into contact with the heat-sensitive element and so preventing its untimely operation.

Institutional

This is used in detention and mental health centres. In such facilities it is important that no part of the sprinkler head is available for the attachment of bed linen, clothing or ligatures.

8.9   Use of hydrant systems to combat fire

Experience has shown that the inclusion of a hydrant system in a multi-storey development for the exclusive purpose of fighting fires is a very worthwhile exercise. In Scotland such provision is mandatory, while in England and Wales it is normal to include such systems voluntarily in the building fire plan. Hydrant systems consist of fire hydrants and rising mains fitted with landing valves, to which the fire service can connect its own hoses, or rising mains fitted with hose reels containing small-diameter hoses, which can easily be used by the occupants of the building. For buildings incorporating oil-storage chambers, the hydrant system will also include provision for foam inlets.

8.9.1   Fire hydrants

There are two kinds of fire hydrant: the underground type, which is placed in a purpose-made pit underground; and the pillar type, which has an outlet fitted to the top of a vertical pillar mounted above ground level.

The hydrants installed in public thoroughfares are statutory. Private fire hydrants are those that have been installed on private land, with the knowledge of the water and fire authorities and constructed according to their specifications. They are usually connected to a ring main system which covers the building site. Ideally, the ring main should be supplied from more than one town main and be so valved that the supply of water to a fire hydrant can come from more than one source. This makes for a reliable supply, and also makes it easier to carry out any alterations and maintenance required.

Fire hydrants should be placed not less than 6 metres from the building so that, in the event of a fire, they can still be used. Generally they should be placed not more than 70 metres from an entry to any building on the site and not more than 150 metres apart. Wherever possible, they are to be sited in positions adjacent to roadways and those hard standings reserved for use by the fire service. Examples of the construction of the pits are given in [8.11]. Note that pillar fire hydrants should be protected from the effects of frost.

8.9.2   *Rising mains with landing valves*

These consist of large-diameter galvanised steel pipes, passing vertically through the building. At each floor level, valved connections (landing valves) are fitted, suitable for the connection of firefighters' hoses.

The installation of a rising main removes the need for firefighters to lay hoses from the ground floor, via the staircase, to the scene of the outbreak. In a multi-storey building, such a process is extremely time-consuming, preventing the blaze from being tackled in its early stages. Also, when hoses are laid it is to be expected that some leaks will occur, causing water damage in parts of the building unconnected with the fire. The provision and use of a rising main obviates such unwanted events.

A rising main can be permanently charged with water from a pressurised supply, in which case it is referred to as a 'wet' riser; or it may be empty but capable of being charged with water by the fire service, when it is referred to as a 'dry' riser. Dry risers should only be installed when the call-out time of the firefighters is known to be a minimal amount, as may be the case when in-house personnel trained in firefighting are involved.

Rising mains should be installed in those buildings that are higher than 18 metres (11 metres for buildings in certain categories in Scotland) and wet or dry risers can be used up to a height of 30 metres – the normal limit at which the fire service turntable ladders can operate.

In London and in cities where high rise buildings are to be found, aerial ladder platforms (ALP) and hydraulic platforms (HP) can extend to heights that approach 60 metres. In the case of buildings more than 60 metres high, it is necessary to use wet rising mains, drawing water from separate storage tanks, reserved exclusively for firefighting purposes. These tanks may be sited at basement, intermediate, or roof level, and in the last two instances the direction of water flow in the mains is downwards and the vertical pipe is then sometimes referred to as a 'down-comer'. Figures 8.7(a), (b), (c) and (d) show several possible arrangements and BS 5306: Part 1: 1976 (amended 1988) is the relevant document dealing with the subject [8.11].

A wet rising main should be installed in such a way that it will not be subjected to freezing conditions, and in the case of both wet and dry risers, the landing valves should be fitted within a ventilated (or pressurised) lobby of a lobby approached stairway, in a stairway enclosure, or

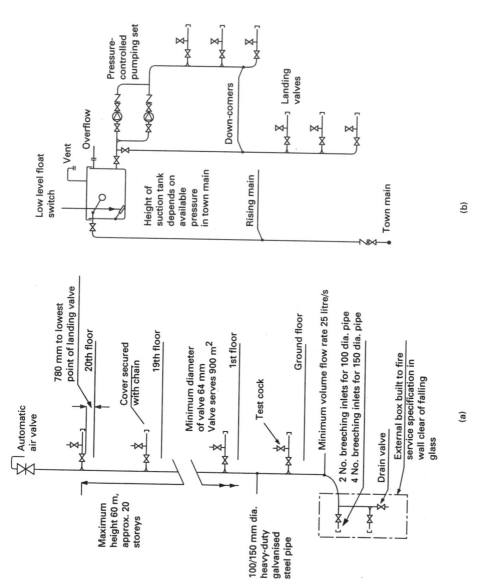

Figure 8.7 (a) Dry riser serving landing valves. (b) Wet riser with down-comers serving landing valves

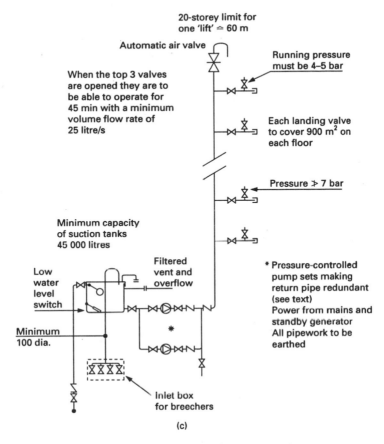

Figure 8.7 (c) Wet riser with landing valves and suction tank, up to 20 storeys

in a position that has been agreed with the fire authority. All parts of rising mains shall be self-draining, including any part of a main positioned in a basement. It is known that hydrant systems are liable to attract the attention of vandals, and for this reason, the landing valves are generally enclosed in boxes, having dimensions, clearance and signage as specified in BS 5041: Part 4. In the case of unenclosed landing valves fitted to a dry riser, they should be strapped shut, with the strap being secured using a padlock.

The British Standard suggests that there should be one rising main for every 900 m² (or part thereof) of the floor area at each level (excluding the ground floor); that they should be spaced no more than 60 metres apart; and, taking into account rises and falls, that the distance from a landing valve to any part of the floor covered by that valve should be within 60 metres. The main is to be constructed of heavy galvanised steel pipes using standard bends or 'easy sweep' large-radius bends, and no elbows are to be used. Where only one landing valve is fitted to

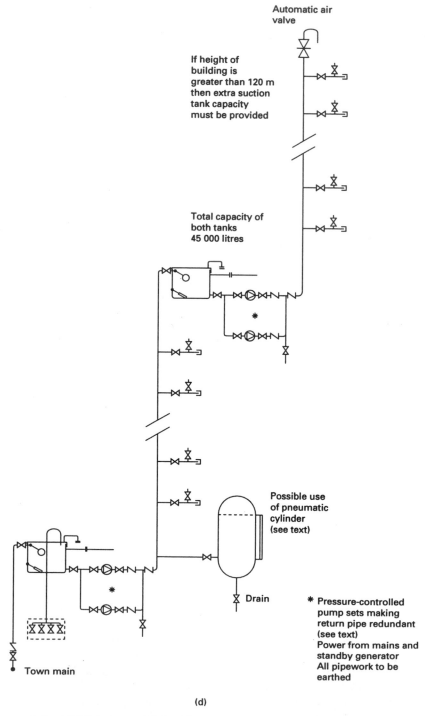

Automatic air
valve

If height of
building is
greater than 120 m
then extra suction
tank capacity
must be provided

Total capacity of
both tanks
45 000 litres

Possible use
of pneumatic
cylinder
(see text)

Drain

* Pressure-controlled
pump sets making
return pipe redundant
(see text)
Power from mains and
standby generator
All pipework to be
earthed

Town main

(d)

Figure 8.7 (d) Wet riser with landing valves and suction tanks, up to 40 storeys

each floor, the diameter of the main is to be 100 mm; where two valves are fitted at any level, the main is to be 150 mm diameter.

At the charging position, a 100 mm diameter dry riser shall be fitted with a two-way inlet breeching, and a 150 mm diameter dry riser with a four-way inlet breeching. As with the landing valves, these are to be installed in boxes having signage complying with the recommendations made in BS 5041.

8.9.3 Adequacy of supply

Generally, the water supply to hydrant systems is to be kept separate from other water services, including other supplies dedicated to other firefighting systems.

Dry risers may be supplied from static or open water supplies, though in urban areas, the supply for dry risers will normally be taken from hydrants connected to the town mains and these must be capable of providing a volume flow rate of at least 25 litre/s. The pressure from the town main will be augmented by that from the portable pumping appliances of the fire service.

Wet risers may be supplied directly from the town main but, as the pressure available diminishes with height, there is a finite limit to the number of floors that may be supplied. In a multi-storey building, the difficulty may be overcome by supplying a low level suction tank having a minimum capacity of 45 000 litres, with water from the town main. A pumping set consisting of two automatic pumps (one acting as a standby), each to be powered from a different source, draws water from the tank and delivers it via non-return valves to the wet riser. Reference [8.11] recommends that the combined output of the town main and the suction tank is to be such that three firefighting jets can be operated for a period of 45 minutes with a total volume flow rate of 25 litre/s.

It is likely that the electric motors of each pump will take their power either from the mains supply, or from the emergency standby generator. The system must be capable of delivering a minimum volume flow rate of 25 litre/s at a running pressure of between 4 and 5 bar, measured at the highest landing valves, when up to three valves are fully open. In a multi-storey building, this means that the pressure in the rising main at the lower levels will be higher than these values, and in order to prevent the bursting of the woven hoses attached to the landing valves on the lower floors, the pressure at the outlet from each landing valve must be restricted to 7 bar.

Traditionally (in older installations), this is achieved by fitting a pressure relief valve on the downstream side of each landing valve, set to operate at a pressure of 7 bar. The water that is released must then be routed to a 75 mm diameter galvanised pipe which returns it to the suction tank. The relief valves can be expected to operate when the nozzles attached to the ends of the hoses are closed and water is being pumped from the suction tank.

Nowadays it is possible to limit the pressure in the rising main by controlling the speed of the pumps electronically. The pumps are said to be 'pressure controlled' and this elegant system makes the 75 mm diameter pipe carrying the spill-over water from each landing valve redundant. An explanation of the principle of operation of this method is given in Chapter 7, section 7.7.5, dealing with the supply of water for domestic purposes to high rise buildings.

Typically, the landing valves have a coupling suitable for a 64 mm diameter woven hose. The hose is usually coiled up close to the landing valve and before it can be used it must be 'run out'. This process requires two people and could be dangerous if the fire is in close proximity to the landing valve itself. Also, when hoses of this type and size are used, the force required to support the nozzle (the reaction) is nearly always larger than expected by an untrained person and consequently the chances of the occurrence of excessive water damage are high. The use of a smaller-diameter rubber hose, permanently coupled to the water outlet and contained in a hose reel, has many attractions, and systems using hose reels in place of woven hose connected to landing valves are now in extensive use.

8.9.4 Rising mains with hose reels

Generally the rising mains supplying hose reels will have a nominal diameter that is not less than 50 mm and the branch points to each hose reel must not be less than 25 mm nominal bore. Hose reels may also be connected to an existing internal hydrant or other special-purpose hose reel main which may be routed over an extensive area of the development.

The hose reels can have internal diameters of 19 or 25 mm and a length that can vary, depending on the diameter, from 18 to 45 m. As it is intended that hose reels will be used by the staff of the building as a first line of defence against an outbreak of fire, it is important that the manipulation of the hose can be accomplished by a person of light physique, so that when planning the number and size of the hose reel units, it is better to choose a slightly higher number of smaller units than the converse.

To operate the unit, the valve is opened and the required amount of rubber hose is pulled off the reel to reach the scene of the fire; when the shut-off nozzle is opened, the water displaces the air from the hose, making the system operational. It is possible to make the water come from the nozzle without delay if the hose reel is always 'charged to the nozzle' after each occasion of use. However, BS 5306: Part 1 specifically states that reels should never be left under pressure and, where practicable, hose should be drained prior to being returned to the drum.

Hose reels are bulky pieces of equipment and to avoid causing an obstruction, they can be fitted into a purpose-made recess, designed in accordance with BS 5274. Also, in order to ensure that the hose will run easily from the reel, the whole assembly can be made to swivel outwards in the direction of the applied force acting on the drum when it is being unwound. Illustrations of the various types of hose reel may be obtained from manufacturers' catalogues.

According to [8.11], one hose reel should be provided to cover each 800 m² of floor space, or part thereof, and hose reels should be sited at each floor level in prominent and accessible positions, adjacent to exits in corridors on exit routes. It should be possible to take the nozzle of each reel into every room and for it to be within 6 metres of every part of that room.

8.9.5 Adequacy of supply for hose reels

As a minimum requirement, the water flow rate at the most hydraulically remote hose reel should be not less than 24 litre/min and, when in use at the same time as the nearest adjacent hose reel, the system should be capable of producing a jet of water 6 metres in length. Depending on the length and diameter of the hose, different pressures will be required at each reel and [8.11] gives the following figures as an example: for the maximum lengths of 25 mm and 19 mm diameter hoses, running pressures at inlet to the reel must be 4 and 5 bar respectively.

In some cases the pressure in the town main will not be sufficient to deal with those hose reels situated on the upper floors, and resort must then be made to providing a pressure-controlled pumping set, drawing water from a suction tank having a minimum capacity of 1125 litres. The tank is to be fed from the town main, via a ball valve having a minimum diameter of 50 mm, see figures 8.8(a) and (b) for further details.

It is normal for industrial developments to have a metered water supply and where this is the case, it has become customary for a meter by-pass to be fitted so that, in an emergency, the pressure drop set up by the meter will not be a factor in reducing the volume flow rate of water available for firefighting purposes.

8.9.6 Provision of foam inlets

It has been found to be beneficial to be able to pump foam directly into oil storage chambers, transformer chambers and inaccessible cable ducts, using a fixed system of pipework, with the inlet positioned outside the building.

One foam inlet connected to one line of delivery pipe is to supply not more than three outlets, having a total area of not more than 3200 mm² and serving one space. The position of the inlet should be such that the act of supplying foam may be carried out without heat and smoke from

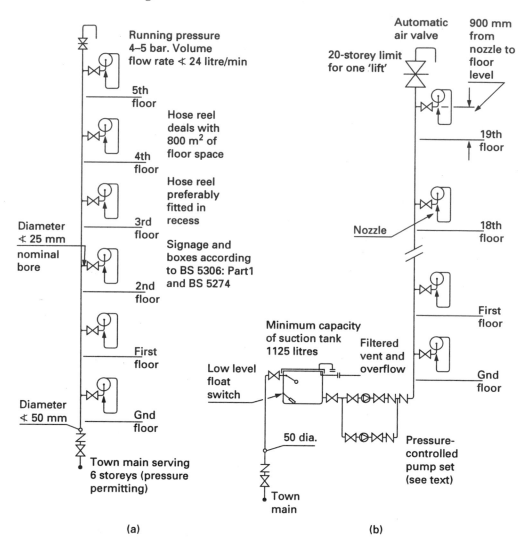

Figure 8.8 (a) Rising main with hose reels, using pressure from town main. (b) Rising main with hose reels, using suction tank up to 20 storeys

the burning oil affecting the process, and the inlet, with its connections, is to be housed in a box complying with the recommendations given in BS 5041: Part 5.

The length of run of the pipework should not exceed 18 metres and is to be constructed of screwed and socketed 65 mm or 80 mm nominal bore galvanised wrought steel and laid to a fall in the direction of the space being protected. Where it is necessary for the piping to change direction, easy sweep (large-radius) bends are to be used and no elbows are to be included in the system.

When protecting an oil storage chamber and tank, the foam outlet(s) is to be positioned approximately 150 mm above the level that the oil would reach if rupture of a full tank had occurred. In the case of one installation with which the author was concerned, the fire prevention officer required that the outlet be directed towards the wall of the chamber, in order that the momentum of the foam issuing from the pipe would spread it over a wide area after being deflected backwards from the surface of the wall. In this way, a blanket of foam would gradually cover the area of burning, so excluding the oxygen and extinguishing the fire.

8.9.7 Recommendations pertaining to the protection of buildings under construction

Reference [8.11] recommends that mains for firefighting purposes should be available for action as soon as the height of a building under construction reaches a point that is 18 metres above the access level of the fire service, and that the mains should be extended as the work progresses. For a building having a finished height less than 60 metres, the mains should be dry. For a building taller than 60 metres, a wet rising main may be installed, or a wet main installation that is initially available for use as a dry main. In the latter case, as soon as a height of 60 metres has been reached, the dry rising main should be converted to a wet rising main.

Such recommendations must be taken into account at the tendering stage and will require a co-ordinated input from the developers, the electricity and water utilities, the subcontractors and the fire service. The authors of [8.11] have set out a series of 'points for consideration' in their Appendix B. A brief résumé of these points is given as follows:

(1) the need for the extension of the mains as work progresses;
(2) the possibility that the fire service will require the system to be tested as the work progresses;
(3) the need for suitable water and electricity supplies and standby apparatus to be provided adjacent to the rising mains;
(4) the need to prevent the water in the mains from freezing;
(5) the possible need to provide for a permanent supply of electricity, in place of the temporary supply, to operate the pumps;
(6) the need to provide a lift for use by the firefighters;
(7) the need to provide suitable access for firefighting appliances;
(8) the need in the case of buildings that have an oil firing system to complete the installation of the foam system, ready for use during the drying out period (if any).

8.9.8   Inspection, checking and testing of hydrant systems

Details of the tests that should be carried out on completion of these systems are given in [8.11]. A permanent record of all the acceptance tests should be kept by the owner or occupier of the building, and these will include information on the following points:

(1) date and time of the inspection or test;
(2) name and identity of the person carrying out the test;
(3) details of the test results;
(4) details of any external conditions affecting the results;
(5) details of any follow-up action required;
(6) details of the work carried out as a result of point (5) with the date and the result of the subsequent retest.

8.9.9   Maintenance of systems and the rectification of defects

Where the services of a competent person are not available 'in house', it is usually possible to take out a service contract with the local water company to carry out maintenance on a regular basis.

(a) Fire hydrants

At least once per year, each hydrant shall be inspected and any leaks around the outlet connection, its flanges and the gland or stuffing box of the isolating valve should be rectified. For underground hydrants, the inspection should take into account the condition of the pit and its frame and cover. The volume flow rate and the pressure at the outlet should also be measured and noted. The inspection should check that the hydrant is not habitually obstructed and that the hydrant indicator plates are in position.

(b) Rising mains

In the case of dry rising mains, a wet test shall be carried out annually, when any leaks may be detected and rectified. At six monthly intervals, checks shall be made on the condition of all parts of the system, including the integrity of the earthing arrangements and the operation of all locks and door hinges attached to the boxes.

In the case of wet rising mains, the six monthly and annual checks should cover those items described above and, in addition, checks should be made on the integrity and cleanliness of all items pertaining to the suction tanks. Checks should be made on the mechanical and electrical equipment associated with the operation and control of the pump set, and the pumps should be serviced and the set tested on completion of the work. If any valve cannot be put into a serviceable condition and no replacement is immediately available, it should be removed and the system left in an operable condition by the strategic insertion of a blank or plug. In those cases where it is not possible to keep the system in a workable condition, the fire authority should be informed and a notice exhibited on site, to the effect that the system is not available for use. When the system is repaired, the fire authority must again be notified, so that it may cancel any arrangements that had been made to cover the situation.

(c) Hose reels

Each hose reel should be regularly checked to ensure that all the associated fittings are leak free, are not blocked with debris and remain fully functional. Once per year the hose should be run out and a check made to see that it does not leak under pressure, and also to check that the volume flow rate is at least 0.4 litre/s. If it is considered impossible to test every hose reel, then the hose reel that is hydraulically furthest from the inlet should be tested. A check should be made to ensure that no hose reel is obstructed and that in the case of a warehouse having merchandise stacked high, the signage referring to the position of each reel is fixed at a level where it is visible.

8.10 Classification of fires

The task of firefighting depends greatly on the nature of the material being burnt, and it has been found that the various types of fire can conveniently be categorised into four classes:

Class A fires – Involve solid material usually of an organic nature, such as wood, cloth and paper.
Class B fires – Involve liquids or liquefiable solids, such as oil, petrol, grease, paint, varnish and fat.
Class C fires – Involve flammable gases.
Class D fires – Involve flammable metals, such as zinc, aluminium, titanium, zirconium, magnesium, uranium and plutonium.

Generally, portable fire extinguishers are provided to deal only with class A and B fires and, because the use of the wrong type of portable extinguisher on class C and D fires can cause dangerous conditions, it is not considered desirable for members of the public to become involved with the latter.

Dangerous conditions can also arise when extinguishers that contain an aqueous extinguishing medium are used on fires occurring in electrical equipment. In this situation, if the stream of extinguishing material is electrically conductive, the operator will receive an electric shock and, even if the material of the stream is itself non-conducting, any moisture appearing on the surrounding surfaces, as a result of the combustion process, may itself present an electrically conductive path, which could include the person using the appliance and any other bystanders.

8.11   Distribution, classification and rating of portable fire extinguishers

Extinguishers are marked with the class of fire for which they are suitable, and this may include more than one class. They are also marked with numbers that indicate the extinguishing capability of the appliance. The numbers quoted refer to the maximum size of test fire that the appliance can be expected to extinguish, when it is being operated by a person who is experienced in the art of firefighting.

In the case of a class A test fire, a wooden crib having a fixed width of 0.5 m and a fixed height of 0.56 m is made up to various prescribed standard lengths according to the details given in BS 5423 [8.12]. If it is found that the maximum standard length of test fire that can be extinguished is 2.1 m, then the appliance is given the rating of 21A. If the maximum standard length of test fire that can be extinguished is 4.3 m, the rating is given as 43A (see BS 5306 [8.13]). A class B test fire may be produced by burning a known volume of flammable liquid in a circular tray manufactured in accordance with the details given in [8.12].

An extinguisher having a classification of 13A/89B is suitable for use on both class A and B fires and has the extinguishing capability as indicated by the numbers that prefix the letters A and B. This method of classification now makes it possible to specify the distribution of fire extinguishers according to their extinguishing capabilities, rather than by their type and size.

For those portable fire extinguishers that comply with earlier British Standards that have been withdrawn, [8.13] gives an equivalent list of ratings for each appliance.

8.11.1   *Selection of portable fire extinguishers for class A fires*

Example

A single-storey building is used as a general office, having a circulation area of 2000 m². Assuming that portable extinguishers having the ratings given are available, specify some possible solutions to the choice of suitable appliances. Rating of extinguishers available: 3A, 5A, 8A, 13A, 21A, 27A, 34A, 43A, 55A and 70A.

Solution

According to [8.13], the total class A rating of the extinguishers must not be less than the area of the floor × 0.065, and in no case less than 26A.

In the case of single occupancy, having an upper floor area not exceeding 100 m², the minimum aggregate rating is 13A and there is no longer a requirement for two extinguishers on each floor. For the example:

$$0.065 \times 2000 = 130A$$

This class A rating may be made up as follows:

$$10 \times 13A \text{ extinguishers} = 130A, \text{ or}$$

$$3 \times 27A + 4 \times 13A \text{ extinguishers} = 133A, \text{ or}$$

$$5 \times 27A \text{ extinguishers} = 135A$$

The British Standard stipulates that the travel distance from any point on the floor to the nearest extinguisher should not exceed 30 m, so this stipulation would probably rule out the possible installation of:

$$2 \times 70A \text{ extinguishers} = 140A$$

The British Standard mentions an upper limit for the mass of each appliance of 23 kg, this being the maximum that it is considered a person can handle effectively.

The appliances chosen should have the same method of operation and, if they are to be used to fight the same class of fire, they should all be similar in shape, appearance and colour.

Appliances should preferably be sited on an escape route and comply with the recommendations given with regard to the height and manner of fixing. If any extinguisher is placed in a position where it is covered from direct view, then attention should be drawn to this, by following the recommendations for signage as given in BS 5499 [8.14].

Appliances intended to deal with a fire in a confined space should be mounted outside that space and in a position where the safety of the operator will not be jeopardised in an emergency.

8.11.2 *Choice of portable fire extinguishers for class A fires*

Generally, the materials that make up class A fires become more difficult to extinguish when there is air movement in the region; under such conditions the use of halon gas as the extinguishing medium has proved to be singularly unsuccessful when compared with the use of water, powder or foam, and the British Standard recommends that appliances using halon should not be installed where there are no other types of appliances available. The use of water as the extinguishing medium is to be preferred to that of powder in hotels, hospitals and homes for the elderly, as the use of powder in a confined space may cause a reduction in visibility, which could produce a state of temporary disorientation in the minds of the occupants and rescuers alike.

8.11.3 Selection of portable fire extinguishers for class B fires

Experience has shown that class B fires are best dealt with using foam as the extinguishing medium. The first application will produce partial extinction and subsequent applications will achieve final extinction. The use of powder is effective provided that every part of the surface area of the burning liquid can be reached. Generally, other types of extinguisher will reduce the flames only for a very short period, after which the full intensity of the fire will be regained.

For those who have to specify the type and rating of portable firefighting extinguishers to be used in a development containing groups of tanks holding flammable liquid, [8.13] gives recommendations (with worked examples) of suitable methods of assessment.

8.11.4 Inspection, maintenance and test discharging of extinguishers

It is the duty of the user of portable firefighting appliances to instigate monthly checks to see whether or not any appliance has been discharged, damaged or removed from its allotted position, and any deficiencies found in the provision of resources should be remedied.

BS 5306 [8.13] gives detailed recommendations with regard to the way in which the annual inspection, service and maintenance of individual types of extinguisher should be carried out. Such annual services are usually performed by a specialist contractor, subject to adequate checks by the occupier.

The recommended intervals of time that should elapse between discharge of the various types of extinguisher are given in the British Standard, some of which are as follows:

Type of extinguisher	Intervals of discharge (years)
Water (stored pressure)	4
Foam (all types)	4
Water (gas cartridge)	5
Powder (gas cartridge)	5
Powder (stored pressure, valve operated	5

It is suggested that on the occasions when test discharging of appliances takes place, this activity could be used to help train personnel. Replacements for the used extinguishers must be available and in order to comply with BS 6643 [8.15], the date of any recharging must be marked on the appliance.

8.12 Types of portable fire extinguisher

There are many different designs of fire extinguisher. Among these, a large group falls naturally into those that use the pressure of a gas contained in a cartridge to expel the extinguishing medium and those that use the pressure of a gas permanently in contact with the extinguishing medium to expel it. The principle of operation of each type may be understood by referring to figures 8.9(a) and (b).

(1) Water (gas cartridge)

Figure 8.9(a) shows the salient points of this type of extinguisher, in which the water is held in a 9 litre container at atmospheric pressure. An antifreeze solution may be added, provided that it is chosen for its non-corrosive properties. It will perform a dual role in preventing the water from freezing and provide some protection from possible electrolytic action between the dissimilar metals used in the construction of the appliance. Fixed inside the container is a steel plastic-coated gas bottle (referred to as a cartridge), containing liquid carbon dioxide (CO_2) under pressure. A protected striking mechanism consists of a spring-loaded plunger, which is capable of puncturing the seal fixed across the top part of the cartridge. When the liquid is released, it flashes off into gaseous form, filling the top of the outer container and forcing the water out through a nozzle, which can be directed on to the fire. Once the process has been started it will continue until all the gas has been expelled. The flow of water from the extinguisher may be halted by inverting the appliance.

(2) Water (stored pressure)

In this case, the water held in the 9 litre container is pressurised to a value of 10 bar by means of dry air, applied by external means through a captive Schrader type valve. Figure 8.9(b) gives the basic details. The appliance is operated by removing a small locking pin from a squeeze type handle; subsequent movement of the handle opens the outlet valve, allowing water to be forced out through the attached hose and nozzle. An advantage of this type of appliance is that by releasing the grip type handle, the flow of water from the nozzle can be stopped and, if required, started again by depressing the handle.

This facility will be available only within the aggregate time allowed for complete discharge of the water from the container. In order to prevent internal corrosion, the steel container is lined with plastic.

(3) Foam (gas cartridge)

This type of extinguisher is similar in construction and operation to the water gas cartridge unit already described. A foam solution takes the place

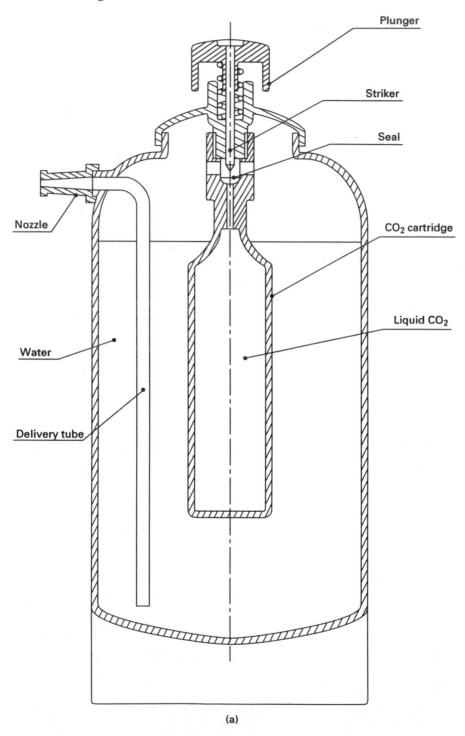

Figure 8.9   (a) Section through a water (gas-cartridge) extinguisher

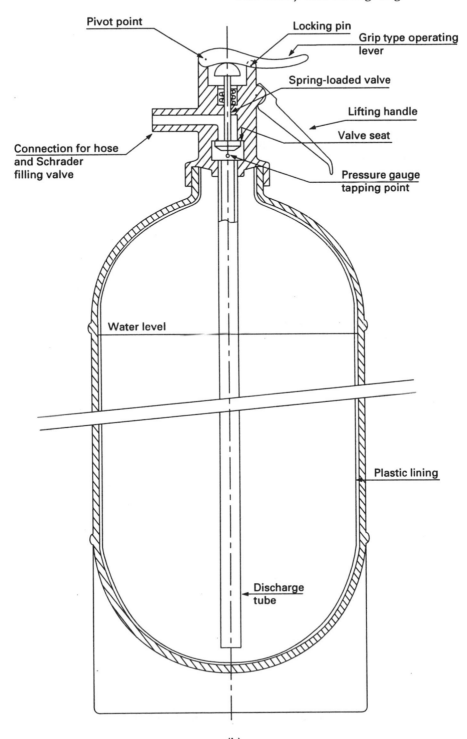

(b)

Figure 8.9 (b) Part section through a water (stored pressure) extinguisher

of the water and it is discharged through a hose and self-aspirating foam branch pipe by means of CO_2 gas, which has been released by the action of a spring-loaded striker knob, breaking the seal on the plastic coated steel cartridge. The discharge of foam can be stopped by inverting the appliance.

(4) Foam (stored pressure)

Again, the construction and operation of this extinguisher are similar to the water stored pressure unit described previously. A foam solution takes the place of the water in the plastic-lined 9 litre container and the space above the foam is pressurised by either air or nitrogen. Operation of the grip type handle opens the outlet valve, and foam is forced up the discharge tube and into the hose and self-aspirating branch pipe, ready for use on the fire.

8.12.1   Other types of portable fire extinguisher

These include powder (gas cartridge), powder (stored pressure) and carbon dioxide, stored under pressure in steel bottles.

Carbon dioxide is electrically non-conductive and therefore ideal for use on fires that occur in electrical apparatus; because it is also a non-contaminant, it is useful for extinguishing cooking oil and other food related fires.

8.13   Automatic/manual firefighting systems using extinguishing media other than water

Such installations using CO_2, halon and other gases, foam and powder are often to be found in places left unattended for long periods and where the prospect of any delay in the arrival of firefighters would be unacceptable. The systems used are nearly always designed by specialist fire safety engineers and are purpose-made to suit an individual client's needs. The methods used to trigger the operation of each system rely variously on electrical, mechanical and pneumatic systems.

8.13.1   Use of carbon dioxide

Most surface fires will be extinguished when the normal concentration of oxygen in air of 21% by volume is reduced to 15%. Carbon dioxide is particularly appropriate for application in closed spaces, where it is likely that there will be no air movement to reduce its concentration. It is suitable to deal with class A and B fires. Class C fires may also be extinguished but in some cases there is a risk of a subsequent explosion occurring. Carbon dioxide is unsuitable for use with chemicals that contain oxygen, such as cellulose nitrate and chlorates. It is also unsuitable for use on metal hydrides and reactive metals such as sodium, potassium, magnesium, titanium and zirconium, as well as in spaces where people are present, because of the risk of asphyxiation.

There are three recognised types of system employing CO_2:

(1) total flooding system;
(2) local application system;
(3) manual hose reel system [8.16].

Maintenance of the system

It is recommended that a weekly check of the installation be carried out, to ensure that all the gas outlets are free from rubbish and that the operating controls are in working order. At least twice yearly a more comprehensive check should be carried out and this is customarily achieved by entering into a service contract with the installer or with a specialist fire safety organisation. Details of the work to be carried out on these occasions are specified in the British Standard.

8.13.2 Use of halon 1301

For many years, halon has been a recognised and effective medium for fighting fires, but is now out of favour. Unlike carbon dioxide, which extinguishes fires by keeping oxygen away from the fire, halon extinguishes fires by inhibiting the chemical reaction between the fuel and the oxygen. Broadly, it deals with the same classes of fire as does CO_2, with the exception (as has previously been mentioned) that it is not as effective against class A 'deep seated' fires as carbon dioxide (see BS 5306 [8.17]).

8.13.3 Composition and suitability of foam as an extinguishing medium

Foam has an extremely low density, enabling it to float on the surface of a burning liquid, cooling it to a limited extent and excluding the oxygen needed to sustain combustion. When this occurs, the layer of foam is referred to as a 'blanket'. When attempting to put out a fire, the foam issuing from the nozzle of the extinguisher should be directed not against the flames, but against a vertical surface so that it flows back on to the surface of the burning area, forming a moving mass of fluid that gradually covers the area of the conflagration.

Foam solution is produced when the correct proportion of water is mixed with foam concentrate. The chemical composition of foam concentrate varies considerably between manufacturers, but the different kinds of foam concentrate may generally be classified as being based upon one of the mixtures now listed.

(a) *Protein (P)*: These are aqueous solutions of hydrolised protein.

(b) *Fluoro-protein (FP)*: These are P concentrates to which fluorinated surface-active agents (surfactants) have been added; the fluid is less viscous than P type foam, and it extinguishes the fire faster and reseals gaps in the foam blanket more readily.

(c) *Film forming fluoro-protein (FFFP)*: These are P concentrates to which other fluorinated surfactants have been added to make the foam less viscous than FP foam; they are resistant to contamination by hydrocarbon liquids and form a film on the surface of some liquid hydrocarbon fuels.

(d) *Synthetic (S)*: These concentrates are solutions of hydrocarbon surfactants; they are used in the USA, but are not generally available in the UK.

(e) *Aqueous film-forming (AFFF)*: These foam concentrates are generally based on a mixture of hydrocarbon and fluorinated hydrocarbon surfactants, and they are film forming on the surface of some liquid hydrocarbon fuels.

(f) *Alcohol resistant (AR)*: These foam concentrates are made for use on liquids that destroy foam; they can have as a base any of the kinds of foam previously listed and, when used on fires of hydrocarbon liquids, they will have an equivalent performance to that of the chosen base.

Foam concentrates can act both as detergents and as degreasing fluids, and skin contact should be avoided. The solutions used are also water pollutants and during exercises should not be allowed to drain away to watercourses. Foam concentrates having a protein base contain salt and produce the same corrosive effects as sea water.

Foam is also categorised in terms of the ratio of the volume of foam produced to the volume of foam solution. This number is greater than 1 and is referred to as the expansion ratio. Expansion ratios fall into three categories: low, medium and high. They have relevance to the type of apparatus needed to produce the foam and also the type of hazard on which it is to be used.

Low expansion foam (LX)

This has an expansion ratio up to 20 and is suitable for extinguishing fires caused by flammable liquids on horizontal surfaces. The relevant British Standard is BS 5306: Section 6.1: 1988.

Medium expansion foam (MX)

This has an expansion ratio that can vary from 21 to 200 and is suitable for use on hydrocarbon fuels and on combustible solids up to a height of about 3 metres. It may be used outdoors and will be effective provided that the air velocity does not exceed 10 m/s. Reference [8.18] gives the relevant British Standard.

High expansion foam (HX)

This has an expansion ratio that can vary from 201 to 1000 and is suitable for use on flammable liquid and combustible solid fires up to a height of at least 10 metres. It is best used indoors for the purpose of submerging the hazard (see [8.18]).

All foams contain water and should not be used on sodium or potassium. Neither should they be used on electrical apparatus that is connected to mains supply. Some flammable compounds are extremely destructive of foam layers, and these compounds are listed in [8.18].

The foam installations dealt with cover fixed and semi-fixed systems, as well as portable and transportable systems. One form of a self-contained fixed foam system is shown in figure 8.10, in which the correct proportions of water and foam concentrate are combined to produce a foam solution that is subsequently passed through a foam 'generator'. This device mixes air with the foam solution, producing foam that is then delivered to the protected area. Figure 8.11 shows an alternative form of fixed system which makes use of a pre-mixed foam solution.

In the case of both systems, the triggering mechanism shown consists of link line control. Other installations may use triggering controls that employ flameproof electrical, pneumatic or hydraulic methods. If electrical methods of triggering are used, an alternative electrical supply having automatic changeover should be available.

Triggering devices are now used to initiate a series of events, resulting in vents being opened and fire doors being closed, and here a note of caution is required. Owing to the possibility of persons being in the wrong place at the wrong time, it must always be possible to override the automatic system, using strategically placed lock-off devices carrying the appropriate signage, as specified in [8.18]. The operation of any lock-off device must not disable the alarm system and its operation must be indicated at the building control centre, where (if required) permits to work may then be issued. There are many kinds of automatic detection and control equipment available having facilities for self-checking, which will warn of the failure of any part of the system, and such devices should comply with the relevant parts of BS 5839: Part 1.

8.13.4 Composition and suitability of powder as an extinguishing medium

Reference [8.19] has classified powders used for firefighting purposes with the letters A, B, C, or D, according to the class of fire for which they are intended. Some powders are classed with more than one letter, such as ABC, and these are capable of dealing with fire risks in all these classes. Powders classed as BC may also be effective when used on class A surface fires.

For general use on BC class fires, sodium hydrogen carbonate, or the

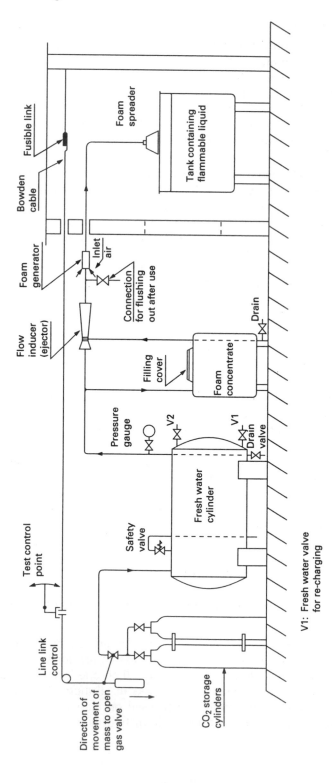

Figure 8.10 Self-contained fixed foam system, using foam concentrate

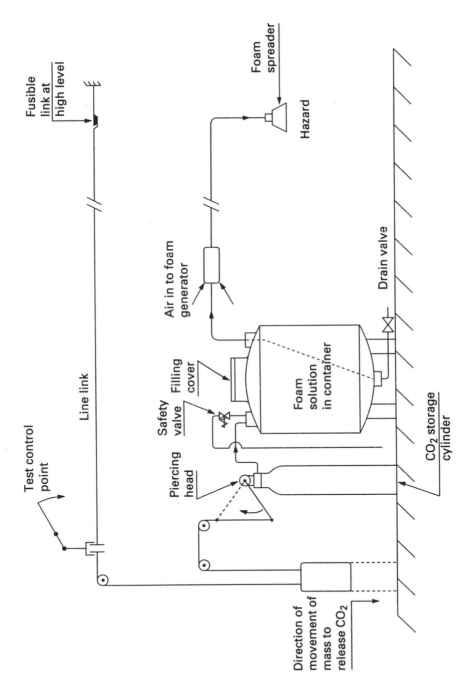

Figure 8.11 Self-contained fixed foam system, using foam solution

more expensive and slightly more effective potassium hydrogen carbonate, can be used. It has also been found that powders consisting of mixtures or compounds of urea and hydrogen carbonate are more effective than the previously mentioned hydrogen carbonates. Mono ammonium phosphate has been found to be effective for general use on ABC class fires because of its fire-retardant properties.

The method of operation is such that in the heat of the fire, a small amount of the chemical decomposes, coating the surface of the remaining particles of powder with a layer, which then acts catalytically, destroying the free radicals that propagate the flame reactions.

The powders used have to be transported to the fire from a storage point and this is done with the aid of a 'propellant'. The propellants used are all gases and the resulting type of flow is referred to as 'two phase', which in this case is made up of a gas and a solid (the powder). The gases used as propellants are air, argon, carbon dioxide, helium and nitrogen. The manufacturers of the powders have found it necessary to include additives to their products which give them properties that in storage make them resistant to caking, packing and moisture absorption, and make them free flowing when being transported.

BS 5306 [8.19] encompasses total flooding systems, local application systems, manual hose reel systems and monitor systems.

The plant needed can take the form of a stored pressure system, in which the propellant is permanently in contact with the powder and occupies the same container, or a gas container system, in which the propellant is not in contact with the powder but is kept in a separate container. This latter type is shown in figure 8.12.

Principles of design for powder systems

Successful operation in the mode of two-phase flow requires a different approach from that required for operating in the modes of either laminar or turbulent flow. Powder manufacturers have data on the flow of gas and powder through pipework and nozzles; before attempting to design a powder system, information on nozzle performance, limiting velocities and mass flow rates for a particular powder, with a particular propellant, must be obtained. Reference [8.19] highlights some interesting design principles concerned with the layout of the pipework, aimed at re-establishing two-phase flow after separation has occurred at bends and tees.

Inspection, testing, maintenance and replenishment of powder systems

The user is expected to put into operation a programme of inspection, to set up a service and maintenance schedule and to keep records in a log book of work done. Guidance on the inspection programme should be obtainable from the installer, the object of the inspections being to detect faults at an early stage so that the functioning of the system in an emergency will not be jeopardised.

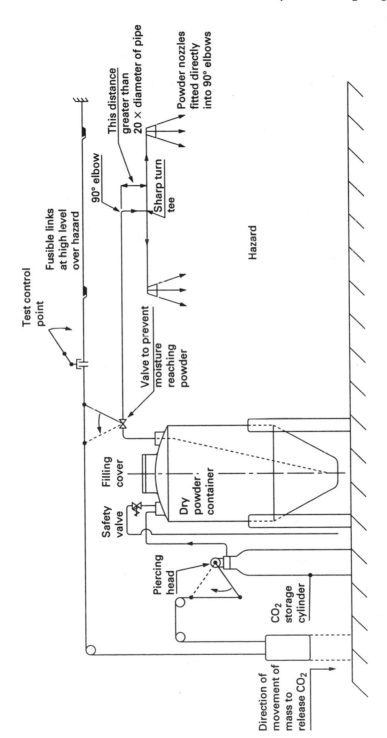

Figure 8.12 Gas container powder system

Inspection programme

Such an inspection programme would involve weekly and monthly checks.
(a) *Weekly*: Check the integrity of all pipework and all control settings. If
the nozzles are fitted with protective seals, check that they are free from
contamination and are sufficient for their purpose. Check (after allowing
for the change in temperature) that the pressure of the propellant has not
fallen by more than 10% of its fully charged value. Rectify any such
deficiency.

Example

The fully charged pressure, p_1, in a stored pressure powder container is
10 bar gauge when the temperature, t_1, is 15°C. Calculate the equivalent
pressure, p_2 gauge, when the temperature, t_2, has fallen to 5°C.

Solution

The combined law PV/T = a constant may be used.
Hence $P_1 V_1/T_1 = P_2 V_2/T_2$.
As the volume of the container does not vary, $V_1 = V_2$, and transposing
for P_2 we have:

$$P_2 = P_1 T_2/T_1 \tag{8.1}$$

Before substituting in any gas law, all pressures and temperatures must
have values that are absolute.

As p_1 = 10 bar gauge,   then P_1 = (10 + 1) bar (absolute), or
                             P_1 = 11 bar.
As t_1 = 15°C,   then T_1 = (15 + 273) K (absolute),
                             or T_1 = 288 K
As t_2 = 5°C,   then T_2 = (5 + 273) K (absolute),
                             or T_2 = 278 K

Substituting in equation (8.1), we have:
P_2 = 11 × 278/288 bar
  or P_2 = 10.62 bar
  hence p_2 = (10.62 − 1) bar gauge
  or p_2 = 9.62 bar gauge is the equivalent pressure of the fully charged
container, allowing for the change in temperature.
(b) *Monthly*: Check that personnel designated to operate the equipment
(including new employees) are qualified for the task and have the necessary
authorisation.

Testing and maintenance schedule

This work is best carried out by specialist fire safety engineers under contract to the user. They will apply the schedule at three monthly intervals, with specific work being carried out every six months and every 12 months. Reference [8.19] gives detailed recommendations.

Replenishment of the system

If it is not possible to obtain fresh powder containers within a period of 24 hours, then spare containers of powder should be kept under dry conditions on the premises, and it is essential that any replacement powder should be of the same type as that for which the system was designed.

8.14 Fire detection for buildings

The reader should be aware that legislation requires that certain categories of buildings must be equipped with an effective means of giving a warning in the event of fire and that, in some cases, consultation with the appropriate authority about the provisions of such an alarm system is obligatory. Also, early contact with the prospective insurer of the building can always be shown to be financially worthwhile to the occupier.

Before an alarm can be given, it is obvious that means must be found to detect any unscheduled fire at the earliest possible moment, and because of the complexity of buildings and their various patterns of occupation, this is most conveniently achieved using detectors that operate automatically.

8.14.1 Stages in the development of a fire

After ignition, there are four distinct stages in the development of a fire:

(1) invisible products of combustion are released;
(2) products of combustion become visible;
(3) the fire produces a visible flame;
(4) there is a rapid increase in the temperature of the resulting gases.

Detectors have been developed that can sense each of the four stages just described and, taken in order, these are:

(1) ionisation detectors;
(2) optical detectors;
(3) radiation detectors;
(4) heat detectors.

There are many different designs of heat detector currently in use and also being developed. The following subsections include descriptions of the most commonly known types that come under the headings previously given.

8.14.2   Fire detectors

(1) Ionisation detector

When a gas is subjected to radioactivity (is irradiated), the atoms of the gas take on negative and positive charges of electricity, and this process is referred to as ionisation. If two electrodes are placed in the gas and connected to form an electrical circuit, a small current will flow between them. The magnitude of the current is affected by the introduction of impurities in the surrounding gas, and it is this effect that in 1941 was first used by Cerberus Ltd of Berkshire, UK, in the detection of visible and invisible smoke particles.

In practice, the detector comprises two chambers, one of which is open to the atmosphere but shielded from it, while the other is sealed; these are shown diagrammatically in figure 8.13(a). Each chamber is irradiated with radioactive foil, having a unit of radioactivity of the order of 0.1 micro-Curie (μCi). Electrodes are provided in each chamber and, when the surrounding air contains no pollutants, a small current of equal magnitude flows in each circuit. The introduction of a pollutant in the form of invisible smoke particles and/or smoke particles from visible smoke reduces the flow of current in the open sampling chamber, producing a voltage difference between the two circuits, which is subsequently amplified. When the magnitude of this voltage difference reaches a value proportional to a prescribed density of smoke, the microprocessor transmits the data to a control panel, where an alarm condition is indicated. This type of detector is described as a 'point' detector, because it monitors events occurring in one position in the building. Ionisation detectors can be designed to deal with smoke particles that vary in size from 0.01 micrometre (μm) to 1 μm and from 0.5 μm to 10 μm.

(2) Optical detectors

There are two types of optical detector. The first is a point detector and relies on the fact that when a beam of light passes through visible smoke particles having a characteristic size of from 0.5 μm to 100 μm, it becomes diffused (see figure 8.13 (b)). Part of the scattered light is detected by a photo-electric cell and after suitable amplification, the signal is then transmitted to the controller which indicates an alarm condition.

The other type is a 'line' detector, so called because it monitors events occurring anywhere along a line that crosses the building, usually at high level (see figure 8.13(c)). This detector functions on the obscuration principle

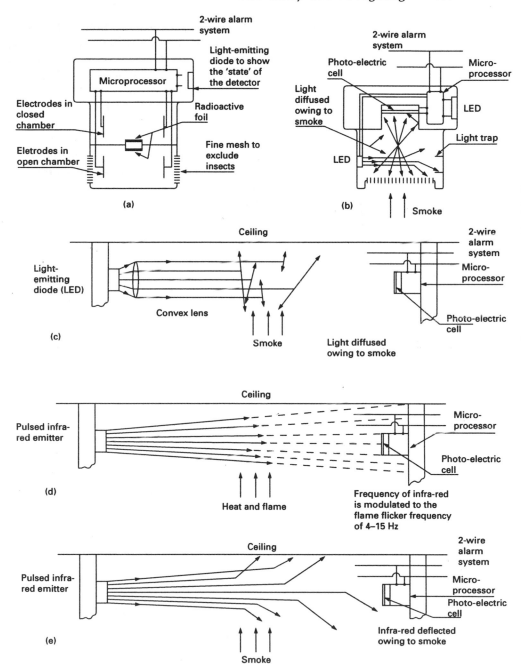

Figure 8.13   (a) Point type ionisation detector. (b) Point type optical detector. (c) Line type optical detector. (d) Line type pulsating infra-red detector, reacting to heat and flame. (e) Line type pulsating infra-red detector, reacting to smoke

– smoke particles occurring anywhere along a beam of light will scatter it sufficiently to prevent it from reaching a photo-electric cell, placed at the opposite end of the line. In this way, the output from the cell is interrupted and the microprocessor then transmits the appropriate data to the controller, which recognises a state of alarm. False alarms may occur, as a result of the light being interrupted by birds, or bats, electing to roost or 'stay awhile' at points lying along the beam, or by stored materials that obstruct the light. The length of the line that can be monitored is limited by the intensity of the light, the quality of the lens used for focusing and the sensitivity of the photo-electric cell. Typically, such a unit will operate up to a distance of about 40 metres.

A different type of detector can increase this distance to 100 metres, by using as the light source a pulsating infra-red beam. This type of detector can operate in two modes. Operation in the first mode (see figure 8.13(d)) relies on the fact that any disturbance in the thermal equilibrium along the line of action of the detector, due to the presence of heat and flame, up to 6 metres on either side of the line, will modulate the frequency at which the photo-electric cell receives the pulsed beam. This change in frequency will be detected by the circuitry of the microprocessor and a state of alarm will then be indicated. Operation in the second mode (see figure 8.13(e)) depends on the production of smoke, anywhere along the line of action of the detector, to deflect the beam from the photo-electric cell and, by default, to indicate a state of alarm.

Smoke detectors are not normally placed in ducts, as the velocity of the air can alter their sensitivity. This difficulty may be overcome by using an air sampling unit. Point detectors of the ionisation type and/or the light diffusing type may be linked to an air sampling unit, which is capable of drawing air from several locations in turn and then of passing it on to a detector. Sampling tubes may be sited in spaces that are normally inaccessible, to give early warning of the presence of smoke.

The use of smoke detectors in an area where process fumes occur, or where the environment is dusty, or steamy, is not recommended. Consideration should be given instead to the use of fixed-temperature heat detectors and/or radiation detectors.

(3) Radiation detectors

A flame emits energy in the form of radiation. This can be identified as consisting of light rays, infra-red radiation and ultra-violet radiation.

Clearly it is possible to design a detector to react to light rays; however, within any building, it is obvious that there will be many sources of light, which may easily be confused with that being emitted from a flame. Hence, radiation detectors are generally designed to be sensitive to ultra-violet radiation or to infra-red radiation.

The ultra-violet component of sunlight is mainly filtered out by the ozone layer in the upper atmosphere and to a certain extent by the glass used in building construction; while the infra-red radiation not associated with an unwanted fire can be eliminated using electronic methods of frequency selection.

During the time of a fire, it is usual for smoke to appear before the flame, and if there is an initial production of thick smoke, it is just possible that a radiation detector, mounted within an enclosure, which relies on 'seeing' the ultra-violet light emitted by the flame, may never detect it. It should be noted that infra-red radiation will pass through smoke but that it is deflected in the process.

(3a) Infra-red detector

An optical filter is used to prevent all electro-magnetic emissions, except those of infra-red, from passing into the detector. The filter also acts as a lens, focusing the beam on to a photo-electric cell.

As mentioned earlier, it is possible that the detector will receive infra-red radiation from other sources in the building and, to overcome this, an electronic filter is used to allow only infra-red radiation, which has a frequency range of between 4 and 15 Hz, to pass into the amplifier. Radiation having these frequencies is characteristic of the radiative output of a flame and it is from such an input that the data required to signal an alarm condition are obtained. This type of detector should be used in situations where a flame occurs early in the process of combustion. Detectors relying on radiation are essential for use where smoke and heat are never likely to build up, as in an outdoor situation.

Note that false alarms have been caused by the frequency of infra-red radiation being modulated on passing through the blades of a fan, and by the movement of foliage fortuitously producing a frequency that is characteristic of the flickering of a flame. In an attempt to minimise the number of false alarms received, the microprocessor has included in its circuitry a timer/integrator, and this delays the transmission of any amplified signal to the controller for between 2 to 15 seconds. Such a detector must be placed where it can 'see' the likely development of a fire and on no account should its 'view' become masked by having material stacked in the vicinity.

In order to deal with a large open plan area, detectors are available that have a motorised lens continuously scanning the area, up to a radius of 90 metres. When an infra-red signal having the characteristic frequency of a flame is received, the motor is stopped and if the first signal is repeated over a period of from 2 to 15 seconds, the data representing an alarm condition are then transmitted to the controller. Alternatively, if the received signal does not persist, the angular motion of the lens is resumed and scanning continues.

A detector for use in an exhibition hall has been developed, having the facility of ignoring infra-red radiation of flame frequency when it emanates from known positions around the 360° scan. These positions may be programmed into the circuitry of the microprocessor and changed when the exhibits are rearranged. Obviously, this type of detector would also be useful in an open plan factory, where processes involving flames are part of the workload. Each detector also provides a visual indication of the state of awareness, either by the use of a flashing light or by the use of coloured lights.

(3b) Ultra-violet detector

This can consist either of a photo-electric cell, or of a gas-filled tube sensitive to ultra-violet light (uvl), having a wavelength that is within the range of 200 nanometres (nm) to 270 nm. Solar radiation does contain uvl having these wavelengths, but fortunately most of the uvl is still being removed by the high-altitude ozone layer, which means that an ultra-violet detector may be used outdoors. Note that detectors requiring a build-up of heat, or smoke are of very little use in an uncovered outdoor situation.

In the case of the gas-filled tube, the uvl has the effect of ionising the gas; this can be recognised by the passing of a small current between a pair of electrodes, fixed inside the tube. The signal is amplified and, in order to cut out false alarms, it is then integrated over a period of from 10 to 15 seconds, after which the alarm condition is established (see figure 8.14(a)). The delay is necessary in order to distinguish between uvl from a flash of lightning and uvl from a flame associated with a fire. Such a detector is obviously not suitable for use in a workplace where electric arc welding is part of the workload. Neither would it be suitable where it is likely that thick black smoke will appear before the instigation of a flame.

(4) Heat detectors

These may take the form of point or line detectors and they may be classified broadly into those devices that rely on a fixed temperature being reached and those that rely on a rate of rise of temperature occurring. Currently, detectors using both these parameters are in use.

A fixed-temperature detector will require a longer time to operate in a cold environment than in a warm one, whereas a rate of rise of temperature detector will require a relatively shorter time and it will be the same in both cases. Figure 8.14(b) shows the relative difference in response rates between these two types of detector and an ionisation detector – provided always that the products of combustion given off by the materials being burnt are suitable for each detector. Note that if the smoke coming from a fire contains large particles, these may not be detected by the ionisation unit.

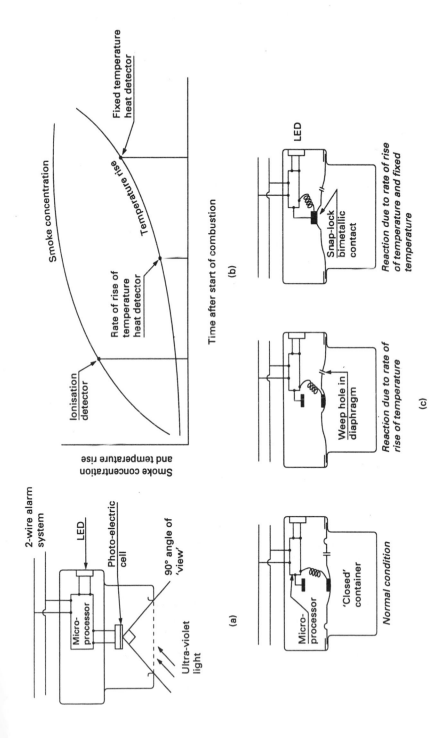

Figure 8.14 (a) Point type ultra-violet light detector. (b) Relative difference in response times, for three types of detector. (c) Dual-action point heat detector

8.14.3 Point heat detectors relying on fixed-temperature changes

An early example of a fixed-temperature, point heat detector depends on the melting of a fusible alloy. Melting of the alloy at temperatures within the range 57°C to 100°C allows an electrical current to flow through the pool of molten metal to the controller, indicating an alarm condition. Obviously, this type of detector is not available for reuse immediately after the alarm has been given.

Another fixed-temperature point heat detector relies on the movement of an electrical contact attached to a snap-lock bimetallic disc, mounted at the centre of a circular flat plate heat collector. When the circuit is closed, the automatic controller indicates the alarm condition. Such a detector is suitable for use where fumes may occasionally be present, such as in a kitchen.

Yet another fixed-temperature point heat detector relies on the analogue output from a temperature-sensitive transistor, referred to as a thermistor. Over the temperature range 20°C to 90°C, this transducer has an output that is approximately linear and when the output corresponding to the set point temperature is reached, the microprocessor triggers the alarm condition. Depending on the design of the microprocessor, the analogue output may also be used to indicate a rate of rise of temperature, giving this type of detector a dual purpose. Yet another variation in the way in which data from a detector can be processed involves the signalling of a 'pre-alarm' condition, which then gives the opportunity of increasing the sensitivity of the detector so that, if required, the alarm state may be given earlier.

8.14.4 Point heat detector relying on a rate of rise of temperature

This detector, sometimes referred to as a 'pneumatic' detector, uses a diaphragm, having a small weep hole, to seal off the end of a chamber containing air. In response to the occurrence of a minimum rate of rise of temperature, the air in the container expands rapidly, deflecting the diaphragm. Attached to the top surface of the diaphragm is a contact, which then completes the alarm circuit. The purpose of the weephole is to allow air that has expanded, or contracted, in response to normal changes in temperature and pressure, to pass through the diaphragm.

According to [8.20], heat detectors containing elements that can react to a fixed maximum temperature should not now be used, as they are unlikely to react to slow-burning fires. This limitation has been overcome by the use of the thermistor analogue type of point heat detector, described in the previous section, and also of those described in the next section.

8.14.5   Two dual-action point heat detectors

These detectors can be installed in a factory situation, where there is likely to be a large variation in the ambient temperature owing to process heating.

One such detector, which will react to both a fixed-temperature rise and to a rate of rise of temperature, consists of the apparatus described in the previous section together with a snap-lock bimetallic disc, fitted so that the centre of the diaphragm can be deflected upwards to make electrical contact when a preset maximum temperature is reached. The operation of this device is shown diagrammatically in figure 8.14(c).

Another dual-purpose heat detector makes use of the action of two bimetallic coils, one of which is in good thermal contact with the air in the room, while the other is shielded from it. There are four ways in which the detector can operate, as shown diagrammatically in figures 8.15(a), (b), (c) and (d). A description of each condition is as follows:

(a) when no extraneous heat is being supplied;
(b) when the temperature change takes place gradually and does not exceed a prescribed maximum value;
(c) when there is a rapid temperature rise due to the start of a fire;
(d) when there is a continuing steady rise in temperature, up to a maximum of either 60°C or 100°C, so that the thermally shielded strip is prevented from further movement by the fixed stop, and the unshielded strip then makes contact.

8.14.6   Line type heat detectors

These detectors take the form of cables whose electrical characteristics change when the temperature changes; or they may consist of fibre optic cables carrying a pulsed laser beam, or of pressurised tubes that fracture at a prescribed temperature. Line type heat detectors (LTHDs) may be classified into three groups:

(1) integrating (analogue) cables;
(2) non-integrating (digital) cables;
(3) semi-rigid sensing elements.

LTHDs belonging to groups (1) and (2) are those most commonly used. At the time of writing, BS 5839: Part 6 *Specification for line type heat exchangers* is in draft form and is being circulated for comment.

According to [8.21], LTHDs have been used over the last 50 years to trigger fire alarms and to initiate firefighting techniques. This type of detector can operate in ambient temperatures between −20°C and 70°C, and they are most likely to be found in an industrial environment fitted in close proximity to the hazard, where the conditions are often extreme

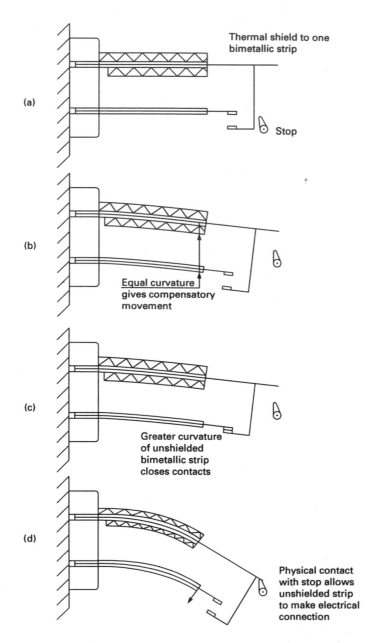

Figure 8.15   (a) to (d) Four operational modes of a dual-action heat detector, using bimetallic strips

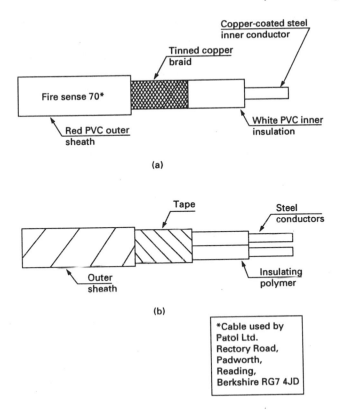

Figure 8.16 (a) Integrating (analogue) linear type heat detector cable.
(b) Non-integrating (digital) linear type heat detector cable

and where other heat detectors would quickly become unserviceable owing to the presence of dirt, dust and moisture. LTHDs can be used in commercial buildings and they are then normally fitted so that the cable crosses the building between 25 and 150 mm below the ceiling. Another useful application is for the detector cable to be routed through cable ducts and walkways.

(1) Integrating LTHD cables

A sectional view of one type of cable is given in figure 8.16(a). The term 'integrating' refers to the summing of a response to temperature change along the whole length of the cable. In this case, the resistance of the white polyvinyl chloride falls, with an increase in the temperature of the air through which the cable is routed, and when a preset (but manually variable) value for the resistance is reached, the alarm is triggered.

The co-axial cable is usually single core with the tinned copper braid acting as both the second conductor and as an effective screen against any stray radio frequency and/or electro-mechanical interference. The response time at 150% of the alarm threshold temperature is 10 s. After an alarm, the resistance of the cable will return to its normal value, provided a temperature of (typically) 150°C has not been exceeded for 10 minutes. This means that the system may be tested non-destructively. The company manufacturing this cable (Patol Ltd, Rectory Road, Padworth, Reading, Berkshire RG7 4JD) also market portable test equipment, which can provide repeatable simulated test conditions. The cable can be bent to a minimum radius of 12 mm several times without damage. Such LTHDs may be used in areas that need to be flameproofed, provided the associated control units are suitably modified.

(2) Non-integrating LTHD cables

This type of cable is shown in figure 8.16(b). The term 'non-integrating' refers to cables that respond to a prescribed temperature at all points along their length. In this case, when the set temperature is reached, the insulating polymer melts, allowing electrical contact to occur between the two adjacent conductors and triggering the alarm. The response time at 150% of the alarm threshold temperature is between 5 and 15 s. The cable cannot be tested non-destructively and clearly, after an alarm, it is always necessary to renew the affected sections. Depending on the type of cable, the minimum recommended radius for bending is between 50 mm and 150 mm, and it may be used in a flameproof area.

(3) Semi-rigid LTHD cables

These cables can be of the integrating or non-integrating type. The integrating type of cable consists of a metal outer sheath with a central conductor. The space between the conductor and the sheath is filled with a ceramic thermistor material, the constitution of which can be varied prior to the manufacture of the cable, so that the change in resistance over a selected temperature band, within a range extending from 20°C to 650°C, may be recognised by a preset controller. Provided the temperature of the cable does not rise above 1000°C, the resistance coefficient of the insulating material remains the same and it follows that the cable may be tested non-destructively.

In cables of the non-integrating type, the insulant consists of an eutectic salt, the composition of which may be varied prior to manufacture, so that melting of the salt will occur at an alarm temperature that lies within the range 120°C to 410°C. As in the previous case, the upper limit of temperature to safeguard the integrity of the cable is 1000°C. Both types of cable may be bent to a radius of 15 mm a limited number of times, and both may also be used in hazardous spaces. Note that these cables should be shielded thermally whenever they are routed in close proximity to a source of radiant heat.

8.14.7   *Choice of fire detectors and their installation*

The choice of detectors will depend on the class of fire that is likely to occur and the combustible materials in the building. Detectors having a high sensitivity (grade 1) are reserved for use on high ceilings, and detectors with a low sensitivity (grade 3) are used on low ceilings. An explanation of this grading system and the methods employed to check it are given in [8.22].

In order to cover as many hazards as possible, it is likely that a combination of different types of detector will be needed and a good deal of useful information on this matter is given in [8.21]. The same reference gives detailed recommendations with regard to the mounting heights, spacing and siting of detectors, for use where they are intended to protect property (system type P) and where they are intended to safeguard life (system type L). In particular it is now recommended that detectors should be provided in rooms that adjoin designated escape routes, especially where sleeping accommodation is to be protected. System types P1 and L1 have the connotations referred to in the previous paragraph and both comprise detectors that are installed throughout the building. System types P2 and L2 both comprise detectors that are installed only in parts of the building. A type M system includes only manual alarms, and a type X system refers to an alarm system designed for use in a multi-occupancy building.

The revised British Standard includes a recommendation that the maximum time that should elapse between the operation of a manual control point and the sounding of the general alarm should be reduced from 8 s to 3 s, and this recommendation operated from January 1990. Details are given of the likely effect of overhead heaters and of the effect of the stratification of smoke, when it occurs, on the performance of detectors.

It is recommended that manual call points are sited in conspicuous positions on exit routes, at the exits and also on the landings or stairways, at a height of 1.4 metres above the floor – in such numbers that no person using the building has to travel more than 30 metres to activate one. In cases where the movement of the occupants is likely to be slower than normal, the travel distance of 30 metres should be reduced considerably and where particular hazards exist, the manual call points should be sited in adjacent positions.

8.15   Alarm systems for buildings

The designers, manufacturers, installers and operators of alarm systems are all very well aware of the effect that recurring false alarms have on personnel. Reference [8.20] lists nine possible causes of false alarms and sets out a course of action that should be taken whenever a false alarm occurs. These are paraphrased as follows:

(1) treat every false alarm as genuine, until it can be shown to be false;
(2) identify the detectors that caused the false alarm and try to establish the cause before they are reset;
(3) keep a note of any activities in the area that may have caused the false alarm;
(4) obtain a record of the false alarm and notify the organisation responsible for servicing the system;
(5) require the service company to carry out an investigation.

As a guide, the British Standard suggests that the number of false alarms should not exceed one/year for every ten detectors connected to the system, and that for individual detectors, the frequency of false alarms should be no more than one every two years.

The continual development work carried out on microprocessors has made it possible for manufacturers to produce elegant alarm systems, which comply easily with the detailed recommendations given in [8.20]. The integrity and output of individual heat detectors can be checked at frequent intervals and comparisons with previous outputs, and those of other units, may be made.

The sensitivity of each detector, and that of the system, may be altered, effectively changing the grade of both the detector and the system. Such changes should be carried out automatically, and when this occurs they are referred to as being 'time related'. This facility produces a lower sensitivity during the hours of occupation, when the start of a fire is more likely to be noticed, and a higher sensitivity outside these hours, when it is not. It is also possible to change from relying on less sensitive heat detectors during the day to a combination of heat detectors and more sensitive smoke detectors at night.

In large buildings, it is customary for the space to be divided into 'zones'; in the event of fire, the control system must distinguish between these zones, sounding the alarms only in those places and in the manner required by the Fire Plan of the building. The British Standard recommends that all alarms (sounders) should have their integrity checked on a frequent and regular basis and, using current technology, this may also be accomplished automatically.

The amount of logging of information recommended in [8.20] is not inconsiderable and this can now be done automatically using a standard computer printout. The all important signal of a state of alarm may be passed automatically to the fire services; ancillary tasks, such as the closing (or opening) of dampers and louvres and the closing down of fans, may all be carried out automatically by the various types of automatic fire detection system currently available. Figure 8.17 shows some of the ways in which alarm detection equipment of different types may be connected.

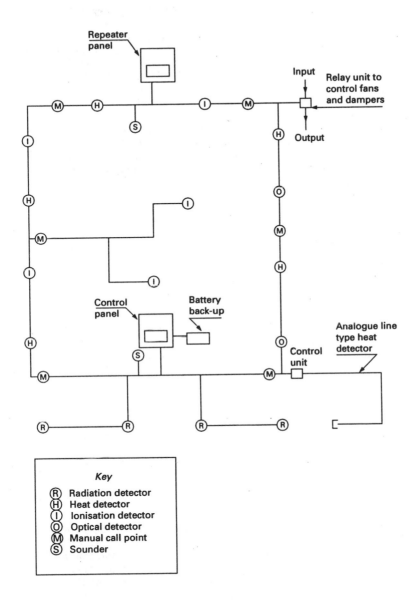

Figure 8.17 Connection of alarm detection equipment to control panel

References

8.1. Fire Protection Association. *Fire Protection on Construction Sites – A Code of Practice*, 1992. FPA, 140 Aldersgate Street, London EC1A 4HX.

8.2. CIBSE Guide, Vol. B. *Installation and Equipment Data*, 1986. Section B5-4. CIBSE, Delta House, 222 Balham High Road, London SW12 9BS.

8.3. BS 5588 *Fire precautions in the design and construction of buildings. Part 2: 1989 Code of practice for shops.* BSI, London.

8.4. BS 5588 *Fire precautions in the design and construction of buildings. Part 3: 1990 Code of practice for office buildings.* BSI, London.

8.5. BS 5588 *Fire precautions in the design and construction of buildings. Part 4: 1986 Code of practice for smoke control in protected escape routes using pressurisation* (amended). BSI, London.

8.6. BS 5588 *Fire precautions in the design and construction of buildings. Part 10: 1991 Code of practice for shopping complexes.* BSI, London.

8.7. Loss Prevention Council. *Code of practice for the construction of buildings*, 1992. LPC, 140 Aldersgate Street, London EC1 4HY.

8.8. BS 5306 *Fire extinguishing installations and equipment on premises. Part 2: 1990 Specification for sprinkler systems* (reviewed annually). BSI, London.

8.9. *Manual of Firemanship. Book 9 Fire Protection of Buildings.* Last update 1990. HMSO, London.

8.10. Nash P. and Young R.A. *Automatic Sprinkler Systems for Fire Protection*, 1991. Paramount Publishing, Borehamwood, Hertfordshire WD6 1RT.

8.11. BS 5306 Part 1: 1988 *Hydrant systems hose reels and foam inlets* (under revision). BSI, London.

8.12. BS 5423: 1987 *Specification for portable fire extinguishers* (amended 1989). BSI, London.

8.13. BS 5306: Part 3: 1985 *Code of practice for selection, installation and maintenance of portable fire extinguishers.* BSI, London.

8.14. BS 5499 *Fire safety signs, notices and graphic symbols. Part 1: 1990 Specification for fire safety signs.* BSI, London.

8.15. BS 6643 *Recharging fire extinguishers. Part 1: 1985 Specification for procedure and materials* (amended 1993). BSI, London.

8.16. BS 5306: Part 4: 1986 *Specification for carbon dioxide systems.* BSI, London.

8.17. BS 5306: Part 5: 1992 *Halon systems.* Section 5.1: Specification for halon 1301 total flooding systems. BSI, London.

8.18. BS 5306: Part 6: 1989 *Foam systems.* Section 6.2: Specification for medium and high expansion foam systems. BSI, London.

8.19. BS 5306: Part 7: 1988 *Specification for powder systems.* BSI, London.

8.20. BS 5839 *Fire detection and alarm systems in buildings. Part 1: 1988 Code of practice for system design, installation and servicing.* BSI, London.

8.21. Willey M. Line type heat detectors – their operation and application. *The Fire Surveyor*, Volume 21, number 2, April 1992. Paramount Publishing, Borehamwood, Hertfordshire WD6 1RT.

8.22. BS 5445 *Components of automatic fire detection systems. Part 5: 1977 Heat sensitive detectors – point detectors containing a static element.* BSI, London.

9 Transportation in Buildings

9.1 Origins of lifts

In England, during the period of the industrial revolution (1750–1850), steam power was applied to the operation of capstans, winches and winding drums used for moving and lifting loads. In the early 1800s, steam power was also used to pump water into a cylinder containing a ram, which was attached to a lifting platform. The stroke of the ram was equal to the height of travel and the cylinder had to be buried to an equivalent depth.

The industrial revolution was a time of intense human activity, requiring good communication systems. In the absence of electronic methods of communication, it was necessary for workplaces to be grouped together, resulting in ever-growing areas of dense population, in which the cost of space became prohibitively high. An obvious solution to the problem was to erect taller buildings, and at this time it became apparent that there was a need not only for goods lifts but also for passenger lifts.

In 1852, an American, Elisha Graves Otis, from Yonkers, New York, designed and produced a safety device that came into operation whenever the lift speed became excessive. The device took the form of wedge type clamps fitted to the car which, when they were activated, gripped the guide rails, bringing the car to a halt.

In 1853, Otis formed a company to manufacture hoists, and the first hydraulic lift using a single ram was produced by the Otis Lift Company in 1878. Later systems coupled the end of the ram to sheaves containing pulleys and ropes which supported the lift. This increased the velocity ratio to 2 and made it unnecessary to sink the hydraulic unit into the ground (see figure 9.3(c) later in the chapter).

It was common practice for the lift ropes to be wound and unwound from a drum, which became progressively larger as the buildings became taller. With this arrangement it was convenient for both the drum and the steam engine powering it to be sited in the basement. In 1889, in New York, the Otis Lift Company fitted an electric motor in the basement of the Demarest building which powered the winding drum of the lift through a reduction system.

431

In 1894, developments in electrical circuitry made it possible to produce the first pushbutton control system for use with electrical traction, and it was also realised that the electric motor (unlike the steam engine) could conveniently be sited at the top of the lift shaft, giving engineers the possibility of being able to use a combined drive and suspension unit. This was demonstrated in England in 1895 and a passenger lift, complete with balancing mass, was successfully operated in this way without the use of a winding drum. This was accomplished by mounting an electrically driven sheave at the top of the building. The sheave had multiple grooves turned in it and ropes were then placed over it, forming an inverted 'U'. The lift was attached to the ropes on one side of the sheave and the balancing mass was fixed to the ropes on the other. It was shown that, provided the angle of contact was large enough, the friction force produced between the ropes and the surface of the sheave was sufficient to transmit the necessary driving torque both to 'hold' and to accelerate the car and its contents. This system made it unnecessary to 'store' rope on a winding drum, making it possible to deal with many more floors; this is now the preferred arrangement for electric traction lifts.

In 1904, developments in the design of electric motors made it possible to couple the armature of the motor directly to the sheave, making transmission through a gearbox unnecessary. Motors using this 'gearless' transmission are used today for passenger lifts that operate at speeds in excess of 2.5 m/s.

Earlier lifts required the personal motor skills of an operator to stop them so that the platform was level with each floor but, in 1915, automatic levelling was introduced, whereby the operator relinquished control of the car just before each stopping place was reached, allowing automatic levelling to take place by means of trip switches installed at each floor level.

9.2 Use of buffers

Readers may have noticed that buffers are provided in the base of the well (the pit). These are solely for the purpose of stopping the car, or the balancing mass, in the event of overrunning. They are not, as is commonly believed, designed to cater for 'free fall' conditions of the lift car. As previously explained, such an exigency is catered for by the automatic application of wedges to the guide rails of the car. Buffers are classified as being either of the energy-absorbing type, or of the energy-dissipating type.

9.2.1 Energy-absorbing buffers

These consist of a compression spring, constrained to move in a vertical direction, when acted upon by a vertical downward force. An alternative to the use of a spring is to use an elastic material such as rubber. As the compression stroke progresses, the reaction to the vertical force increases and the deceleration of the car becomes progressively greater. Energy-absorbing buffers are suitable for rated car speeds up to 1.6 m/s.

9.2.2 Energy-dissipating buffers

These take the form of an oil-operated 'dashpot'. The vertical force is applied to the end of a plunger, which displaces a volume of oil through a series of small orifices. The reaction force remains largely constant throughout the travel of the plunger, producing a near uniform deceleration of the car. Energy-dissipating buffers are suitable for all rated car speeds. When the vertical force is removed, the dashpot plunger is returned automatically to its original position and BS 5655 [9.1] requires an electrically operated device to prevent the car from being started until the plunger has reached the fully extended position.

9.3 Development of automatically operated doors

Another development concerned the automatic opening and closing of the inner and outer doors. This was accomplished by providing the operator with an override button, so that in the case of a passenger requiring more time to negotiate the entrance, or in the case of the jamming of an object between the doors, automatic operation could be discontinued.

Subsequent development work produced other forms of 'fail safe' automatic door controls. These can be divided into those that require contact with an obstruction to cause the doors to reverse and those that do not require physical contact to establish door reversal.

9.3.1 Fail safe doors requiring physical contact to initiate reversal

In one type, spring-loaded pressure-sensitive edges are fitted to the doors so that whenever they come into contact with an object, the edges are deflected. This deflection operates reversal switches, which check the closing motion of the doors and then reverse them. Experience shows that frequent maintenance work is required on this type of system.

Where small power, variable-voltage, ac motors are used to operate the doors, another system employs an electronic logic control, which compares the magnitude of feedback signals with standard preset values. These signals are made to be representative of the speed and position of the doors and the torque exerted by the motor. When an obstruction is encountered, the feedback signals change rapidly from the norm and, when this change is detected, the motor is stopped and then reversed to an intermediate position, so giving time for the obstruction to be cleared.

9.3.2 Fail safe doors not requiring physical contact to initiate reversal

One such development utilises electronic proximity devices, fitted adjacent to the edges of the doors; this system is capable of stopping the closing motion of the doors and of opening them partially, before actual contact takes place. This facility is invaluable for use in hospitals and in homes for the elderly.

Another method that will operate without physical contact occurring involves the use of an infra-red 'curtain', simulated by the diagonal projection from both sides of the car of about 60 individual infra-red beams. When an object interrupts the beams, the closing motion of the doors is checked and they are then opened a limited amount. This action usually gives sufficient time for the obstruction to be removed without encroaching too much on the expected 'round trip time' of the lift.

9.4 Handling capacities of various circulatory methods

The CIBSE Guide [9.2] discusses in detail the different handling capacities of corridors and ramps, staircases and portals, conveyors, elevators and lifts, and the figures given in table 9.1 are quoted from that reference. Except for lifts, a width of 1 m has been assumed and the figures for the escalator assume 1 person/step.

It should be noted that the value for the handling capacity of lifts of 50 persons/minute is the lowest of all the means of transport, notwithstanding the fact that the density of 5 persons/m^2 is also the highest. The Guide also points out that the value for the handling capacity of the lift of 50 persons/minute was obtained by evaluation of the performance of the (currently) most efficient 8-car group, each car having a capacity of 21 persons and serving 14 floors. When considering the movement of people from floor to floor, the table shows the overall superiority of an escalator system over the other methods, whereas when moving people horizontally, it is clear that the conveyor system has the highest handling capacity of 120 persons/minute.

There are four reasons why the carrying capacity of escalators is superior to that of lifts:

(1) escalators operate in a continuous handling mode, so that there is no waiting, whereas lifts handle their loads in batches;
(2) the direction in which escalators operate may be reversed to suit the traffic flow;
(3) the mechanism is less complicated in escalators and there are likely to be fewer unscheduled stoppages than is the case with lifts;
(4) escalators can be used as a stairway even when they are not operating; however, most insurance companies discourage such use because of the effect of the variable step height on the progression of pedestrians.

Table 9.1 Handling capacities of various circulatory methods

Circulatory mode	Circulatory method	Handling capacity (persons/min)	Density (persons/m²)
Horizontal	Corridor	80	1.4
	Portal	60	1.4
	Conveyor	120	2.0
Incline	Stairs	60	2.0
	Ramp	85	2.2
	Escalator	90	2.5
Vertical	Lift	50	5.0

Data drawn from CIBSE Guide D by permission of the Chartered Institution of Building Services Engineers.

The preferred configurations for escalators are given in figures 9.1(a), (b) and (c). The relevant British Standard is BS 5655 [9.3].

9.5 Planning a lift system

The cost of a lift system can be in the region of 10% of the total building cost and ideally the design of the lift installation and lobby should be contemporaneous with that of the building.

The preferred position for the lift machinery room is on the top of the lift staff. It should be possible for town planners, architects and their assistants to be able to plan the skyline of their buildings in such a way that the normal expected projections caused by plant rooms, tank rooms and lift machinery rooms can easily be accommodated with the maximum of aesthetic appeal and the minimum of capital cost. Lift machinery rooms can be placed alongside the well at the bottom and/or the top, but at considerably increased overall cost.

Provided that the building is less than five storeys, the switch from an electric traction lift to a hydraulically operated one will make the task of positioning the machine room much less onerous. However, if the projected number of lift starts/hour is likely to be excessive, cooling of the hydraulic oil may become a problem unless a closed-loop logic control microprocessor is linked to the control valve. This system detects a change in the pressure and temperature of the oil, and opens the valve a few seconds later or earlier to allow for the perceived changes, so preserving the levelling capabilities of the car.

It is obviously essential that the dimensions of the openings in the lift well and those governing the verticality of the well itself should be given tolerances, so that the subsequent vertical movement of the lift will be

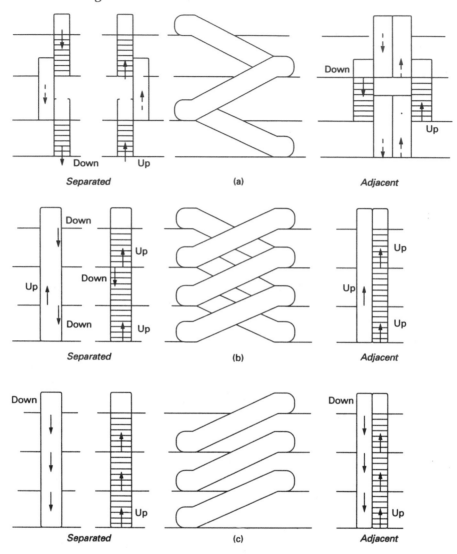

Figure 9.1 (a) Parallel escalators. (b) Cross-over escalators. (c) Walk around escalators

facilitated. Typical of the tolerances given in [9.3] for dimensions governing verticality are as follows:

Well height (m)	*Tolerance on plumb dimension (mm)*
Up to 30	+ 25, −0
Between 30 and 60	+ 35, −0
Between 60 and 90	+ 50, −0

The size of lift well depends on the configuration of the doors and it should be noted that, where a group of lifts is to be installed, it is good practice to arrange for a common well to be constructed. This not only saves on the cost of the dividing walls but avoids the 'piston effect' of air being forced through the crackage at each landing, producing noise whenever a lift approaches or leaves it.

It must also be expected that the machinery used to drive the lift(s), in spite of being supported on anti-vibration mountings, will still produce a noise spectrum that is characteristic of the particular plant installed. Consequently, any building facilities requiring a noise-free environment should not be sited adjacent to the machine room.

For those who have to design and specify a suitable lift system, BS 5655 [9.3] gives two worked examples. The CIBSE Guide D [9.2] also gives a series of worked examples and demonstrates the use of tables to provide equivalent solutions. The authors of the Guide have also included an example that gives help with the complicated task of comparing two tenders for a proposed lift system.

Choosing the best lift system is clearly an involved task and most lift manufacturers use computer programs to produce the most satisfactory solution to their potential customers' needs. This involves the consideration and manipulation of many factors, all of which are interdependent. In some cases, the quantities obey mathematical rules, as is the case with the physical motion of the lift, which includes values for velocity, distance travelled, acceleration and time of arrival, but other factors, which involve assumptions about the behaviour of the passengers, are imponderables. For instance, it is not possible to predict accurately the way in which the chosen traffic control system will be operated, to predict the time it takes for passengers to transfer into or out of the car, or to decide whether the direction of traffic flow at a given moment will be uni-directional or multi-directional.

Because of the existence of these unknown parameters, the operation of designing a lift system can never be an exact science and whenever problems of this nature have to be solved, engineers and others have to rely on empirical information modified by the results of past experience. The formulae given in [9.4] are now universally accepted and the lift manufacturers have incorporated these into tables and computer programs, to provide information on the handling capacities and round trip times for their own lifts.

9.6   Quality of service of lifts

The CIBSE Guide D [9.2] suggests that the quality of service should be evaluated for journeys starting from the main terminal, when the up-peak travel profile is occurring. An indication of the quality of service is obtained by evaluating the ratio of (the average round trip time for one car)/

(the number of cars in a group). This is called the 'up-peak interval' (seconds) and it is also linked to the handling capacity, or 'quantity' of service.

The Guide recommends that all lift installations should be sized to deal with the number of passengers requesting service during the heaviest 5 minutes of the up-peak traffic profile, in the (usually) well-founded belief that service in all the other operational modes of the lift will then be satisfied. There are always exceptions to any rule and these are known to occur in hotels at meal times, in hospitals that enforce strict visiting times and in those finance houses that are expected to be operational at set times of the day. Clearly, such cases should be evaluated in a different way.

As a general guide to the level of service that can be provided by a single lift serving several floors, the Guide offers the following information:

• one lift per 3 floors gives excellent service
• one lift per 4 floors gives average service
• one lift per 5 floors gives poor service.

This rule of thumb must, of course, be qualified by the possible need to achieve a particular interval and handling capacity.

In recent years, there has been a change in the expectations of those who use lifts, and these are reflected in up-peak interval times which are now shorter than those recommended in BS 5655: Part 6. For different types of buildings, the Guide gives some suggested values for up-peak intervals for lifts, loaded to 80% of their capacity, and for the likely percentage arrival rate of the personnel into the main terminal. It also includes a table giving the estimated density of population for each type of building.

In the particular case of an office block having a percentage arrival rate of 11 to 15%, the figures for the up-peak intervals with the qualifying comments are as follows:

• 20 s or less, indicates an excellent system
• 25 s indicates a good system
• 30 s indicates a satisfactory system
• 40 s indicates a poor system and
• 50 s or greater, indicates an unsatisfactory system.

9.7   Basic specification for an office block lift system

Example

A prestige multiple tenancy office block has six floors above the main terminal, each having a net usable area of 1500 m². The density of population may be taken as 20 m²/person and the percentage arrival rate of the total population of the building as 17%. Calculate the number of persons likely to use the lift system over the busiest 5 minutes of the up-peak period, when the building is 80% occupied.

If the recommended interval range extends from 20 to 25 s, specify an acceptable interval for the proposed system. Also determine the rated capacity of the cars, when the average number of passengers occupying a car during the up-peak period is assumed to be 80% of the rated capacity. Take the standard car sizes over the range required as 6, 8, 10 and 13 persons.

Solution

Number of persons/floor = 1500/20
                        = 75 persons/floor
Total population   = 6 × 75
                        = 450 persons
Expected population   = 0.8 × 450
                        = 360 persons
Expected peak arrival   = 0.17 × 360
                        = 61.2 persons
                        say 62 persons

As the building is classified as a prestige building, the shorter interval (waiting time) of 20 seconds should be chosen, even though the implementation may prove to be more expensive than if the interval of 25 seconds was used. Hence the lift system should be capable of handling 62 persons, with a 20 second interval, over the up-peak period of 5 minutes.

The capacity of the cars may be determined as follows:

Number of trips in 5 minutes for all cars = 300/20
                                          = 15 trips
Number of persons per trip   = 62/15
                                          = 4.1 persons
Size of one car   = 4.1/0.8
                                          = 5.125 persons

The nearest standard car size is 6 persons.

9.8 Arrangement of lifts for tall buildings

The CIBSE Guide D categorises tall buildings as having a maximum of 40 floors and very tall buildings as having more than 40 floors. Details of the traffic design methods used for very tall buildings are given in [9.5].

For any building it can be shown that for a given car size, the number of stops increases as the number of floors being served increases and that, as a result, both the quality and quantity of service become downgraded. The problem may be overcome by dividing the building into zones, each of which is served exclusively by a group of lifts and, as a general rule, the Guide suggests that such zoning should extend to a maximum of 16 floors.

Traditionally there are two kinds of zoning, one of which is referred to as 'stacked' and the other as 'skipped' (UK), or 'interleaved' (USA). Figures 9.2(a) and (b) show the two systems. In the stacked system all cars within a zone can call at all floors, whereas in the skip floor system each car can only call at every other floor.

In the stacked system it is convenient to try to make the round trip time approximately the same for all the lifts. This means that the number of floors served by the lifts in the upper and medium zones will be progressively fewer than those served by the lifts in the medium and lower zones. In this way, time is allowed for the lifts in the upper zones to make the express 'jump' from the main terminal to the commencement of the appropriate zone.

In the UK, the skip floor system is commonly to be found in local authority medium to high rise housing. Such a system is cheaper in first cost than having two lifts that call at all floors. Each lift has to call at half the number of floors and, accordingly, only half the number of openings and landing doors is required. However, the service to the tenants is poorer than if there were two cars to take them to the floor on which they live (a duplex system). It is only to be expected that tenants attempting to reach their flats from the main terminal will summon both cars and, if the car that does not serve their floors happens to arrive first, this would be taken to an adjacent floor. This practice results in a greater number of redundant calls for both lifts to the main terminal and also increased maintenance and running costs. For these and other reasons, the Guide does not recommend the use of skip floor (interleaved) zones.

9.9   Further recommendations on planning in the CIBSE Guide D

(1) It is suggested that where basement levels exist, these should be served by a traffic system that is separate from that serving the floors above ground.
(2) Multiple entrances to buildings at different levels should be designed to converge to a single lift lobby forming the main terminal for the whole building.
(3) The design of lifts for use in a hotel should take into account that the pattern of traffic produces two peak load times, one at check-in and the other at check-out, with a considerable amount of interfloor movement occurring at the same time. It is suggested that one lift per 90 to 100 keys might form the basis for a preliminary evaluation and that a separate 'service' lift system should be dedicated to the transport of baggage, goods and service personnel.
(4) Hospital bed lifts should be kept separate from the lifts serving staff and visitors.
(5) Most of the traffic in shopping centres will be accommodated by the

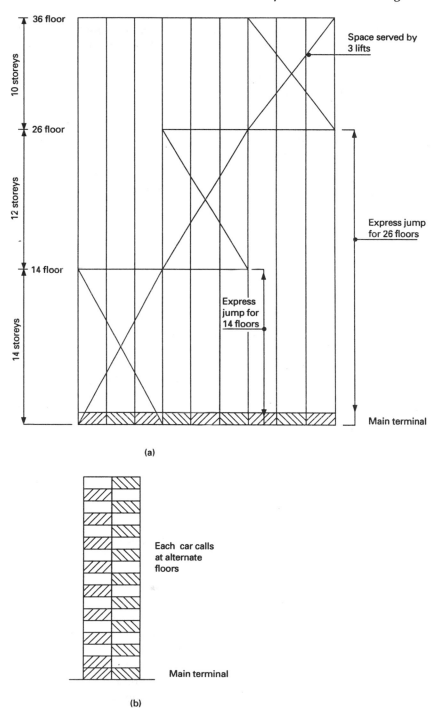

Figure 9.2 (a) Stacked zones. (b) Interleaved zones

use of escalators, however it is suggested that, as a guide, the conventional lift capacity should be based on one person per 100 m² of gross lettable retail area. The lifts should be duplex and have an interval of 40 to 60 seconds. It is recognised that observation lifts, while contributing to the traffic flow, will have a much reduced handling capacity when compared with that of conventional lifts.

(6) All mid to high rise office buildings should be provided with at least one goods lift having a minimum capacity in the range of 1600 to 2000 kg. It is also suggested that when the gross total floor area approaches 30 000 m² one dedicated goods lift should be provided, and that for each additional 40 000 m² of floor space one extra goods lift should be installed.

Details of vertical and inclined transport systems follow.

9.10   Lifts for the disabled

The most popular size of lift to cater for wheelchairs is a 630 kg, eight-person car, having internal dimensions of 1100 mm wide by 1400 mm deep. This size of car satisfies the recommendations given in [9.6] and [9.7].

The Guide suggests that internal mirrors could be used in an attempt to limit the claustrophobic effect associated with small cars. Toughened glass mirrors may also be used to good effect in an attempt to combat the vandalising of lift interiors. This was first tried out by the Otis Lift Company upon the recommendations of an environmental psychologist, who asserted that acts of vandalism would be lessened if the perpetrators could see themselves committing the crime. Subsequently, the advice was shown to be good and mindless destruction was no longer wreaked on the car and its fittings. The use of closed circuit television is also known to produce an inhibiting influence on potential vandals and thieves.

In the UK, where a lift for the disabled is installed in a nursing home, or warden assisted accommodation, it is current practice to fit a two-way communication system to the car, linked to the warden. In the relatively unlikely event of a lift stopping between floors, it is usually possible for the car to be manually wound down or up to the nearest floor, whichever is the easier.

The Otis Lift Company have a system that is referred to as 'remote elevator monitoring' (REM), whereby in an emergency it is possible for a representative of the company to have a two-way conversation with passengers who may be of a nervous disposition and who find themselves 'inadvertently detained' in the lift. The system is such that an engineer can be despatched to the site with an estimated time of arrival (ETA), which information can then be communicated to the passengers.

Lift speeds in nursing homes and homes for the elderly should be within the range of 0.4 m/s to 0.6 m/s, although for heights no greater than 12 metres a lower speed of 0.25 m/s is considered acceptable. In those cases where a stretcher has to be accommodated, a 1000 kg, 13-person car, having internal dimensions of 1100 mm wide by 2100 mm deep, will be satisfactory.

Whenever people enter a building, they do so with the subconscious expectation that in an emergency they may have to make a quick exit from it and in the case of fire, it is known that the most effective way of evacuating a building is via stairwells that have the appropriate fire resistance. When people who are disabled enter a building, they will probably do so by means of a lift, and in the case of a fire they will require to be evacuated in the same way.

Reference [9.8] gives detailed recommendations about the design and construction of escape lifts that may be considered suitable for use by the disabled during a conflagration. It is recommended that lifts should be operated by authorised persons, following an agreed evacuation procedure.

Under certain circumstances, subject to the agreement of the fire authority, it may be permissible for the firefighting lift to be used prior to the arrival of the fire service. In some buildings, it may be possible to use a passenger lift, provided that it has been suitably upgraded. The additional specification will require the car to:

(1) have the same fire resistance as a protected stairway;
(2) be of a suitable size;
(3) have a secondary power supply;
(4) have a switch that will put the movement of the car under the exclusive control of the operator;
(5) have a simple communication system that is able to identify the floor on which help is required to the operator of the lift.

It will also be necessary for the fire authority to give permission for this relaxation, and to help it come to a satisfactory conclusion; it must be made aware of the existence of an agreed management procedure covering the proposed use of the lift in the event of a fire.

9.11 Passenger lifts using electric traction

All new lifts and service lifts must conform to the recommendations made in BS 5655 [9.1], which deals with the safety of the equipment utilised in the lift installations.

It is comforting to note that when considering the tensile stresses that occur in the suspension ropes, the safety factor must be at least 12 in the case of a traction drive having three ropes or more, and 16 in the case of a traction drive having two ropes. The factor of safety is defined as the ratio between the minimum breaking load in one rope to the maximum force in this rope, when the car is stationary and carrying its rated load at its lowest level.

Ropes can also fail if the bending stress incurred by the separate fibres becomes too great. Such a situation will arise if the rope is made to pass around a pulley or a drum that has a limited radius of curvature. The maximum value of the bending stress occurs in the outer fibres of the rope, and the fibres will fail if it becomes too great. This type of failure can often be observed when a small-diameter winding drum forms part of a winch (often attached to a motor vehicle).

In the case of ropes used for lifts, BS 5655 [9.1] requires that the ratio of the pitch circle diameter of the drum to the nominal diameter of the suspension ropes should not be less than 40.

Details of the various methods of grouping lifts in a building are given in [9.3]. Reference [9.9] gives the recommended sizes for the machine room which, for electric traction lifts should be situated vertically above the lift shaft. As mentioned earlier, other positions for the machine room are possible but only at considerably increased expense.

9.12 Passenger lifts using a hydraulic system

The preferred position of the machine room for a lift powered by hydraulic means is immediately adjacent to the lift shaft at the lowest level served. If it is suspected that cooling of the oil will be necessary, then this will obviously be facilitated if the machine room can be sited adjacent to an external wall.

The vertical travel for passenger type hydraulic lifts is limited to approximately 18 metres and it should be noted that they are not capable of achieving the same flight times or of providing the same intensity of service as can be obtained from electric traction lifts. In general, they also require a higher starting current and, even though it is not necessary to apply power on the downward flight, they require more power than an electric traction lift, mainly because hydraulic lifts are not normally equipped with a balancing mass. Hydraulic lifts have a lower capital cost than electric traction lifts but cost three or four times as much to run. However, they may be used to good effect when powering observation lifts and goods lifts, and when used in establishments that cater for the needs of the elderly and the disabled.

Hydraulic lifts have good levelling characteristics and do not require as much headroom as those powered by electric traction, although both types require the provision of a lifting beam sited vertically above the well. Also, because hydraulic lifts do not normally operate with a balancing mass, the dimensions of the well can be made smaller. It is also worth noting that with a hydraulic lift, because the load carried by the building is restricted to forces and bending moments acting at low level, the structural demands on the building are much less stringent; therefore, in the case of an existing development undergoing a retro-fit, this structural consideration may make the choice of a hydraulically operated lift essential.

9.13 Goods lifts

Goods lifts (like passenger lifts) are manufactured in standard sizes, as are specified in BS 5655 [9.9].

Where it is known that the loading and unloading will be carried out using a fork lift truck, the floor of the lift and the door sills will be subjected to point loads acting through the wheels of the truck. These forces will be made up of the weight of the truck and the weight of the load; when the platform is designed, such forces must be treated as rolling loads.

In those cases where it is necessary for the load handlers to accompany the load, the Guide suggests that the internal dimensions of the lift should, as a minimum, be made equal to the size of the proposed load plus 600 mm and that, whenever possible, the next higher standard size of lift should then be chosen. In this way, when internal bumper bars are fitted to protect the walls, there will still be sufficient internal working space. If corrosive liquids are to be carried, then in the case of spillage it must be possible to wash out the inside of the car without the liquid spilling into the lift well. It is customary to fit hand-operated folding shutter gates, as these are easily adapted to fit varying car widths and also because they require the least space in the lift well.

9.13.1 Goods lifts using a hydraulic system

If the height of travel of the goods lift is no more than about 12 metres then hydraulic means of traction can be used.

If the mass being carried is 1250 kg or less, it is normal for the car to be suspended from a cantilever secured to the top of the hydraulic ram, and this form of fixing is referred to as being of the single side-acting direct type, see figure 9.3(a). For loads in excess of 1250 kg it is normal for the top of the ram to be fixed to the underside of the platform, and this form of attachment is referred to as being of the single, central direct type, see figure 9.3(b).

The speed and height of travel of the lift can both be increased by a factor of 2 by introducing the arrangements shown in figures 9.3(c) and (d), referred to respectively as the single side-acting indirect type and the twin side-acting indirect type. Because both these configurations use suspension cables, it is a requirement that safety gear be fitted. Figure 9.3(e) shows a twin side-acting direct type of coupling. Whenever hydraulic rams are used, the reaction to the load being carried is provided by the ground floor and is transmitted directly to the foundations. However, in the case of electric traction having an overhead machine room, the load and the balancing mass must be supported by structural means from the top of the building.

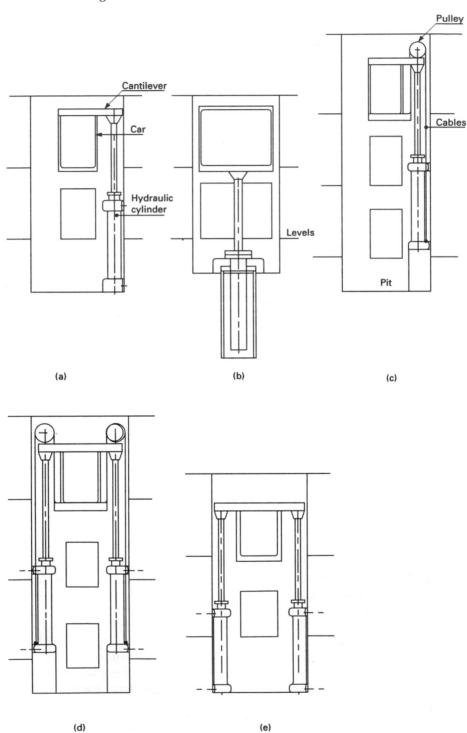

Figure 9.3 (a) Single side-acting direct. (b) Single central direct. (c) Single side-acting indirect. (d) Twin side-acting indirect. (e) Twin side-acting direct

9.14   Goods and passenger lifts using rack and pinion drives

In this system, the car levers itself up the face of the building with a recommended maximum speed of 1.5 m/s, using the gear teeth of two pinions that mesh with a gear wheel having an infinitely large radius (a rack).

Each pinion is driven by an identical motor-gear unit, both of which are mounted on the car roof – making the provision of a machine room unnecessary. It is possible for the motors powering the lift to be either electric or hydraulic. The safety gear consists of a self-contained centrifugal governor and braking mechanism, mounted on the roof of the car and coupled to a third pinion, also meshing with the rack. When the governor reacts to an excessive increase in the speed of the pinion, the brake is applied and the car is brought to a halt, usually within a distance of 1 metre. In order to ensure that the pinions and the rack are always correctly in mesh, a sub-assembly attached to the car consists of two guide rollers that make contact with the plane (reverse) side of the rack.

The car is cantilevered from a framework to which four contoured guide rollers are attached. These move over four hollow circular section 'rails' attached to the four corners of a hollow rectangular structural framework, referred to as the 'mast'. Within the framework of the mast, smaller diameter hollow circular section rails are fitted, and these act as guides for the balancing mass. The mast, which is of modular construction, is anchored to the wall of the building. The modules are 1.5 metres in length and are bolted together. In the case of a building under construction, the mast may be extended, ready for work scheduled to commence on the next storey; for buildings where 'change of use' is envisaged, it is possible either to extend or reduce the travel of the lift as required. These operations can be carried out from the roof of the car, making it unnecessary to erect scaffolding for the purpose.

In this system, the structural load is carried by the mast, which transfers it to pressure pads at the base of the building. The bending moment produced on the framework, owing to the car being suspended from cantilevers, is resisted by the bolts anchoring the mast to the wall. The system is structurally self-contained and a well is not required. Note that, with some alterations to the way in which the car is suspended and the way in which the balancing mass is supported, this particular form of transport may be adapted to operate both in the vertical mode and that of gently curving inclines.

9.15   Service lifts

These are normally powered by an electric traction drive which operates the car at speeds of between 0.2 m/s and 0.5 m/s.

The loads carried by these lifts are generally limited to a maximum of

500 kg and the dimensions are such that they are unsuitable for carrying people. The appropriate British Standard for this type of lift is given in [9.10] from which standard sizes and many other details may be obtained. If the service lift is intended to carry documents, it will require power-operated doors and may be fitted with automatic tilting trays, having collection boxes at each landing. Purpose-made fitments are available for handling all types of goods, and if the service lift is handling food, the compartment must be constructed with smooth corners (for ease of washing-out) and it may be heated.

A service lift is normally supplied to the customer complete with the structural framework in which it operates, and this ensures that any builder's work required is kept to a minimum. Obviously, there are many different designs of lifts but as a guide to the size of lift well required, the Guide suggests that the dimensions of the well should be:

well width = width of car + 500 mm

well depth = depth of car + 300 mm

The height of the soffit of the well above the highest floor served is obtained by evaluating the height of the hatch, plus the height of the car, plus an allowance of between 500 mm and 1000 mm (depending on the requirements of the supplier). Also, depending on the arrangements for the landing doors, the depth of the pit needed for service should be between 150 mm and 1000 mm. Normally the machine room will be of the same plan area as the well and should ideally be placed above it, having a minimum internal height of 600 mm and a clear access that is at least 600 mm square.

9.16 Motor vehicle lifts

The design of these lifts will almost certainly be individual to the development. Such lifts are often used to achieve access to restricted parking areas that are sited within a building. The entrance will usually require to be fire rated and when this is the case, the supplier of the equipment should be notified.

Whenever the entrance is open to the atmosphere, the provision of ramps will help in preventing rainwater from entering the lift well. If automatic doors are provided, special care should be exercised to see that the control system has a back-up device to prevent the door acting as a self-appointed motor vehicle crusher. The most reliable and inexpensive pushbutton-operated doors are the four-panel centre opening type, which however do require extra space in the well. Those who have to specify such a system will find details of the recommended sizes for the lift well, lift and machine room in the Guide [9.2].

Where the height of travel is no more than approximately 12 metres, it is usual to power the lift by means of twin acting hydraulic cylinders, and speeds of between 0.2 m/s and 0.3 m/s are considered to be satisfactory. For heights in excess of this, electric traction may be used, when a slightly higher speed of about 0.5 m/s is acceptable. In order to limit the magnitude of the starting current, owing to the relatively high payload, it is customary to employ a rope factor (velocity ratio) of up to 4:1. The control of the lift from within the motor vehicle may be facilitated by the use of strip-type pushes fitted on both sides of the lift car.

9.17   Firefighting lifts

Not every development needs a firefighting lift. The circumstances that prescribe such a need are given in [9.11] and [9.12], which also give detailed guidance on the orientation of firefighting lifts and stairways. Some of the recommendations made are paraphrased as follows:

(1) the firefighting lift must have a capacity of at least 630 kg, a width of 1100 mm, a depth of 1400 mm and a height of 2000 mm;
(2) the speed should be sufficient to reach the top of the building in less than 60 s;
(3) operating doors must be provided giving a clear opening 800 mm wide × 2000 mm high, and they must be operable in response to the use of buttons inside the car;
(4) lift position indicators should be provided in the car and at the fire service access level (FSAL), which (level) should also be delineated inside the car;
(5) a two-way intercommunications unit, consisting of a microphone and speaker fitted in the car, should be provided;
(6) the apparatus must link the car to the machine room and to the FSAL, where handsets may be provided;
(7) the walls and roof of the car should be constructed of non-combustible material;
(8) the signage on the lift must be executed and displayed in accordance with the recommendations made in [9.12];
(9) an emergency hatch complying with [9.11] should be provided;
(10) the floor of the fire service access level should be provided with drains and contoured in such a way that water will not readily find its way into the lift shaft;
(11) an attempt should be made to keep the electrical control equipment free from the effects of the splashing of water, by the use of suitable shrouds and covers fixed to the equipment inside and outside the car and in the well;
(12) audible and visual indications that the lift is in firefighting mode should be provided within the well and machine room;

(13) the mains power supply to the firefighting lift is to be derived from a dedicated sub-main, as shown in Chapter 6, figure 6.10, and the run of the cables must be suitably fire protected;
(14) the interval between the loss of the main power supply and the provision of power from the secondary supply should not be greater than 30 s and on restoration of the power, the lift must re-establish its position within 10 s, without returning to the FSAL;
(15) a fire brigade switch should be positioned in a locked metal cabinet marked 'FB' adjacent to the lift entrance at the FSAL.

9.17.1   Control of the firefighting lift

Activation of the fire brigade switch renders the firefighting lift operational and simultaneously configures the way in which all the lifts in the group subsequently operate, as follows:

(1) A car sign is displayed in each lift within the group, indicating that it is 'under fire service control', and all lifts are then returned to the FSAL without stopping.
(2) After the passengers have left, all the doors except those of the firefighting lift are closed and the lifts are disabled.
(3) Control of the firefighting lift and doors is achieved manually from within the car.
(4) Continued application of pressure is required on the appropriate door buttons for the doors to open or close completely. If pressure on the button is removed before the opening or closing cycle has been completed, then the doors revert to the previous open or closed position. The movement of any button must also operate indicator lights within the car.
(5) The two-way communications system inside the car of the firefighting lift is activated.

For firefighting purposes, the capacity of secondary power supplies should be such that the firefighting lift may be operated for at least 3 hours, and in the event of the main power supply failing before the lifts have been returned to the FSAL, it must be possible subsequently to recover at least one lift at a time.

9.17.2   Inspection and testing of firefighting lifts

Upon taking over a completed development, the user can expect that the lift installer will have carried out operational tests in accordance with the recommendations given in [9.13] and a test certificate should form part of the documentation. The Guide recommends that specific checks should be carried out at the following times:

Weekly: The fire brigade lift switch should be operated, when the functions outlined in points (1) to (5) given in the previous section should be realised.

Monthly: Failure of the main power supply should be simulated and the firefighting lift should also be brought into action.

Six monthly: The maintenance company of the lift manufacturer should carry out inspection and testing according to the recommendations given in [9.13].

Yearly: Appendix C1 of [9.13] describes the tests to be carried out, and successful completion of these tests should be recognised by the issue of a certificate.

9.18 Control systems for lift installations

The control of lifts may be carried out using any one of three technologies: (1) electromechanical, (2) solid-state logic, or (3) computer-based. The divisions between the methods of control are not clear cut, as electromechanical switching technology is also used to implement the technologies used in methods (2) and (3).

(1) Electromechanical methods

These methods of controlling lifts were the first to be developed and involve the use of electromagnetic relays and mechanically operated switching devices.

Electromagnetic relays are used to switch small electric currents; these smaller currents are sometimes used to operate electromagnetic contactors, which are used to switch larger currents. Both devices use electrical contacts which, in the case of contactors, need to be replaced when the combined effects of spark erosion, dust and insects from the atmosphere destroy the electrical conductivity of the surface of the contacts. The Guide gives some interesting facts and figures about the number of operations that contactors are likely to perform before breakdown occurs, and those interested in the application of statistics to the probability of the occurrence of breakdowns will find further details in [9.14].

The satisfactory functioning of mechanically operated switching devices relies upon the accurate juxtaposition of levers and sliding contacts, and when the effects of wear and possible distortion combine to alter the relative positions of the members outside the acceptable tolerances, this method of control becomes impaired.

It is true to say that the failure of relays and the frequent need to carry out adjustments to the mechanical switching devices have resulted in a relatively large amount of downtime for those lift installations that are entering the 'wear-out phase' of their operating life. The exclusive use of such methods of control now tends to be restricted to passenger and goods lifts operating at speeds of up to 1 m/s in buildings having no more than three storeys.

(2) Solid-state logic methods

Solid-state devices are known to be extremely reliable and use less power than electromechanical devices. They are used to produce integrated circuit boards, designed to operate at 12 to 15 V, which provide high immunity to the effects of extraneous electrical signals.

Input signals relating to the landing and car call points, the automatic doors and the lift spatial transducers are all routed to the solid-state logic boards. The output from each logic board is amplified and passed to conventional relays, which are energised and then operate the motor controller, the direction indicators and the call indicators. Reference [9.3] recommends that while solid-state switching methods may be employed to control logic circuits and the drive motor, they should not replace safety contacts, for which conventional electromechanical devices should still be employed.

Note that the operating temperature of the passive integrated circuits should be kept at a steady temperature, which is above 5°C for electric traction and above 15°C for hydraulic traction [9.3].

The Guide [9.2] makes the point that by operating the logic circuitry at between 70 and 80% of its rated capacity, the reliability of the equipment is doubled. A common source of failure with this type of circuitry occurs because of the way in which the separate boards are joined together. Edge connectors are known to be inferior to DIN type connectors and, when high reliability is important, they should be avoided.

Manufacturers are now able to 'tailor' a lift control system to suit the needs of a particular client by using microprocessor technology, and the circuits used in this particular approach are described as 'application-specific integrated circuits' (ASIC).

Solid-state logic technology is well suited to controlling goods lifts, single passenger lifts and up to two lifts in a group, with the last two applications operating at speeds of up to 2 m/s. Provided the density of traffic is low, the technology may be used in small hotels and residential buildings up to 12 storeys in height.

(3) Computer-based methods

This method of controlling lifts is the last of three methods to be developed. In theory, the computer can be programmed to control the lifts in an infinite number of ways, depending on the elegance of the program being used. In practice, however, suppliers of the equipment tend to use programs that have been developed for one particular company and subsequently field-tested for its specific installations. Use of a computer makes it possible to diagnose faults by recording information on the performance of the installation, and it can be programmed to give an indication of the presence (or otherwise) of both input and output signals across a chosen component. The program used can include options to vary the velocity of opening and closing of the lift doors and also the position at which a lift may be parked, to suit the density of traffic at a particular time and place.

The monitoring of the way in which a lift installation is operating may be undertaken by the lift company with the use of a modem link, and in the case of the Otis Lift Company this system is referred to us 'remote elevator monitoring' (REM). The client may also wish to receive this information and to display it on his business management system. There is a great deal of development work currently being undertaken, both nationally and internationally, to try to produce a rationalisation of lift monitoring standards. For those working in this field, the Guide [9.2] devotes two chapters to this and other related topics, including the evaluation of computer programs for use with lifts.

Computer-based methods of control rely on the proper functioning of solid-state devices and on the use of electromechanical switching, and the reliability of the system will depend on the way in which the various components of the different technologies are matched. According to [9.2], a successful application of this technology can produce an improvement in the traffic handling capabilities of a lift group of between 30 and 40%. The technology may be applied to all types of lift installations at speeds that vary from 0.5 m/s to 10 m/s, and it is particularly suitable for application to groups of lifts.

9.19 Processing of landing calls

Landing calls may be dealt with by using any one of the following four collection methods.

Single automatic pushbutton control

In this system, the lift may be called by only one person (or group of persons) at a time. At this stage, the sign 'lift in use' is displayed on all the other floors and the lift will not respond to, or memorise, any other landing call until it has discharged its passengers.

This method is suitable for goods lifts and for passenger lifts containing up to six persons, and it may be used in residential accommodation up to three storeys high.

Non-directional collective control

In this system, a landing call received by a lift when it is moving up or down will make it stop at the landing. If the lift subsequently moves in the opposite direction from that required by the new occupant, it will be in answer to the needs of a passenger who called it earlier. The lift will continue to discharge its passengers in the order in which the buttons are pressed, until the memory contains no more information. This system produces long waiting times, giving poor traffic handling characteristics.

Down-collective control

This system is only suitable for buildings where the amount of interfloor movement is expected to be minimal, such as multi-storey car parks and residential buildings.

All landing calls are accepted as down calls, except calls coming from the ground floor which are accepted as up calls. When a number of landing calls are made, the lift moves to the position of the highest call as a priority, subsequently collecting passengers on the downward flight.

Full-collective control

In this system, two pushbuttons are provided at each landing position, one dealing with and signifying upward calls and the other with downward calls. These are registered separately and answered only in the direction in which the car is moving at that time. When the direction of the car changes, the calls requiring the movement of passengers in the new direction are then processed. If a call is received when the car is stationary, that call is then answered.

The system is eminently suitable for use in office blocks, where it can deal with both interfloor traffic and peak directional traffic.

9.20   Application of computer control to a group of lifts

A computer can be programmed to collect data from each landing and from each lift in a group. That data may then be compared with information that is already stored and, depending on the results of the comparisons, external actions may then be initiated. Such a computational process is referred to as an 'algorithm' and with the use of suitable algorithms, it is possible to change the way in which a group of lifts operate, without changing or manually adjusting any hardware.

It is possible to prioritise certain landing calls and, in order to prevent the interval for the group becoming unacceptably long when the number of landing calls in one part of the building increases above a programmed 'norm', idle lifts can be sent to the landing. Lifts can also be 'parked' at strategic positions ready to deal with a perceived sudden increase in traffic, or in response to a landing call, the lift that is nearest in flight time will be sent to take the call. To help deal with traffic peaks, the rate at which car doors open and close may also be changed temporarily. Frequently, in a multiple tenancy building one or two floors may undergo a change in use, so that the traffic profile for the group changes; adjustments needed for the group to deal with this may be made automatically by the controller.

Back-up for the system can be provided by the use of more than one computer, and the second (and subsequent) computers can be programmed to check the operations of the control computer. In this way, if a failure occurs, the functioning of the group will be taken over by a second control computer and this arrangement has resulted in group control reliability approaching 100%.

9.21 Temperature control in machine rooms

According to the recommendations given in [9.3], the temperature of the air inside machine rooms for traction lifts should be kept between 40°C and 5°C, and for hydraulic lifts between 35°C and 15°C. All electric motors produce between 0.25 and 0.33 of their rated capacity as waste heat, with the exception of those that operate using variable voltage and variable frequency type controls.

In the UK, it is usually possible to ensure that the recommended conditions are achieved by using a combination of natural and forced ventilation to dissipate the heat and a thermostatically controlled bank of electric tubular heaters to provide the heat input. Figure 9.4 shows a possible arrangement, in which the cooling fan attached to the lift motor discharges the heated air directly to the atmosphere.

For an installation using hydraulic traction, the electric motor and pump are often of the submersible type and the heat dissipated from the motor serves to warm the oil contained in the reservoir. In installations where it is likely that the oil in the reservoir will be cooled below 15°C overnight, or at weekends, the lift manufacturer will install a thermostatically controlled electric immersion heater. Where the number of starts/hour exceeds the rated number, the temperature of the oil will rise and cooling will be required. The lift manufacturer may then install an oil-to-air radiant fin cooler.

Note that even when an oil cooler is fitted, it is still necessary to keep the machine room at a temperature that lies between the limits already quoted. Normally, in the UK, effective cooling can be obtained by circulating outside air through the space; however, if the environment is such that the temperature differential becomes too small to make this method viable, the use of a direct expansion cooling module complete with a remote operating alarm, to indicate an unnacceptably high rise in room temperature, will overcome the difficulty.

9.22 Modernisation of lift systems

When a lift installation has been operating for 15 to 20 years, it will almost certainly be possible to improve its performance by fitting computerised

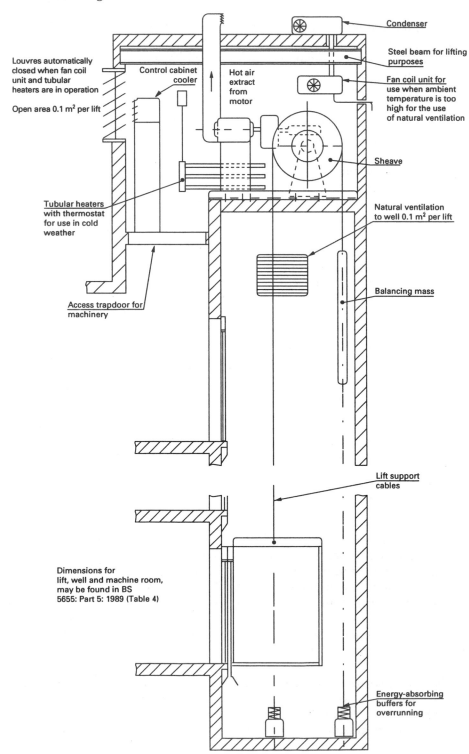

Condenser

Steel beam for lifting purposes

Fan coil unit for use when ambient temperature is too high for the use of natural ventilation

Louvres automatically closed when fan coil unit and tubular heaters are in operation

Open area 0.1 m² per lift

Control cabinet cooler

Hot air extract from motor

Sheave

Tubular heaters with thermostat for use in cold weather

Natural ventilation to well 0.1 m² per lift

Balancing mass

Access trapdoor for machinery

Lift support cables

Dimensions for lift, well and machine room, may be found in BS 5655: Part 5: 1989 (Table 4)

Energy-absorbing buffers for overrunning

Figure 9.4   View of lift, well and machine room, showing methods of

controllers and new variable-voltage variable-frequency motors, which also improve the power factor. In some cases, the owner may be advised to continue to use the earlier motor generator dc motor set (Ward Leonard set). This is because some manufacturers' products were over-engineered, making them capable of giving many more years of useful service. However, the addition of a new controller incorporating the use of feedback techniques would produce advantages in the smooth running of the lift and should be fitted to the motor control system.

Lift consultants and lift manufacturers will carry out an on-site survey of the performance of the existing lift installation, and by using suitable computer programs, they will then be in a position to predict the likely improvements to be achieved by the proposed installation over that of the existing one.

Where a group of lifts is being refurbished, the likely cost of the kind of work being suggested may be of the order of £100,000 and it is often the case that the owner is more likely to have the work carried out if it is done (and paid for) over a period of, say, 5 years. This has encouraged lift manufacturers to adopt what is sometimes referred to as a modular approach to the refurbishment of lifts; this involves the updating of parts of the system, usually on a yearly basis.

When the building is being refurbished, the owner has the opportunity to bring the lift installation up to the requirements of the latest British Standards and/or the latest Health and Safety at Work legislation, as well as changing the interior decor of the cars. As the lift installation represents about 10% of the capital cost of the building, it makes good financial sense to improve the effective operation and enhance the aesthetic appeal of the existing investment. Owners carrying out refurbishment should be aware of the contents of [9.15].

9.23 Paternosters

In the UK, this form of vertical transportation is now considered to be virtually obsolete.

The paternoster transportation system consists of a number of doorless, two-passenger capacity lift cars, which are attached at intervals to an endless chain, moving continuously in a common vertical shaft. The velocity of the car is approximately 0.4 m/s and a car passes each floor at roughly 10 second intervals, one going up and the other going down. If the cars are attached to the chain at intervals of 4 metres, the system is capable of handling 720 persons/h in both directions. Because the cars move at a constant speed, power is not required for acceleration purposes and the same number of people may be transported for less power than that which would be required in an equivalent electric traction lift system.

Such an installation is useful where there is continuous interfloor traffic, as might be the case in a university library. Boarding the lift car requires a minimal amount of manual dexterity.

Recommendations regarding paternosters are given in BS 2655, which has now been superseded by BS 5655, although the latter does not include any references pertaining to paternosters. However, information is available in *Passenger-carrying paternosters*, HSE Plant and Machinery Guidance Note PM 8 (HMSO, London, 1977).

9.24 Escalators and passenger conveyors

In the 1890s, inclined moving belts or stairways to transport passengers from one floor to another were developed; these subsequently acquired the name 'escalator' around 1900. The modern escalator uses a succession of steps, each of which is mounted on open roller chain links, which are constrained to follow the contours of a fixed track. The track is initially horizontal and the steps form a horizontal platform on which the passengers can stand. After a distance of about 1 metre, the track begins to take up the angle of slope of the escalator which is normally 30°. However, where the rise does not exceed 6 metres and the rated speed does not exceed 0.5 m/s, this angle can be 35° to the horizontal, see figure 9.5.

During the journey, the steps remain in the horizontal position but as the links on which they are mounted are inclined, each step rises above the one behind it to form a staircase. Towards the top of the stairs, the roller tracks again become horizontal and the steps form a horizontal platform, extending for about 1 metre to enable the passengers to make a smooth transition from the moving staircase to their own subsequent personal velocity on a level surface.

Below floor level, the track guiding the rollers curves around the main driving wheel and then follows a path that is parallel to the line of the staircase and the lower horizontal section, before following the curve of the carriage wheel to complete the closed loop. The linear velocity of the handrail is meant to be the same as a point on the steps, and [9.16] gives a tolerance for this of 0 to +2.0% of the speed of the steps. The driving motor and gearbox are located in a pit, which can be positioned either at the top or bottom of the escalator.

Safety measures are detailed in [9.16] and include:

(1) a 'no-volt and overload release mechanism', forming part of the motor starter, an earth leakage relay and a direction detector incorporating a 'fail-safe' shutdown mechanism;
(2) the provision of a speed governor;
(3) the provision of a brake which [9.16] recommends should produce a uniform deceleration and be activated in the event of power failure, or mechanical failure of the roller chain or of the handrail;

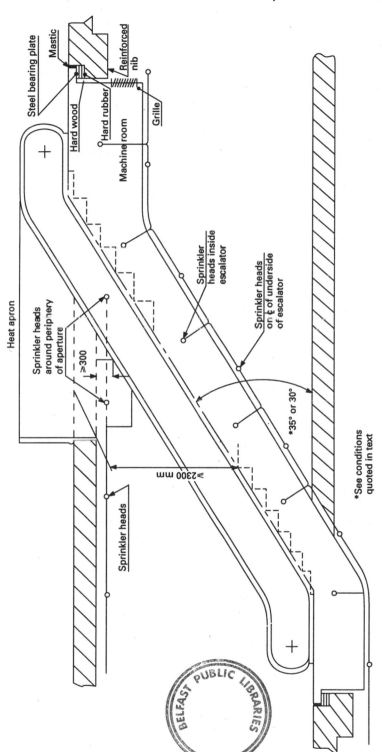

Figure 9.5 Elevation of typical escalator, showing heat apron and positions of sprinkler heads

(4) the provision of emergency stop buttons at the top and bottom of the escalator;
(5) the provision of balustrades, handrails and side panels, made from non-snag materials and constructed and manufactured according to the recommendations given in [9.16];
(6) the illumination at the landings of indoor escalators at a minimum of 50 lx, with other lighting being related to that of the general lighting in the area;
(7) in the machine space, the lighting and socket outlet circuits are to be supplied from separate circuits from that which serves the drive motor.

See also [9.17].

9.24.1 Passenger conveyors

These consist of an endless moving walkway, which may be laid in a horizontal direction, or along inclines where the angle of inclination is no more than 12° to the horizontal. The passenger conveyor has been found to be useful in preventing congestion from occurring in small concourses and in delivering people and their luggage over relatively long distances to board an aircraft, or to visit a customs hall.

Both escalators and passenger conveyors must conform to the safety and operating requirements given in BS 5656 [9.16]. Where the local fire authority requires the provision of smoke detectors, sprinkler heads and fire shutters, these and any interfacing required will normally be supplied by the supplier of the escalator or passenger conveyor to a specification provided by the mechanical/electrical services consultant.

9.24.2 Installation details

Escalators can be shipped to the site in three separate parts which, when they are assembled, produce a statically determinate structure. Generally, however, smaller escalators can be delivered to the site already assembled.

Some early planning is required between the architect and the contractors, which should start by taking a decision as to the precise stage in the construction process at which the escalator(s) should be delivered.

If it is decided to take early delivery, the task of manoeuvring the escalator(s) into position would be relatively straightforward, provided there is a suitable hard-standing with no mud, which extends from the off-loading position to the point of destination. Subsequently, however, there is a real possibility that the escalator might be damaged by later construction processes, such as the erection and later dismantling of shuttering.

If a later delivery is planned, there may be an access problem due to a lack of clearance or to columns being in the way; or the point loads from load handling trolleys may exceed the permissible floor loadings. This latter problem is likely to arise where manoeuvring is required on intermediate floors. It may be overcome by the provision of props placed in suitable positions and by arranging to spread the load from the trolley(s) over a larger floor area, by supporting the trolley wheels on rigid steel plates. Lifting eyes will be required and should be placed at suitable points along the access route and at the destination point. If it is intended to lift the escalator into position using hydraulic jacks, these may require special-purpose thrust pads to be incorporated in the floor(s). Architects and builders should also note that all machine rooms must be fireproofed, and whenever they are sited on the ground or lower floors, they must also be waterproofed.

9.25 Use of an installation quality plan for escalators, passenger conveyors and lifts

It makes good sense for the manufacturers of the equipment that is subsequently to be installed in a development to ensure that the building is constructed in such a way that the installation can proceed smoothly towards the successful commissioning of that equipment.

For lifts in particular, checks on the vertical alignment of the well and the suitable positioning of fixings for the guide rails and the subsequent correct alignment of the rails should be carried out throughout the construction process. Also, checks should be made on the correct positioning of landing door frames and the associated equipment.

For escalators, checks should be made on the positions of the bearing plates, which should conform to the tolerances given, and on the suitable construction of the machine rooms. The results of the checks should be recorded on an 'installation quality' plan, which may then be used as a record of the (hopefully) satisfactory progress of all the systems towards an ultimately successful commissioning.

9.26 Tests on completion

These are normally carried out by the manufacturer of the equipment. The conventional 'snagging' list is then compiled by the architect and the owner's representative for presentation to the parties concerned. For lifts in particular, [9.1] and [9.18] give the recommendations for carrying out completion tests and further suggest that, before any lift is put into service, a thorough examination should be carried out and a report should be issued.

For those lifts that are installed in premises covered either by the Factories Act 1961, or by the Offices, Shops and Railway Premises Act 1963, there is a statutory obligation for an examination followed by a report to be commissioned and carried out at least every six months.

There is no statutory obligation for escalators to be tested in this way but [9.16] recommends that escalators should be examined and a report made following the format suggested in [9.19].

As well as the Factories Act 1961, there is a set of regulations referred to as the 'Offices, Shops and Railway Premises (Hoists and Lifts) Regulations 1967', which details similar obligations that must be observed by owners; however, in this case all reports mentioning defects are to be submitted to the local authority environmental health inspectors.

The Health and Safety Executive (HSE) has a duty to advise industry on safety matters, and this is done by the issuing of Guidance Notes, some of which have been mentioned in the text and appear in the references given at the end of the chapter. In particular, HSE Guidance Note PM7 refers to 'the thorough examination and testing of lifts' and this gives details of the time intervals between carrying out inspections on the salient pieces of equipment used in lift installations. Also, HSE Guidance Note PM26 refers to 'Safety at lift landings'. A summary of the contents of these Guidance Notes is given in the CIBSE Guide D [9.2].

9.27 Maintenance of lifts and escalators

The age old adage 'If it ain't broke, don't fix it' has no place in the philosophy of lift maintenance. The statutory obligations vested in the owner relating to the testing and examining of lifts also extend to their maintenance and, to this end, preventive maintenance must be employed.

The manufacturers of lifts and escalators generally offer two types of contract. The first, referred to as a 'service' contract, covers checking and lubrication, with repairs, replacements and subsequent adjustments forming the subject of extra costs. The second form of contract is comprehensive and includes a long-term programme of preventive maintenance and encompasses the supplying and fixing of replacement parts, together with all call-out charges.

With the increasing use of computer control and data-logging, it is now possible to be more circumspect in the replacement of those parts of the system to which preventive maintenance applies and statistics may also be applied to the performance of the installation, giving a view on how well it and the many individual parts are operating.

9.28 Quality assurance applied to lifts and escalators

Quality assurance is a management inspection tool designed specifically to maintain a state of excellence in all aspects of the activities associated with the design, specification, manufacture, installation and maintenance of a product. The general philosophy is presented in BS 5750 *Quality systems.*

A mechanism for applying the recommendations to all aspects of the UK lift and escalator industry has been set up, drawing on ideas presented in a paper by P.H. Day [9.20]. Purchasers of lift and escalator equipment can be assured that when they deal with companies who have been granted certification under the scheme, all goods and activities associated with their purchase will have been subject to quality control. This will have been carried out according to the recommendations made by an independent group of people, representing all sections of the industry and including purchasers, insurers and government bodies. The scheme can be extended to the scrutiny of specifications and drawings by an authorised inspection body, who will also carry out surveillance of the subsequent installation and observation of the testing of the equipment. On completion of the project, a technical report that includes recommendations for not less than 5 years' maintenance, together with a users' manual and the relevant detail drawings, will be provided.

References

9.1. BS 5655 *Lifts and service lifts. Part 1: 1986 Safety rules for the construction and installation of electric lifts.* BSI, London.
9.2. Transportation systems in buildings. *CIBSE Guide*, Vol. D, 1993. CIBSE, Delta House, 222 Balham High Road, London SW12 9BS.
9.3. BS 5655 *Lifts and service lifts. Part 6: 1990 Code of practice for selection and installation.* BSI, London.
9.4. Barney G.C. and Dos Santos S.M. Improved traffic design methods for lift systems. *Building Science*, Volume 10, 1975, pp. 272–285.
9.5. Fortune J.W. Top-down sky lobby lift design. *Elevatori*, Volume 19, number 2, February 1990, pp. 25–31.
9.6. BS 5810: 1979 *Code of practice for access for the disabled to buildings.* BSI, London.
9.7. *The Building Regulations 1991.* HMSO, London.
9.8. BS 5588: *Part 8: 1988 Code of practice for means of escape for disabled people.* BSI, London.
9.9. BS 5655: *Part 5: 1989 Specification for dimensions of standard electric lift arrangements.* BSI, London.
9.10. BS 5655: *Part 3: 1989 Specification for electric service lifts.* BSI, London.
9.11. *Building Regulations 1985: Mandatory rules for means of escape in case of fire.* HMSO, London.
9.12. BS 5588 *Fire precautions in the design, construction and use of buildings: Part 5: 1986 Code of practice for firefighting stairways and lifts.* BSI, London.
9.13. BS 5655: *Part 10: 1986 Specification for testing and inspection of electric and hydraulic lifts.* BSI, London.
9.14. O'Connor P.D.T. *Practical Reliability Engineering*, 1991. J. Wiley, Chichester.
9.15. BS 5655: *Part 11: 1979 Specification for modernisation and reconstruction.* BSI, London.
9.16. BS 5656 9EN 115: 1983 *Safety rules for the construction and installation of escalators and passenger conveyors.* BSI, London.
9.17. Safety in the use of escalators. *HSE Plant and Machinery Guidance Note PM34*, 1983. HMSO, London.
9.18. BS 5655: *Part 2: 1988 Hydraulic lifts.* BSI, London.
9.19. Escalators: periodic thorough examination. *HSE Plant and Machinery Guidance Note PM45*, 1984. HMSO, London.
9.20. Day P.H. New innovations from the UK elevator. *Technology 3*, 1990. IAEE Publications, Manchester.

Index